# CERTIFIED IN RISK AND INFORMATION SYSTEMS CONTROL™

CRISC™ Review Manual 2012

**ISACA®**

With 95,000 constituents in 160 countries, ISACA (*www.isaca.org*) is a leading global provider of knowledge, certifications, community, advocacy and education on information systems (IS) assurance and security, enterprise governance and management of IT, and IT-related risk and compliance. Founded in 1969, the nonprofit, independent ISACA hosts international conferences, publishes the *ISACA® Journal*, and develops international IS auditing and control standards, which help its constituents ensure trust in, and value from, information systems. It also advances and attests IT skills and knowledge through the globally respected Certified Information Systems Auditor® (CISA®), Certified Information Security Manager® (CISM®), Certified in the Governance of Enterprise IT® (CGEIT®) and Certified in Risk and Information Systems Control™ (CRISC™) designations. ISACA continually updates COBIT®, which helps IT professionals and enterprise leaders fulfill their IT governance and management responsibilities, particularly in the areas of assurance, security, risk and control, and deliver value to the business.

**Disclaimer**

ISACA has designed and created *CRISC™ Review Manual 2012* primarily as an educational resource to assist individuals preparing to take the CRISC certification exam. It was produced independently from the CRISC exam and the CRISC Certification Committee, which has had no responsibility for its content. Copies of past exams are not released to the public and were not made available to ISACA for preparation of this publication. ISACA makes no representations or warranties whatsoever with regard to these or other ISACA publications assuring candidates' passage of the CRISC exam.



**ISACA**
3701 Algonquin Road, Suite 1010
Rolling Meadows, IL 60008 USA
Phone: +1.847.253.1445
Fax: +1.847.253.1443
E-mail: *info@isaca.org*
Web site: *www.isaca.org*

ISBN 978-1-60420-226-7
*CRISC™ Review Manual 2012*
Printed in the United States of America

# Acknowledgments

This manual is the result of contributions from volunteers across the globe who are actively involved in risk management and information systems control design, implementation, monitoring and maintenance and who generously contributed their time and expertise. This international team exhibited a spirit and selflessness that has become the hallmark of contributors to ISACA manuals. Their participation and insight are truly appreciated.

**CRISC Developer**
Kevin M. Henry, CISA, CISM, CRISC, CISSP, SCF, KM Henry & Affiliates Management, Inc., Canada

**CRISC Subject Matter Expert Reviewers**
Sandy Fadale, CISM, CGEIT, CRISC, Bell Aliant Regional Communications, Canada
Shawna M. Flanders, CISA, CISM, CRISC, ACS, CSSGB, PSCU Financial Services, USA
David Githira, CRISC, Nairobi
Danny Ha, CISA, CISM, CGEIT, CRISC, CISSP, CSSLP, OneNet Company, Hong Kong
Pokit Lok, CISA, Hong Kong Productivity Council, Hong Kong
Jambunaathan Ramani, CISA, Capgemini Business Services Ltd., India
Darron Sun, CISA, CRISC, CISSP, ACS, AIAA, FLMI, Manulife Financial, Hong Kong
Gijo Varghese, CISM, CRISC, SA Health, Australia

ISACA has begun planning the 2013 edition of the *CRISC™ Review Manual*. Volunteer participation drives the success of the manual. If you are interested in becoming a member of the select group of professionals involved in this global project, we want to hear from you. Please e-mail us at *criscqae@isaca.org*.

Susan M. Caldwell
Chief Executive Officer
ISACA

Terence J. Trsar
Chief Professional Development Officer
ISACA

# Table of Contents

Starting Page

**Part II—Risk Management and Information Systems Control in Practice**

CRISC
Certified in Risk and Information Systems Control™
An ISACA® Certification

# Introduction

## A. Overview

**Contents**

The introduction contains the following topics:

# B. About This Manual

## Purpose of This Manual

ISACA is pleased to offer the second edition of the *CRISC™ Review Manual*.

The purpose of the manual is to provide CRISC candidates with information and references to assist in the preparation and study for the Certified in Risk and Information Systems Control (CRISC) exam.

Certification has resulted in a positive impact on many careers, including worldwide recognition for professional experience and enhanced knowledge and skills. The Certified in Risk and Information Systems Control™ certification (CRISC™, pronounced "see-risk") is designed for IT and business professionals who have hands-on experience with risk identification, assessment and evaluation; risk response; risk monitoring; IS control design and implementation; and IS control monitoring and maintenance. We wish you success with the CRISC exam.

**Note:** The *CRISC Review Manual 2012* is intended to assist candidates in preparing for the CRISC exam. The manual is **one source of preparation for the exam and should not be thought of as the only source nor should it be viewed as a comprehensive collection of all the information and experience that is required to pass the exam.** No single publication offers such coverage and detail.

## Basis for the Content

The content in the manual is based on the current CRISC job practice found at *www.isaca.org/criscjobpractice.*

This manual is the result of contributions from volunteers across the globe who are actively involved in risk management and IS control design, implementation, monitoring and maintenance and who generously contributed their time and expertise.

## Organization of This Manual

The *CRISC™ Review Manual 2012* is organized into two parts:
Part I—Risk Management and Information Systems Control Theory and Concepts
Part II—Risk Management and Information Systems Control in Practice

## Part I

*Part I—Risk Management and Information Systems Control Theory and Concepts* consists of five chapters, each dedicated to one of the five CRISC domains, as described in the CRISC job practice:

1. Domain 1—Risk Identification, Assessment and Evaluation
2. Domain 2—Risk Response
3. Domain 3—Risk Monitoring
4. Domain 4—Information Systems Control Design and Implementation
5. Domain 5—Information Systems Control Monitoring and Maintenance

## Exhibit B.1: High-level Relationship Between CRISC Domains

**Exhibit B.1** describes the high-level relationship between CRISC domains:

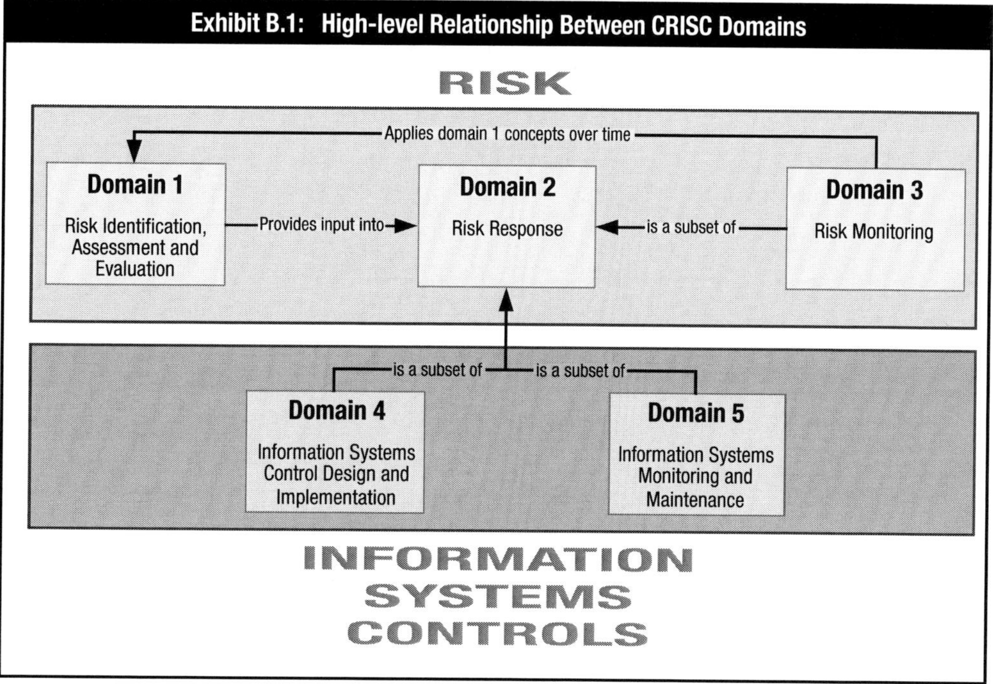


## Part I Chapter Structure

Each of the Part I chapters:
- Depicts the tasks performed by individuals who have a management, advisory or assurance role related to risk and IS control
- Describes the knowledge required to perform these tasks
- Serves as a definition of the roles and responsibilities of the professionals performing risk and IS control work

Domain 1 describes how risk management ties into risk governance and introduces essential risk governance concepts, such as risk appetite and tolerance. While not established by the CRISC, these concepts are an essential input into the risk practitioner's activities and help the CRISC candidate better understand the environment to which the risk practitioner contributes.

**Note:** The knowledge statements from domains 1, 4 and 5 that relate to business or IT process-specific risk, controls, control objectives, activities and metrics are addressed in Part II of this publication.

## Exhibit B.2: Knowledge Statements Addressed in Part II

**Exhibit B.2** highlights the process-specific knowledge statements **(in bold)** addressed in Part II:

### Exhibit B.2: Process-specific Knowledge Statements Addressed in Part II

***Domain 1—Task Statements***

1.1 Collect information and review documentation to ensure that risk scenarios are identified and evaluated.
1.2 Identify legal, regulatory and contractual requirements and organizational policies and standards related to information systems to determine their potential impact on the business objectives.
1.3 Identify potential threats and vulnerabilities for business processes, associated data and supporting capabilities to assist in the evaluation of enterprise risk.
1.4 Create and maintain a risk register to ensure that all identified risk factors are accounted for.
1.5 Assemble risk scenarios to estimate the likelihood and impact of significant events to the organization.
1.6 Analyze risk scenarios to determine their impact on business objectives.
1.7 Develop a risk awareness program and conduct training to ensure that stakeholders understand risk and contribute to the risk management process and to promote a risk-aware culture.
1.8 Correlate identified risk scenarios to relevant business processes to assist in identifying risk ownership.
1.9 Validate risk appetite and tolerance with senior leadership and key stakeholders to ensure alignment.

***Domain 2—Task Statements***

2.1 Identify and evaluate risk response options and provide management with information to enable risk response decisions.
2.2 Review risk responses with the relevant stakeholders for validation of efficiency, effectiveness and economy.
2.3 Apply risk criteria to assist in the development of the risk profile for management approval.
2.4 Assist in the development of risk response action plans to address risk factors identified in the organizational risk profile.
2.5 Assist in the development of business cases supporting the investment plan to ensure that risk responses are aligned with the identified business objectives.

***Domain 3—Task Statements***

3.1 Collect and validate data that measure key risk indicators (KRIs) to monitor and communicate their status to relevant stakeholders.
3.2 Monitor and communicate key risk indicators (KRIs) and management activities to assist relevant stakeholders in their decision-making process.
3.3 Facilitate independent risk assessments and risk management process reviews to ensure that they are performed efficiently and effectively.
3.4 Identify and report on risk, including compliance, to initiate corrective action and meet business and regulatory requirements.

***Domain 4—Task Statements***

4.1 Interview process owners and review process design documentation to gain an understanding of the business process objectives.
4.2 Analyze and document business process objectives and design to identify required information systems controls.
4.3 Design information systems controls in consultation with process owners to ensure alignment with business needs and objectives.
4.4 Facilitate the identification of resources (e.g., people, infrastructure, information, architecture) required to implement and operate information systems controls at an optimal level.
4.5 Monitor the information systems control design and implementation process to ensure that it is implemented effectively and within time, budget and scope.
4.6 Provide progress reports on the implementation of information systems controls to inform stakeholders and to ensure that deviations are promptly addressed.
4.7 Test information systems controls to verify effectiveness and efficiency prior to implementation.
4.8 Implement information systems controls to mitigate risk.
4.9 Facilitate the identification of metrics and key performance indicators (KPIs) to enable the measurement of information systems control performance in meeting business objectives.
4.10 Assess and recommend tools to automate information systems control processes.
4.11 Provide documentation and training to ensure that information systems controls are effectively performed.
4.12 Ensure that all controls are assigned control owners to establish accountability.
4.13 Establish control criteria to enable control life cycle management.

***Domain 5—Task Statements***

5.1 Plan, supervise and conduct testing to confirm continuous efficiency and effectiveness of information systems controls.
5.2 Collect information and review documentation to identify information systems control deficiencies.
5.3 Review information systems policies, standards and procedures to verify that they address the organization's internal and external requirements.
5.4 Assess and recommend tools and techniques to automate information systems control verification processes.
5.5 Evaluate the current state of information systems processes using a maturity model to identify the gaps between current and targeted process maturity.
5.6 Determine approach to correct information systems control deficiencies and maturity gaps to ensure that deficiencies are appropriately considered and remediated.
5.7 Maintain sufficient, adequate evidence to support conclusions on the existence and operating effectiveness of information systems controls.
5.8 Provide information systems control status reporting to relevant stakeholders to enable informed decision making.

***Domain 1—Knowledge Statements—Knowledge of…***

1.1 standards, frameworks and leading practices related to risk identification, assessment and evaluation
1.2 techniques for risk identification, classification, assessment and evaluation
1.3 quantitative and qualitative risk evaluation methods
1.4 business goals and objectives
1.5 organizational structures
**1.6 risk scenarios related to business processes and initiatives**
1.7 business information criteria
**1.8 threats and vulnerabilities related to business processes and initiatives**
**1.9 information systems architecture (e.g., platforms, networks, applications, databases and operating systems)**
**1.10 information security concepts**
**1.11 threats and vulnerabilities related to third-party management**
**1.12 threats and vulnerabilities related to data management**
**1.13 threats and vulnerabilities related to the system development life cycle**
**1.14 threats and vulnerabilities related to project and program management**
**1.15 threats and vulnerabilities related to business continuity and disaster recovery management**
**1.16 threats and vulnerabilities related to management of IT operations**
1.17 the elements of a risk register
1.18 risk scenario development tools and techniques
1.19 risk awareness training tools and techniques
1.20 principles of risk ownership
1.21 current and forthcoming laws, regulations and standards
**1.22 threats and vulnerabilities associated with emerging technologies**

***Domain 2—Knowledge Statements—Knowledge of…***

2.1 standards, frameworks and leading practices related to risk response
2.2 risk response options
2.3 cost-benefit analysis and return on investment (ROI)
2.4 risk appetite and tolerance
2.5 organizational risk management policies
2.6 parameters for risk response selection
2.7 project management tools and techniques
2.8 portfolio, investment and value management
2.9 exception management
2.10 residual risk

***Domain 3—Knowledge Statements—Knowledge of…***

3.1 standards, frameworks and leading practices related to risk monitoring
3.2 principles of risk ownership
3.3 risk and compliance reporting requirements, tools and techniques
3.4 key performance indicators (KPIs) and key risk indicators (KRIs)
3.5 risk assessment methodologies
3.6 data extraction, validation, aggregation and analysis tools and techniques
3.7 various types of reviews of the organization's risk monitoring process (e.g., internal and external audits, peer reviews, regulatory reviews, quality reviews)

***Domain 4—Knowledge Statements—Knowledge of…***

4.1 standards, frameworks and leading practices related to information systems control design and implementation
4.2 business process review tools and techniques
4.3 testing methodologies and practices related to information systems control design and implementation
**4.4 control practices related to business processes and initiatives**
**4.5 the information systems architecture (e.g., platforms, networks, applications, databases and operating systems)**
**4.6 controls related to information security**
**4.7 controls related to third-party management**
**4.8 controls related to data management**
**4.9 controls related to the system development life cycle**
**4.10 controls related to project and program management**
**4.11 controls related to business continuity and disaster recovery management**
**4.12 controls related to management of IT operations**
4.13 software and hardware certification and accreditation practices
4.14 the concept of control objectives
4.15 governance, risk and compliance (GRC) tools
4.16 tools and techniques to educate and train users

***Domain 5—Knowledge Statements—Knowledge of…***

5.1 standards, frameworks and leading practices related to information systems control monitoring and maintenance
5.2 enterprise security architecture
5.3 monitoring tools and techniques
5.4 maturity models
5.5 control objectives, activities and metrics related to IT operations and business processes and initiatives
**5.6 control objectives, activities and metrics related to incident and problem management**
5.7 security testing and assessment tools and techniques
5.8 control objectives, activities and metrics related to information systems architecture (platforms, networks, applications, databases and operating systems)
**5.9 control objectives, activities and metrics related to information security**
**5.10 control objectives, activities and metrics related to third-party management**
**5.11 control objectives, activities and metrics related to data management**
**5.12 control objectives, activities and metrics related to the system development life cycle**
**5.13 control objectives, activities and metrics related to project and program management**
**5.14 control objectives, activities and metrics related to software and hardware certification and accreditation practices**
**5.15 control objectives, activities and metrics related to business continuity and disaster recovery management**
5.16 applicable laws and regulations

## Part II

*Part II—Risk Management and Information Systems Control in Practice* contains selected process-specific chapters:

1. Determining the IT Strategy
2. Project and Program Management
3. Change Management
4. Third Party Management
5. Continuous Service Assurance
6. Information Security Management
7. Configuration Management
8. Problem Management
9. Data Management
10. Physical Environment Management
11. IT Operations

## Part II Chapter Structure

Each chapter introduces one IT or business process at a high level and address the following, as they relate to that specific process:

- Common risk
- Key risk indicators (KRIs)
- Controls
- Control metrics
- Control monitoring practices

This knowledge is necessary to perform the tasks discussed in Part I of the manual and is required for passage of the CRISC exam.

The learning objectives for each of these domains appear at the beginning of each chapter with the corresponding task and knowledge statements that are tested on the exam. Exam candidates should evaluate their strengths, based on knowledge and experience, in each of these processes.

## Additional Resources

The manual also contains the following sections:

- **Study Questions, Answers and Explanations**—To familiarize candidates with question structure and general content, 25 study questions, sorted by domain with explanations for the correct and incorrect answers, are provided.
- **Glossary**—The definitions of key terms and acronyms are provided for concepts in which CRISC practitioners felt it necessary to ensure a common understanding.
- **Suggested Resources for Further Study**—As candidates read through the manual and encounter topics that are new to them or ones in which they feel their knowledge and experience are limited, additional references should be sought. Suggested resources for further study are provided in Parts I and II, and a comprehensive list appears at the end of the manual.
- **List of Exhibits**— A comprehensive listing of all of the exhibits in each chapter is provided at the end of the manual.

## Written Material Is Not a Substitute for Experience

The CRISC exam is a practice-based exam. Simply reading the material in this manual will not properly prepare candidates for the exam.

The study questions included in this manual are designed to provide further clarity to the content presented in the manual and to depict the type of questions typically found on the CRISC exam. The practice questions and answers:

- Should *not* be used independently as a source of knowledge
- Should *not* be considered a measure of one's ability to answer questions correctly on the exam for any domain
- *Are* intended to familiarize candidates with question structure and general content
- *May or may not* be similar to questions that will appear on the actual exam

The reference material includes other publications that could be used to further acquire and better understand detailed information on the topics addressed in the manual.

## Disclaimer

No representations or warranties are made by ISACA in regard to this or other ISACA publications assuring candidates' passage of the CRISC exam. This publication was produced independently of the CRISC Certification Committee, which has no responsibility for the content of this manual.

## Annual Updates to This Manual

The *CRISC Review Manual* will be updated annually to keep pace with rapid changes in the field of IT-related business risk management. As such, your comments and suggestions regarding this manual are welcomed.

After the exam is over, please take a moment to complete the online questionnaire. Your observations will be extremely valuable for the preparation of the 2013 edition of the manual.

A link to an online feedback questionnaire is posted at *www.isaca.org/studyaidsevaluation*.

## Other Study Aids

The CRISC candidate may find it useful to study the *CRISC™ Review Questions, Answers & Explanations Manual 2011* and the *CRISC™ Review Questions, Answers & Explanations Manual 2012 Supplement*. Each consists of 100 multiple-choice study questions.

# C. About the CRISC Exam

## Introduction

The CRISC certification is designed to meet the growing demand for professionals who can integrate enterprise risk management (ERM) with discrete IS control skills. The technical skills and practices the CRISC certification promotes and evaluates are the building blocks of success in this growing field, and the CRISC designation demonstrates proficiency in this role.

## Development of the CRISC Exam

The CRISC job practice, based on the results of extensive research and feedback from subject matter experts (SMEs) around the world, serves as the basis for the exam and the experience requirements to earn the CRISC certification.

The CRISC Certification Committee oversees the development of the exam and ensures the currency of its content. Questions for the CRISC exam are developed through a comprehensive process designed to enhance the ultimate quality of the exam.

The process includes a Test Enhancement Subcommittee (TES) that works with item writers to develop and review questions before they are submitted to the CRISC Certification Committee for review.

## Exam Scope

There are five domains in the CRISC job practice. Each domain is accompanied by tasks and knowledge statements that depict the tasks performed by CRISCs and the knowledge required to perform these tasks. Exam candidates will be tested based on their practical knowledge associated with performing the tasks in the percentages listed in the following table.

| Domain No. | Domain Name | Percentage |
|---|---|---|
| 1 | Risk Identification, Assessment and Evaluation | 31% |
| 2 | Risk Response | 17% |
| 3 | Risk Monitoring | 17% |
| 4 | Information Systems Control Design and Implementation | 17% |
| 5 | Information Systems Control Monitoring and Maintenance | 18% |

**Note:** The percentages listed with the domains indicate the emphasis or percentage of questions that will appear on the exam from each domain. For a description of each domain's task and knowledge statements, visit *www.isaca.org/criscjobpractice*.

## Exam Design

The exam consists of 200 multiple-choice questions and is administered biannually in June and December during a four-hour session. At this time, the exam is offered in English only.

## Types of Questions

CRISC exam questions are developed with the intent of measuring and testing practical knowledge and the application of general concepts and standards. All questions are designed with one best answer. The candidate is asked to choose the correct or best answer from the options.

Every CRISC question has a stem (question) and four choices (possible answers).

| Question Component | Description |
|---|---|
| Stem | The stem may be in the form of a question or incomplete statement. In some instances, a scenario may also be included. The questions with a scenario normally include a description of a situation and require the candidate to answer two or more questions based on the information provided. |
| Answers | Each question has four answer choices. The candidate has to select the single most appropriate answer. Some questions may require the candidate to choose the appropriate answer based on a qualifier, such as **MOST** or **BEST**. |

The candidate is required to read the question carefully, eliminate known incorrect answers and then make the best choice possible.

## Preparing for the Exam

Good preparation for the CRISC exam can be achieved through an organized plan of study. To assist individuals with the development of a successful study plan, ISACA offers study aids and review courses to exam candidates. See *www.isaca.org/criscbooks* to view the ISACA study aids that can help prepare for the exam.

## Suggested Resources for Further Study

The following resources are available from ISACA in addition to the *CRISC™ Review Manual 2012*.

<table>
<tr><th>Resource</th><th>Function</th></tr>
<tr><td>The Risk IT Framework</td><td rowspan="3">Although knowledge of the Risk IT framework and COBIT is not specifically tested on the CRISC exam, the Risk IT and COBIT processes are reflected in the CRISC job practice knowledge statements. As such, a thorough review of the Risk IT framework and COBIT is recommended for candidates preparing for the CRISC certification.<br><br>Note: The COBIT 4.1 framework is available at no charge from ISACA and can be downloaded at www.isaca.org/cobit. The new COBIT 5 framework will be available in 2012.</td></tr>
<tr><td>The Risk IT Practitioner Guide</td></tr>
<tr><td>COBIT® 4.1</td></tr>
<tr><td>CRISC™ Review Questions, Answers & Explanations Manual 2011</td><td>Designed to provide CRISC candidates with an understanding of the type and structure of questions and content that will appear on the CRISC exam, the CRISC™ Review Questions, Answers & Explanations Manual 2011 consists of 100 multiple-choice study questions.<br><br>To help candidates maximize study efforts, questions are presented in the following two ways:<br>• Sorted by job practice area, allowing CRISC candidates to focus on particular topics<br>• Scrambled as a sample 100-question exam, enabling candidates to effectively determine their strengths and weaknesses and allowing them to simulate an actual exam</td></tr>
<tr><td>CRISC™ Review Questions, Answers & Explanations Manual 2012 Supplement</td><td>The CRISC supplement offers an additional 100 study questions in the same structure as the CRISC™ Review Questions, Answers & Explanations Manual 2011</td></tr>
</table>

## Exam Schedule

**The CRISC exam is offered twice a year, in June and December.**

## Exam Registration

Exam registration information, including dates, deadlines and fees for registration opening, early bird and closing dates, is available in the Exam Registration Information (*Bulletin of Information* [BOI]) at *www.isaca.org/criscexam.*

You may register online or complete the registration form within the BOI and fax or mail it to ISACA for processing.

When available for each exam, the Exam Registration Information (BOI) is published online at *www.isaca.org/criscboi.*

## Additional General Information

Additional information regarding the CRISC certification and exam is provided at the end of this manual.

# D. Earning the CRISC Certification

### Earning the CRISC Certification

Candidates who pass the CRISC exam are not automatically CRISC-certified and cannot use the CRISC designation until the completed application is received and formally approved.

Candidates will be awarded the CRISC certification once they have met (and continue to meet) the following requirements:
1. Successful completion of the CRISC exam
2. Submission of an application
3. Experience in risk management and IS control
4. Adherence to the ISACA Code of Professional Ethics
5. Adherence to the CRISC continuing professional education (CPE) policy

### Successful Completion of the CRISC Exam

CRISC exam details are addressed in section C (About the CRISC Exam) of this introduction.

### Submission of an Application

Once a candidate passes the CRISC exam, the individual has five years from the date of the exam to apply for certification. To be certified, successful candidates must complete the application and submit it for approval. All work experience must be verified using the appropriate forms included in the application. The form can be downloaded at *www.isaca.org/criscapp*. A processing fee of $50 must accompany all applications.

Once certified, the new CRISC will receive a CRISC certificate and a pin. At the time of application, individuals must also acknowledge that ISACA reserves the right, but is not obligated, to publish or otherwise disclose their CRISC status.

Please note that certification application decisions are not final as there is an appeal process for certification application denials. Inquiries regarding denials of certification can be sent to *certification@isaca.org*.

### Experience in Risk Management and IS Control

Certification is granted initially to individuals who have completed the CRISC exam successfully and meet the following work experience requirements:
1. A minimum of at least three years of cumulative work experience performing the tasks of a CRISC professional across at least three CRISC domains. There will be no substitutions or experience waivers.
2. Experience must have been gained within the 10-year period preceding the application date for certification or within five years from the date of initially passing the exam.
3. All experience must be verified independently with employers.

### Adherence to the ISACA Code of Professional Ethics

Members of ISACA and/or holders of the CRISC certification agree to the ISACA Code of Professional Ethics to guide professional and personal conduct.

Failure to adhere to this Code of Professional Ethics can result in an investigation into a member's and/or certification holder's conduct and, ultimately, in disciplinary measures. The ISACA Code of Professional Ethics can be viewed online at *www.isaca.org/ethics*.

## Adherence to the CRISC CPE Policy

To maintain certification, the CRISC must:
- Pay an annual maintenance fee
- Attain and report an annual minimum of 20 CPE hours per cycle year
- Attain and report a minimum of 120 CPE hours for the three-year reporting period

The objectives of the continuing professional education program are to:
- Maintain an individual's competency by requiring the update of existing knowledge and skills in the areas of risk and IS control.
- Provide a means to differentiate between qualified CRISCs and those who have not met the requirements for continuation of their certification.
- Provide a mechanism for monitoring risk and IS control professionals' maintenance of their competency.
- Aid management in developing sound risk and IS control functions by providing criteria for personnel selection and development.

The CRISC CPE policy can be viewed at *www.isaca.org/crisccpepolicy*.

## Revocation of CRISC Certification

The CRISC Certification Committee may, at its discretion and after due and thorough consideration, revoke an individual's CRISC certification for any of the following reasons:
- Failing to comply with the CRISC CPE policy
- Violating any provision of the ISACA Code of Professional Ethics
- Falsifying or deliberately failing to provide relevant information
- Intentionally misstating a material fact
- Engaging or assisting others in dishonest, unauthorized or inappropriate behavior at any time in connection with the CRISC exam or the certification process

**Page intentionally left blank**

# Part I—Risk Management and Information Systems Control Theory and Concepts

## Domain 1—Risk Identification, Assessment and Evaluation

### A. Chapter Overview

**Introduction**

This chapter provides the core practices of risk identification, assessment and evaluation with which the risk practitioner should be familiar.

The chapter also provides information about "The Big Picture" or risk governance to ensure that the risk practitioner can effectively differentiate between governing and managing IT-related business risk.

**Inputs (Tie-back)**

The risk governance function is the responsibility of senior management. It shapes the risk culture within which the risk practitioner functions, and provides inputs for many of the tasks performed during risk identification, assessment and evaluation. The risk practitioner is expected to provide risk response recommendations in alignment with such inputs, particularly regarding risk appetite and risk tolerance levels.

**Process Objectives**

Risk identification, assessment and evaluation is concerned with correctly determining the risk faced by the enterprise and providing recommendations to senior management on how to effectively maintain risk at an acceptable level, including, but not limited to:
- Identifying risk, including emerging risk and risk associated with people, processes, technology, architecture, applications, information, natural factors and physical threats
- Assessing the risk levels associated with each threat, including anticipated risk likelihood and impact and the effectiveness of current and planned controls
- Calculating the risk levels using both quantitative and qualitative metrics and determining the impact of the risk on the ability of the business to meet its goals and objectives

**Outputs (Tie-forward)**

The output of the risk assessment process helps to identify appropriate controls for reducing or eliminating risk during the risk response process. To determine the likelihood of a future adverse event, threats to the enterprise must be analyzed in conjunction with the potential vulnerabilities and the controls in place for the system.

This process includes the creation of a comprehensive, prioritized inventory of relevant risk—often in the form of a risk register—as well as risk response recommendations. Other deliverables may include risk awareness programs and training.

**Learning Objectives**

As a result of completing this chapter, the risk practitioner should be able to:
- Differentiate between risk management and risk governance.
- Identify the roles and responsibilities for risk management.
- Identify relevant standards, frameworks and practices.
- Explain the meaning of key risk management concepts, including "risk appetite" and "risk tolerance."
- Differentiate between threats and vulnerabilities.
- Apply risk identification, classification, quantitative/qualitative assessment and evaluation techniques.
- Describe the key elements of a risk register.
- Describe risk scenario development tools and techniques.
- Help develop and support risk awareness training tools and techniques.
- Relate security concepts to risk assessment.

## Contents

This chapter contains the following sections:

## B. Task and Knowledge Statements

### Introduction

This section describes the task and knowledge statements for Domain 1, which focuses on identifying, assessing and evaluating risk to enable the execution of the enterprise risk management (ERM) strategy.

### Domain 1 Task Statements

The following table describes the task statements for Domain 1.

| No. | Task Statement (TS) |
|---|---|
| TS1.1 | Collect information and review documentation to ensure that risk scenarios are identified and evaluated. |
| TS1.2 | Identify legal, regulatory and contractual requirements and organizational policies and standards related to information systems to determine their potential impact on the business objectives. |
| TS1.3 | Identify potential threats and vulnerabilities for business processes, associated data and supporting capabilities to assist in the evaluation of enterprise risk. |
| TS1.4 | Create and maintain a risk register to ensure that all identified risk factors are accounted for. |
| TS1.5 | Assemble risk scenarios to estimate the likelihood and impact of significant events to the enterprise. |
| TS1.6 | Analyze risk scenarios to determine their impact on business objectives. |
| TS1.7 | Develop a risk awareness program and conduct training to ensure that stakeholders understand risk and contribute to the risk management process and to promote a risk-aware culture. |
| TS1.8 | Correlate identified risk scenarios to relevant business processes to assist in identifying risk ownership. |
| TS1.9 | Validate risk appetite and tolerance with senior leadership and key stakeholders to ensure alignment. |

## Domain 1 Knowledge Statements

The following table describes the knowledge statements for Domain 1.

| No. | Knowledge Statement (KS)<br>Knowledge of: |
|---|---|
| KS1.1 | Standards, frameworks and leading practices related to risk identification, assessment and evaluation |
| KS1.2 | Techniques for risk identification, classification, assessment and evaluation |
| KS1.3 | Quantitative and qualitative risk evaluation methods |
| KS1.4 | Business goals and objectives |
| KS1.5 | Organizational structures |
| KS1.6 | Risk scenarios related to business processes and initiatives |
| KS1.7 | Business information criteria |
| KS1.8 | Threats and vulnerabilities related to business processes and initiatives |
| KS1.9 | Information systems architecture (e.g., platforms, networks, applications, databases and operating systems) |
| KS1.10 | Information security concepts |
| KS1.11 | Threats and vulnerabilities related to third-party management |
| KS1.12 | Threats and vulnerabilities related to data management |
| KS1.13 | Threats and vulnerabilities related to the system development life cycle |
| KS1.14 | Threats and vulnerabilities related to project and program management |
| KS1.15 | Threats and vulnerabilities related to business continuity and disaster recovery management |
| KS1.16 | Threats and vulnerabilities related to management of IT operations |
| KS1.17 | The elements of a risk register |
| KS1.18 | Risk scenario development tools and techniques |
| KS1.19 | Risk awareness training tools and techniques |
| KS1.20 | Principles of risk ownership |
| KS1.21 | Current and forthcoming laws, regulations and standards |
| KS1.22 | Threats and vulnerabilities associated with emerging technologies |

**Note:** Knowledge statements 1.8, 1.11-1.16 and 1.22 are process specific and are addressed in Part II of this manual.

# C. The Big Picture—Risk Management and Risk Governance

## 1. Section Overview

### Introduction

Enterprises continuously plan, operate and deploy business activities and processes to achieve business objectives. The risk practitioner is actively involved in ensuring that the operational risk of each business activity is assessed; monitored; and, if necessary, addressed.

Each business activity carries both risk and opportunity, and the risk practitioner must be aware of the need to balance business needs and productivity with effective controls. Some controls that will be specifically considered are IS controls; however, the risk practitioner must also be familiar with other risk response methods as seen in *Domain Two—Risk Response*.

The risk practitioner must be capable of evaluating risk across the entire enterprise, not just on a systemwide or departmentwide basis. Having a grasp of the "Big Picture" will allow risk management efforts to be better integrated with business priorities and culture; provide a more efficient, interoperable, measureable and effective risk response; and support governance requirements more thoroughly.

## 2. Key Terms and Principles

### Definition of Risk

Risk reflects the combination of the likelihood of events occurring and the impact those events have on the enterprise.

Risk—the potential for events and their consequences—contains both:
- Opportunities for benefit (upside)
- Threats to success (downside)

### Guiding Principles for Effective Risk Management

The following are guiding principles for effective risk management:
- Maintain focus on the business mission, goals and objectives.
- Integrate IT risk management into enterprise risk management (ERM).
- Balance the costs and benefits of managing risk.
- Promote fair and open communication.
- Establish tone at the top and assign personal accountability.
- Promote continuous improvement as part of daily activities.

The following table provides further detail.

| Guiding Principles for Effective Risk Management | |
|---|---|
| **Principle** | **Description** |
| Maintain focus on the business mission, goals and objectives. | • All risk is treated as a business risk, and the risk management approach must be comprehensive and cross-functional.<br>• The focus is on business outcome. Each business function supports the achievement of business objectives; IT-related risk is expressed as the impact it can have on the achievement of business objectives or strategy.<br>• Every risk analysis considers business and IT-process resilience and contains a dependency analysis of how the business process depends on IT-related resources, such as:<br>– People<br>– Information<br>– Applications<br>– Infrastructure<br>• IT-related business risk is viewed from two angles:<br>– Protection against value destruction<br>– Enablement of value generation |

## Guiding Principles for Effective Risk Management *(cont.)*

| Guiding Principles for Effective Risk Management *(cont.)* | |
|---|---|
| **Principle** | **Description** |
| Integrate IT risk management into enterprise risk management (ERM). | Most enterprises rely so heavily on information systems that a systems failure could have devastating consequences. For that reason, IT risk must be considered as a part of overall business risk, and risk assessments must be built into all aspects of the enterprise's business model. This includes ensuring that:<br>• Business objectives and the amount of risk that the enterprise is prepared to take are clearly defined and documented.<br>• The entity's risk appetite reflects its risk management philosophy and influences the culture and operating style (as stated in the Committee of Sponsoring Organizations of the Treadway Commission [COSO] *Enterprise Risk Management—Integrated Framework*).<br>• Risk issues are integrated for each business unit, i.e., the risk view is consolidated across the overall enterprise.<br>• Attestation of/sign-off on control environment is provided. |
| Balance the costs and benefits of managing risk. | • Risk is prioritized and addressed in line with risk appetite and tolerance.<br>• Controls are implemented to prevent or minimize impact and are based on a cost-benefit analysis. In other words, controls are not implemented simply for the sake of implementing controls.<br>• Existing controls are leveraged to address multiple risk factors or to address risk more efficiently. |
| Promote fair and open communication. | • Open, accurate, timely and transparent information on IT risk is exchanged and serves as the basis for all risk-related decisions.<br>• Risk issues, principles and risk management methods are integrated across the enterprise.<br>• Technical findings are translated into relevant and understandable business terms. |
| Establish tone at the top and assign personal accountability. | • Key personnel, i.e., influencers, business owners and the board of directors, is engaged in risk management.<br>• There is clear assignment and acceptance of risk ownership.<br>• Top management provides direction by means of policies, procedures and the right level of enforcement.<br>• Enterprise leadership actively promotes a risk-aware culture.<br>• Authorized individuals make risk decisions, including business-focused IT risk, e.g., for IT investment decisions, project funding, major IT environment changes, risk assessments, and the monitoring and testing of controls. |
| Promote continuous improvement as part of daily activities. | • Because of the dynamic nature of risk, risk management is an iterative, perpetual and ongoing process.<br>• The enterprise pays attention to consistent risk assessment methods, roles and responsibilities, tools, techniques, and criteria across the enterprise, noting especially:<br>– Identification of key processes and associated risk<br>– Understanding of impacts on achieving business objectives<br>– Identification of triggers that indicate when an update of the framework is required<br>• Risk management practices are appropriately prioritized and embedded in enterprise decision-making processes that enable *risk-return*-aware business decisions.<br>• Risk management practices are straightforward and easy to use and contain practices to detect, prevent and mitigate threat and potential risk. |

## 3. Essentials of Risk Management

| | |
|---|---|
| **Definition of Management** | Often differentiated from governance as the distinction between being "committed" (governance) and "involved" (management), management entails the judicious use of means (resources, people, processes, practices, etc.) to achieve an identified end.<br><br>Management is responsible for execution within the direction set by the guiding body or unit. Management is about planning, building, organizing and controlling operational activities to align with the direction set by the governance body. It is a means or instrument by which the governance body achieves a result or objective. |
| **Definition of Risk Management** | Risk management is the identification, assessment and prioritization of risk followed by coordinated and economical application of resources to minimize, monitor and control the probability and/or impact of adverse events or to maximize the realization of opportunities. |
| **Holistic Risk and Opportunity Management** | Risk and opportunity go hand in hand. To provide business value to stakeholders, enterprises must engage in various activities and initiatives, all of which carry various degrees of uncertainty and, therefore, risk.<br><br>Managing risk and opportunity is a key strategic activity for enterprise success.<br><br>Risk management must be integrated across all business processes. Risk must be carefully considered whenever strategic plans are being considered and whenever a change to an enterprise's processes or systems is being implemented. Performing prudent risk management early in each business activity will allow the enterprise to take steps to minimize risk and avoid situations where an unnecessary level of risk would be encountered. |
| **Responsibilities and Accountability for Risk Management** | Various members of the enterprise have responsibility or accountability for risk management. In this context:<br>• *Responsibility* belongs to those who must ensure that the activities are completed successfully.<br>• *Accountability* applies to those individuals, groups or entities that are ultimately responsible for the subject matter, process or scope. |

## 4. Essentials of Risk Governance

| | |
|---|---|
| **Topic Overview** | This topic contains a brief introduction to risk governance to provide the risk practitioner with a baseline understanding of the holistic environment in which the risk practitioner functions. |
| **Relevance** | Risk governance addresses the oversight of the business risk management strategy of the enterprise.<br><br>Risk governance is the domain of senior management and the shareholders of the enterprise. They establish the enterprise's risk culture and the acceptable levels of risk; set up the management framework; and ensure that the risk management function is operating effectively to identify, manage, monitor and report on current and potential risk facing the enterprise. |

## 4.1 Key Terms and Principles

**Definition of Governance**

Governance is derived from the Greek verb meaning "to steer." A governance system refers to all the means and mechanisms that enable multiple stakeholders in an enterprise to have an organized say in evaluating conditions and options; setting direction; and monitoring compliance, performance and progress against plans to satisfy specific enterprise objectives.

> **Note:** Means and mechanisms set direction and monitor compliance and performance aligned with the overall objectives, and include frameworks, principles, policies, sponsorship, structures and decision mechanisms, roles and responsibilities, and processes and practices. In most enterprises, this is the responsibility of the board of directors under the leadership of the chief executive officer (CEO) and chairman.

---

**Definition of Risk Governance**

Risk governance is a strategic business function that helps ensure that:

- Risk management activities align with the enterprise's loss capacity and leadership's subjective tolerance of it.
- The risk management strategy is aligned with the overall business strategy.

Enterprise decisions consider the full range of (risk) opportunities and consequences.

---

**Responsibility for Risk Governance**

Risk governance is ultimately the responsibility of the board of directors and senior management. They establish the enterprise's risk culture and the acceptable levels of risk; set up the management framework; and ensure that the risk management function is operating effectively to identify, manage, monitor and report on current and potential risk facing the enterprise.

> **Note:** While risk governance and the decisions made in the execution of risk governance ultimately are not the responsibility of the risk practitioner, the practitioner must nevertheless contribute to and enable sound risk management decisions through the execution of many underlying tasks associated with the risk governance process.

## 4.2 Risk Governance Objectives

### Risk Governance Objectives

Effective risk governance helps ensure that risk management practices are embedded in the enterprise, enabling it to secure optimal risk-adjusted return. Risk governance has three main objectives:

1. Establish and maintain a common risk view.
2. Integrate risk management into the enterprise.
3. Make risk-aware business decisions.

| Risk Governance Objective | Description |
|---|---|
| 1. Establish and maintain a common risk view. | Effective risk governance establishes the common view of risk for the enterprise. This determines which controls are necessary to mitigate risk and how risk-based controls are integrated into business processes and IS.<br><br>The risk governance function sets the tone of the business in how to determine an acceptable level of risk tolerance.<br><br>Risk governance is a continuous life cycle that requires regular reporting and ongoing review. The risk governance function must oversee the operations of the risk management team. |
| 2. Integrate risk management into the enterprise. | Integrating risk management into the enterprise enforces a holistic enterprise risk management (ERM) approach across the entire enterprise. It requires the integration of risk management into every department, function, system and geographic location. Understanding that risk in one department or system may pose an unacceptable risk to another department or system requires that all business processes be compliant with a baseline level of risk management.<br><br>The objective of ERM is to establish the authority to require all business processes to undergo a risk analysis on a periodic basis or when there is a significant change to the internal or external environment. |
| 3. Make risk-aware business decisions. | To make risk-aware business decisions, the risk governance function must consider the full range of opportunities and consequences of each such decision and its impact on the enterprise, its place in society and the environment. |

### Foundation for Effective Risk Governance

To effectively govern enterprise and IT risk, there must be an:
- Understanding and consensus with respect to the risk appetite and risk tolerance of the enterprise
- Awareness of risk and the need for effective communication about risk throughout the enterprise
- Understanding of the elements of risk culture

## 4.3 Risk Appetite and Tolerance

### Introduction

"Risk appetite" and "risk tolerance" are concepts that are frequently used, but the potential for misunderstanding is high. Some people use the concepts interchangeably; others see a clear difference.

### Definition of Risk Appetite

The broad-based amount of risk that a company or other entity is willing to accept in pursuit of its mission (or vision)

## Definition of Risk Tolerance

The acceptable variation relative to the achievement of an objective (often best measured in the same units as those used to measure the related objective)

> **Note:** These definitions are compatible with the Committee of Sponsoring Organizations of the Treadway Commission (COSO) ERM definitions, which are equivalent to the ISO 31000 definition in Guide 73:2009, Risk Management Vocabulary.

## Major Factors Influencing Risk Appetite

When considering the risk appetite for the enterprise, the following two major factors are important:

- The enterprise's objective capacity to absorb loss, e.g., financial loss, reputation damage
- The (management) culture or predisposition toward risk taking—cautious or aggressive. (What is the amount of loss that the enterprise wants to accept to pursue a return?)

> **Note:** Risk appetite and risk tolerance should be applied not only to risk assessments, but also to all risk decision making.

## Risk Appetite Variations Between Enterprises

Risk appetite can and will be different among enterprises—there is no absolute norm or standard of what constitutes acceptable and unacceptable risk. Every enterprise has to define its own risk appetite levels and should:

- Ensure that such definitions/levels are:
  - In line with the overall risk culture that the enterprise wants to express (i.e., ranging from very risk averse to risk taking/opportunity seeking)
  - Well defined, understood and communicated
  - Reviewed on a regular basis

## Exhibit 1.1: Risk Map Indicating Risk Appetite Bands

In practice, risk appetite can be defined in terms of combinations of frequency and magnitude of a risk, using risk maps. **Exhibit 1.1** and the following table depict and describe different bands of risk significance, based on frequency and magnitude of risk.

**Exhibit 1.1: Risk Map Indicating Risk Appetite Bands**

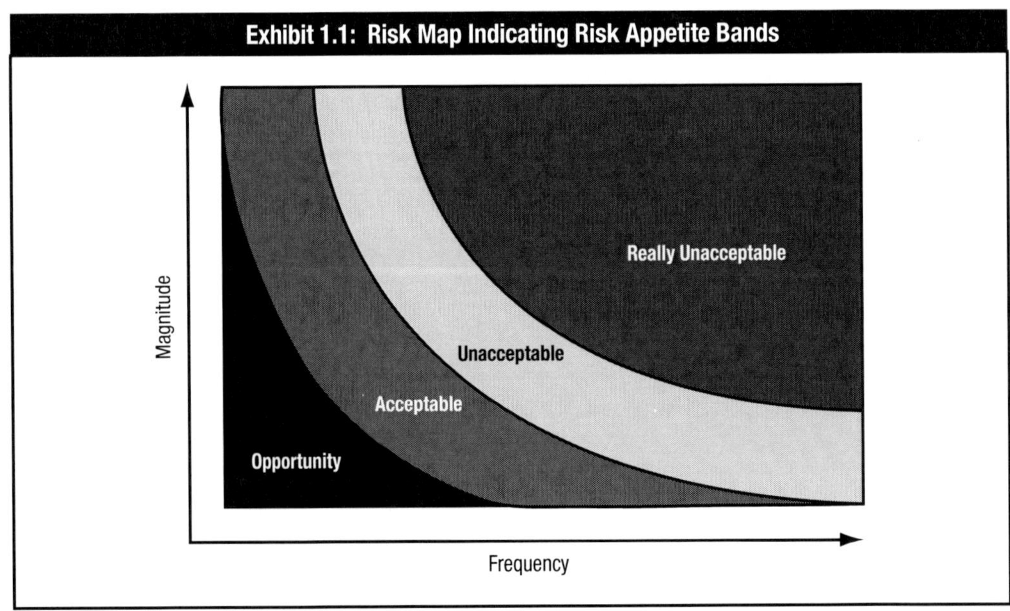

**Exhibit 1.1: Risk Map Indicating Risk Appetite Bands *(cont.)***

| Risk Level | Description |
|---|---|
| Really unacceptable | Indicates really unacceptable risk. The enterprise estimates that this level of risk is far beyond its normal risk appetite. Any risk found to be in this band may trigger an immediate risk response. |
| Unacceptable | Indicates elevated risk, i.e., also above acceptable risk appetite. The enterprise may, as a matter of policy, require mitigation or another adequate response to be defined within certain time boundaries. |
| Acceptable | Indicates a normal, acceptable level of risk, usually with no special action required, except for maintaining the current controls or other responses. |
| Opportunity | Indicates very low risk, in which cost-saving opportunities may be found by decreasing the degree of control or in which opportunities for assuming more risk may arise. |

**Note:** This risk appetite scheme is an example. Each enterprise needs to define its own risk appetite levels and review them regularly.

## Risk Tolerance Example

Risk tolerance is the acceptable deviation from the level set by the risk appetite and business objectives.

**Example:** Standards require projects to be completed within the estimated budgets and time, but overruns of 10 percent of budget or 20 percent of time are tolerated.

## Guidelines for Risk Appetite and Risk Tolerance

The guidelines listed in the following table apply to risk appetite and risk tolerance.

| Guidelines for Risk Appetite and Risk Tolerance | |
|---|---|
| **Guideline** | **Description** |
| Risk appetite and risk tolerance must connect. | Risk appetite and risk tolerance go hand in hand. Risk tolerance is defined at the enterprise level and is reflected in policies set by the executives. At lower (tactical) levels of the enterprise, or in some entities of the enterprise, exceptions can be tolerated (or different thresholds defined) as long as the overall exposure does not exceed the risk appetite set at the enterprise level. Any business initiative includes a risk component, so management should have the discretion to pursue new opportunities of risk.<br><br>Enterprises in which policies are cast in stone rather than as "lines in the sand," could lack the agility and innovation to exploit new business opportunities. Conversely, there are situations in which policies are based on specific legal, regulatory or industry requirements in which it is appropriate to have no risk tolerance for failure to comply. |
| Exceptions to risk tolerance standards must be reviewed and approved. | Risk tolerance is defined at the enterprise level by the board and clearly communicated to all stakeholders. A process should be in place to review and approve any exceptions to such standards. |

## Guidelines for Risk Appetite and Risk Tolerance *(cont.)*

| Guidelines for Risk Appetite and Risk Tolerance *(cont.)* | |
|---|---|
| **Guideline** | **Description** |
| Risk appetite and tolerance change over time. | Risk appetite and tolerance change due to:<br>• New technology<br>• New organizational structures<br>• New market conditions<br>• New business strategy<br>• Many other factors<br><br>Such factors require an enterprise to reassess its risk portfolio at regular intervals and also require the enterprise to reconfirm its risk appetite at regular intervals, triggering risk policy reviews.<br><br>In this respect, an enterprise also needs to understand that the better risk management it has in place, the more risk can be taken in pursuit of return. |
| Cost of risk mitigation options can affect risk tolerance. | There may be circumstances in which the cost/business impact of risk mitigation options exceeds an enterprise's capabilities/resources, thus forcing higher tolerance for one or more risk conditions.<br><br>**Example:** If a regulation states that sensitive data at rest must be encrypted, yet there is no feasible encryption solution or the cost of implementing a solution would have a large negative impact, the enterprise may choose to accept the risk associated with regulatory noncompliance, which is a risk trade-off. |

## 4.4 Risk Culture

### Definition of Risk Culture

Risk culture is the shared values and beliefs that govern the attitudes and behaviors toward risk taking, care and integrity, and determines how openly risk and losses are reported and discussed.

### Relevance

Risk management efforts are directly and indirectly affected by the risk culture.
A risk-aware culture:
- Characteristically offers a setting in which components of risk are discussed openly and acceptable levels of risk are understood and maintained

Risk awareness also implies that all levels within an enterprise are aware of why a response is needed and how to respond to adverse IT events.

### Responsibility for Risk Culture

In a risk-aware culture, the tone is set at the top with board and business executives who:
- Set direction.
- Communicate risk-aware decision making.
- Reward effective risk management behaviors.

The risk governance function is the body that most significantly affects the risk culture of an enterprise since organizational beliefs are created and shaped at the leadership level.

## Exhibit 1.2: Elements of a Risk Culture

Risk culture is a concept that is not easy to describe. **Exhibit 1.2** and the following table depict and describe the series of behaviors that are elements of a risk culture.

**Exhibit 1.2: Elements of a Risk Culture**

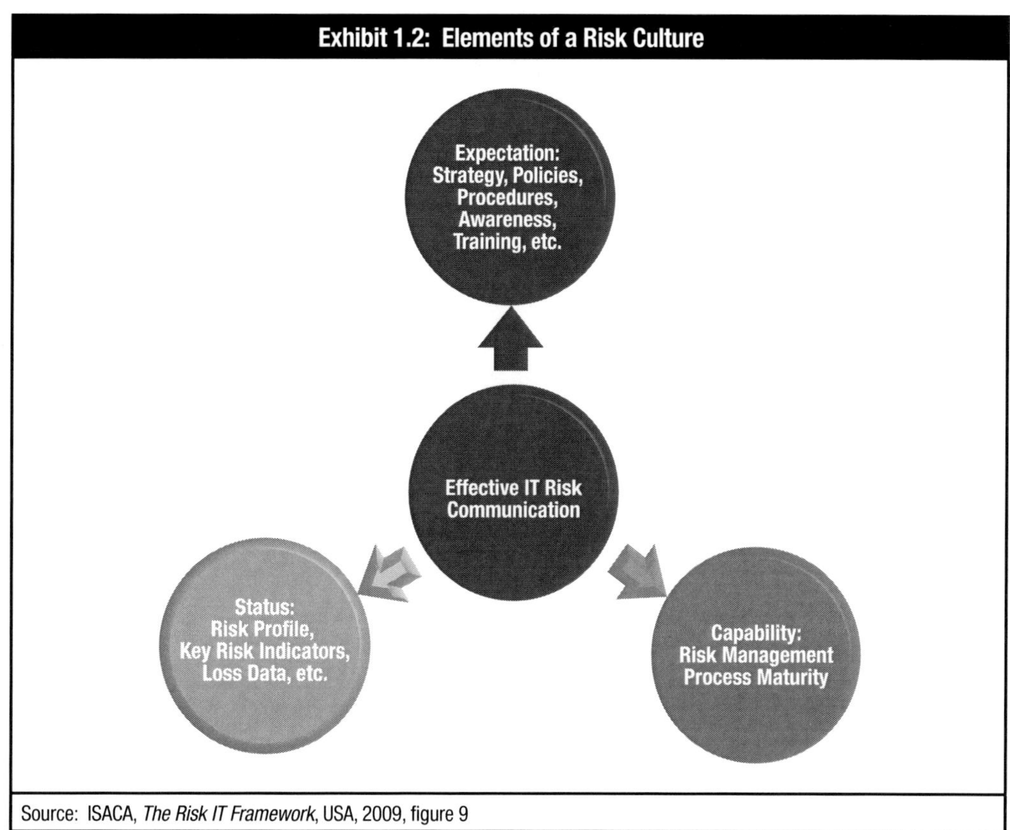


Source: ISACA, *The Risk IT Framework*, USA, 2009, figure 9

| Elements of a Risk Culture | |
|---|---|
| Behavior toward taking risk | How much risk does the enterprise feel it can absorb, and what specific risk is it willing to take? |
| Behavior toward policy compliance | To what extent will people embrace and/or comply with policy? |
| Behavior toward negative outcomes | How does the enterprise deal with negative outcomes, i.e., loss events or missed opportunities? Will it learn from them and try to adjust, or will blame be assigned without treating the root cause? |
| **Symptoms of an Inadequate or Problematic Risk Culture** | |
| Misalignment between real risk appetite and translation into policies | Management's real position toward risk can be reasonably aggressive and risk taking, whereas the policies that are created reflect a much stricter attitude. |

## Elements of a Risk Culture *(cont.)*

| Symptoms of an Inadequate or Problematic Risk Culture *(cont.)* | |
|---|---|
| Existence of a "blame culture" | This type of culture should, by all means, be avoided; it is the most effective inhibitor of relevant and efficient communication.<br><br>In a blame culture, business units tend to point the finger at IT when projects are not delivered on time or do not meet expectations. In doing so, they fail to realize how the business unit's involvement up front affects project success.<br><br>In extreme cases, the business unit may assign blame for a failure to meet the expectations that the unit never clearly communicated. The "blame game" only detracts from effective communication across units, further fueling delays. Executive leadership must identify and quickly control a blame culture if collaboration is to be fostered throughout the enterprise. |
| Benefits of a Risk-aware Culture | A risk-aware culture offers a setting in which components of risk are discussed openly and acceptable levels of risk are understood and maintained.'<br>Risk awareness also implies that all levels within an enterprise understand why a response is needed and how to respond to adverse IT events. |

# D. Risk Management Frameworks, Standards and Practices

## 1. Key Terms and Principles

### Definition of Framework

A framework is a generally accepted, business-process-oriented structure that establishes a common language and enables repeatable business processes.

**Note:** This term may be defined differently in different disciplines. This definition suits the purposes of this manual.

### Definition of Standard

A standard establishes mandatory rules, specifications and metrics used to measure compliance against quality, value, etc.

Standards are usually intended for compliance purposes and to provide assurance to others who interact with a process or outputs of a process (e.g., food and drug quality).

Standards are intended to be implemented in a rigid way and to minimize the number of deviations based on a cost-benefit analysis. Deviations from the standard should only be granted on an "exception" basis and should follow a defined approval process.

### Definition of Leading Practice

A practice is a frequent or usual action performed as an application of knowledge

A **leading practice** is defined as an action that optimally applies knowledge in a particular area.

Practices are issued by a "recognized authority" that is appropriate to the subject matter. Issuing bodies may include professional associations and academic institutions or commercial entities such as software vendors. They are generally based on a combination of research, expert insight and peer review.

**Note:** Practices usually are derived from and supplement/support standards and frameworks and are the least formal of the three.

### Relevance of Risk Management Frameworks, Standards and Practices

Frameworks, standards and practices matter to the risk practitioner because they:
- Provide a systematic view of "things to watch" that could result in harm to customers or an enterprise
- Act as a guide to focus efforts of diverse teams
- Save time and costs, such as training costs, operational costs and performance improvement costs
- Help achieve business objectives more quickly and easily
- Provide credibility to engage functional (e.g., chief financial officer [CFO]) and C-suite leadership

## 2. Examples of Frameworks Related to Risk Management and IS Control

### Examples of Risk Management Frameworks

The following table provides examples of frameworks related to risk management.

| Issuing Body | Publication |
|---|---|
| ISACA | *The Risk IT Framework* |
| ISACA | *Enterprise Value: Governance of IT Investments, The Val IT Framework 2.0* |
| ISACA | COBIT® 4.1<br><br>**Note:** The COBIT 4.1 framework is available at no charge from ISACA and can be downloaded at *www.isaca.org/cobit.* The new COBIT 5 framework will be available in 2012. |
| Committee of Sponsoring Organizations of the Treadway Commission (COSO) | *Enterprise Risk Management—Integrated Framework* |
| US National Institute of Standards and Technology (NIST) | Risk Management Framework (RMF) |

**Reminder:** Frameworks can be applied flexibly within an enterprise.

## 3. Examples of Standards Related to Risk Management and IS Control

### Examples of Risk Management Standards

Standards related to risk management include, but are not limited to, those in the following table.

| Issuing Body | Publication |
|---|---|
| ISACA | IT Audit and Assurance Standards |
| International Organization for Standardization (ISO) | ISO 31000:2009<br><br>**Note:** Unlike other "standards," this was not intended to be used for certification. |
| ISO/International Electrotechnical Commission (IEC) | ISO/IEC 2700x (for information security management systems [ISMSs]) |
| British Standards Institution (BSI) | BS 25999-x (for business continuity)<br><br>BS 25999 comprises two parts:<br>• Part 1, the Code of Practice, provides business continuity management (BCM) best practice recommendations. Please note that this is a guidance document only.<br>• Part 2, the Specification, provides the requirements for a BCM system (BCMS) based on BCM best practice. This is the part of the standard that can be used to demonstrate compliance via an auditing and certification process. |
| Payment Card Industry (PCI) Security Standards Council | PCI Data Security Standard (PCI DSS) |

**Reminder:** Standards—including corporate standards, which are not addressed here—ideally define measurable objectives to enable compliance assessments. Standards are intended to be implemented in a rigid way with variations only as allowed in the standard.

## 4. Examples of Leading Practices Related to Risk Management and IS Control

### Examples of Risk Management or Control Leading Practices

The following table provides examples leading practices related to risk management or control.

| Issuing Body | Publication |
|---|---|
| ISACA | *The Risk IT Practitioner Guide* |
| ISO/IEC | ISO/IEC 2700x (for Information Security Management Systems) |
| NIST | NIST Special Publication (SP) 800-37, Revision 1, Guide for Applying the Risk Management Framework to Federal Information Systems |
| Carnegie Mellon University (CMU) Software Engineering Institute (SEI) | Operationally Critical Threat, Asset, and Vulnerability Evaluation℠ (OCTAVE®) |
| Spanish Ministry for Public Administrations | Methodology for Information Systems Risk Analysis and Management (MAGERIT version 2) |

# E. The Risk Identification, Assessment and Evaluation Process

**Introduction**

Risk identification, evaluation, and evaluation process is the first step in the risk management process. It is concerned with the determination of risk levels and providing guidance for the later steps of risk response, monitoring and maintenance.

**Risk Identification**

Risk identification is the process of determining and documenting the risk that an enterprise faces. The identification of risk is based on the recognition of threats, vulnerabilities, assets and controls in the enterprise's operational environment.

**Risk Assessment**

A process used to identify and evaluate risk and its potential effects

**Scope Note:** Risk assessment includes assessing the critical functions necessary for an enterprise to continue business operations, defining the controls in place to reduce exposure and evaluating the cost for such controls. Risk analysis often involves an evaluation of the probabilities of a particular event.

**Risk Evaluation**

The process of comparing the estimated risk against given risk criteria to determine the significance of the risk [ISO/IEC Guide 73:2002]

**Inputs From Other Domains**

Risk management is a never-ending process. Therefore, effective risk identification, assessment and evaluation includes the review of the enterprise's risk response decisions, risk monitoring results, the existing information systems control architecture and the results of information systems control monitoring efforts.

*Domain 2—Risk Response* provides risk management action plans that affect the enterprise's overall risk profile.
*Domain 3—Risk Monitoring* provides key risk indicators (KRIs) that enable tracking of specific risk over time.
*Domain 4—Information Systems Control Design and Implementation* provides the enterprise's current information systems control inventory.
*Domain 5— Information Systems Control Monitoring and Maintenance* provides timely information on the actual design and operating effectiveness of existing controls.

**Process Objective**

The purpose of the Risk Identification, Assessment and Evaluation process is to document the risk to the enterprise, prioritize the various risk factors and provide recommendations on what risk responses should be implemented.

The risk practitioner needs to ensure that all risk to the enterprise is considered, documented and evaluated. This ensures that the enterprise is aware of all relevant risk and can make risk-aware business decisions.

If the risk practitioner fails to consider both the positive and negative angles of risk during assessment, the credibility of the risk practitioner and the value of the risk identification, assessment and evaluation will be incomplete.

**Outputs to Other Domains**

The primary output of the risk identification, assessment and evaluation process includes a comprehensive, prioritized inventory of relevant risk—often in a form of a risk register—as well as risk response recommendations. These are core inputs for *Domain 2—Risk Response* and *Domain 4—Information Systems Control Design and Implementation.*

## 1. Key Terms and Principles

**Definition of Frequency**

A measure of the rate by which events occur over a certain period of time

**Definition of Magnitude**

A measure of the potential severity of loss or the potential gain from a realized IT event/scenario

**Definition of Risk Aggregation**

The process of integrating risk assessments at a corporate level to obtain a complete view of the overall risk for the enterprise

**Definition of Risk Analysis**

A process by which frequency and magnitude of IT risk scenarios are estimated

**Definition of IT Risk**

The business risk associated with the use, ownership, operation, involvement, influence and adoption of IT within an enterprise

## 2. Introduction to Risk Identification, Assessment and Evaluation

**Risk Identification, Assessment and Evaluation Activities**

Effective risk identification, assessment and evaluation involves:
- Collecting data on:
  - The enterprise's operating environment (both internal and external)
  - Risk events
- Identifying risk factors (both internal and external)
- Analyzing and estimating risk (through the use of techniques such as risk scenarios)
- Identifying business process resilience level, including the related IT services and supporting assets

**Reference:** For more information on:
- Risk scenarios, see section F in this chapter
- Risk factors, see section in G in this chapter
- IT risk identification and assessment, see section I in this chapter

**Exhibit 1.3: IT Risk in the Risk Hierarchy**

IT risk can be categorized in different ways, as depicted in **exhibit 1.3**.

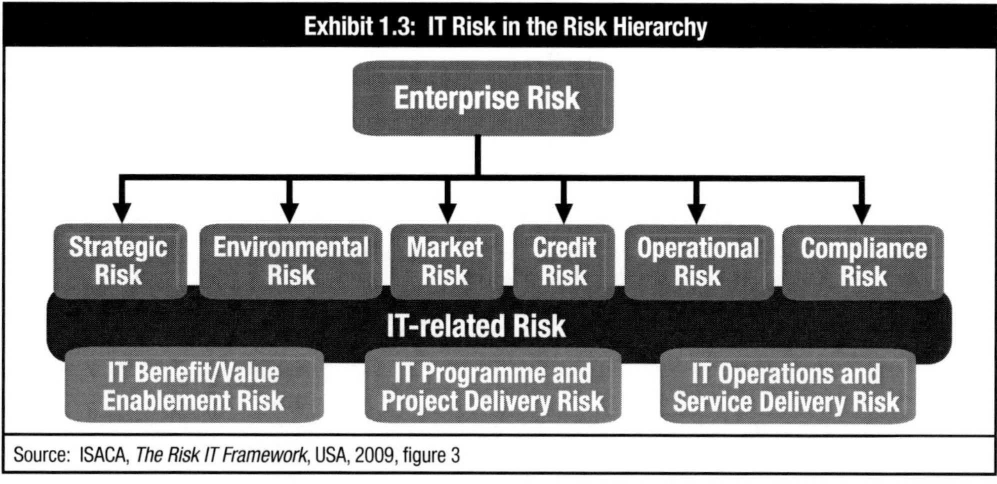

Source: ISACA, *The Risk IT Framework*, USA, 2009, figure 3

**IT Risk Categories**

IT risk can be described as follows:

- **IT benefit/value enablement risk**—Associated with (missed) opportunities to use technology to improve efficiency or effectiveness of business processes or as an enabler for new business initiatives
- **IT program and project delivery risk**—Associated with the contribution of IT to new or improved business solutions, usually in the form of projects and programs
- **IT operations and service delivery risk**—Associated with the performance of IT systems and services, which can bring destruction or reduction of value to the enterprise

**Exhibit 1.4: IT Risk Categories**

**Exhibit 1.4** shows that there is an equivalent upside for all risk; for example, successful project delivery brings new business functionality.

**Exhibit 1.4: IT Risk Categories**

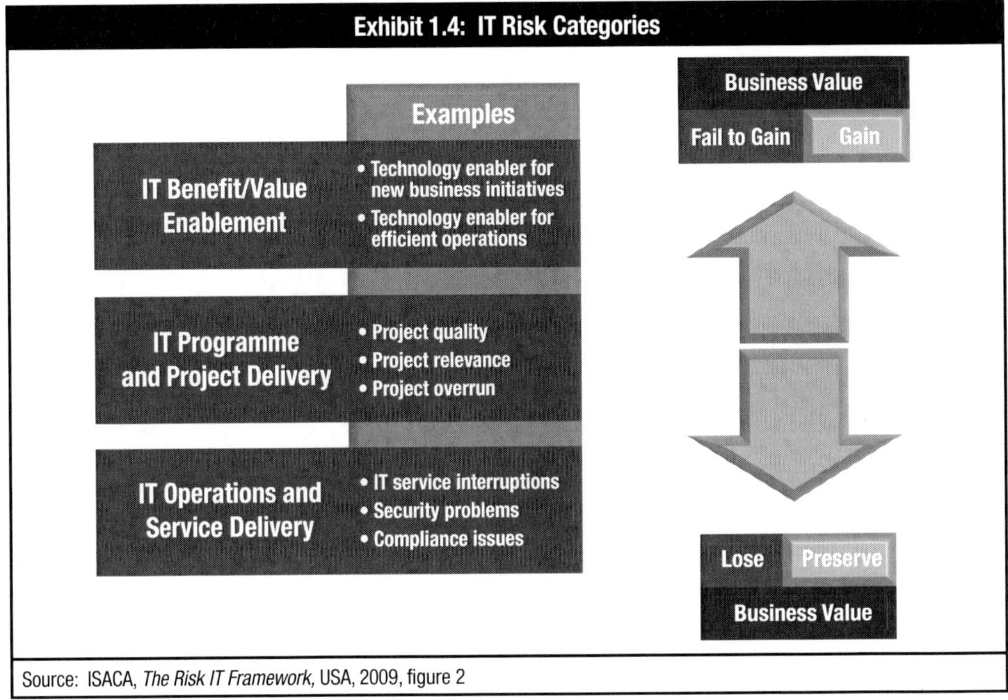


Source: ISACA, *The Risk IT Framework*, USA, 2009, figure 2

## 3. The Risk Identification, Assessment and Evaluation Process

**Exhibit 1.5: Domain 1 High-level Process Phases**

**Exhibit 1.5** depicts the phases of the risk identification, assessment and evaluation process.

**Exhibit 1.5: Domain 1 High-level Process Phases**

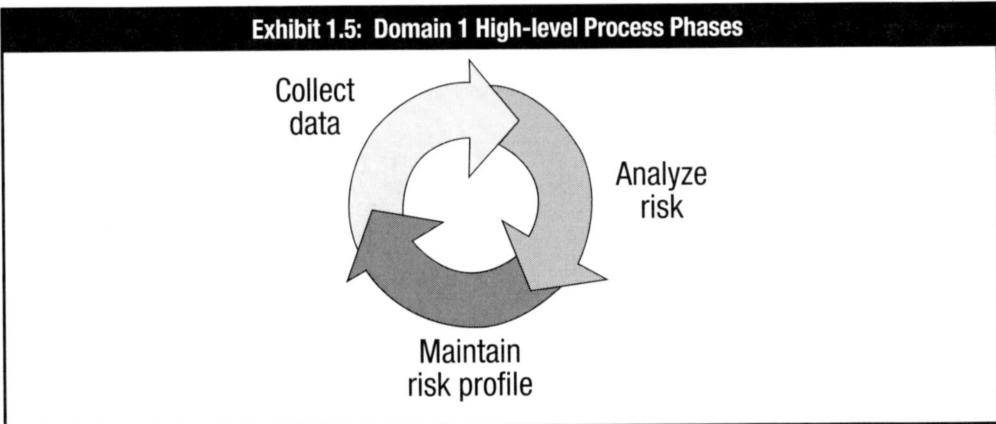

## High-level Phases

The following table describes the phases of the risk identification, assessment and evaluation process.

**Note:** For more information on these phases, see ISACA, The Risk IT Framework, USA, 2009.

| Risk Identification, Assessment and Evaluation High-level Phases | |
|---|---|
| **Phase** | **Description** |
| 1. Collect data. | Requires collecting data on the business environment, types of events, risk categories, etc., to identify relevant data to enable effective risk identification, analysis and reporting |
| 2. Analyze risk. | Requires analyzing risk to develop information to support risk decisions that take into account the business relevance of risk factors. |
| 3. Maintain a risk profile. | Requires maintaining an up-to-date and complete inventory (in a risk register or similar tool) of known threats and their attributes (e.g., expected likelihood, potential impact, disposition), IT, resources, capabilities and controls as understood in the context of business products, services and processes to effectively monitor risk over time |

## Collecting Data

Risk assessment will only be effective if the correct data has been gathered prior to conducting the risk analysis.

Some of the methods used to collect risk data include:
- Interviews
- Questionnaires and surveys
- Facilitated workshops
- Observation
- Testing

## Describing the Business Impact of IT Risk

As most enterprises are fully dependent on working information systems to support the business processes, the coverage of IT-related business risk (IT risk) is a main task of the risk practitioner.

Meaningful IT risk assessments and risk-based decisions require IT risk to be expressed in clear, business-relevant terms. Effective risk management requires mutual understanding between IT and the business over which risk needs to be managed, risk mitigation priorities, and supporting rationales.

IT risk is a risk to the business—specifically, the business risk associated with the use, ownership, operation, involvement, influence and adoption of IT within an enterprise. It consists of IT-related events and conditions that could potentially impact the business. It can occur with both uncertain likelihood and magnitude of impact, and it creates challenges in meeting operational and strategic goals and objectives.

# F. Risk Scenarios

## Introduction

One of the challenges for risk management is to identify relevant risk. One of the techniques to overcome this challenge is the development and use of risk scenarios. It is a core approach to bring realism, insight, organizational engagement, improved analysis and structure to the complex matter of enterprise risk.

Once these scenarios are developed, they are used during the risk evaluation, in which likelihood of the risk and its business impacts are estimated.

## 1. Introduction to Risk Scenarios

### Definition of Risk Scenario

A risk scenario is a description of an event that can lead to a business impact, when, and if, it should occur.

### Purpose of Risk Scenario Analysis

Risk scenario analysis is a technique used to:
- Describe risk in a more concrete and tangible manner
- Allow for proper risk assessment and analysis

### Exhibit 1.6: Risk Scenario Development

**Exhibit 1.6** describes both the top-down and bottom-up scenario development approaches and the different categories of risk factors.

**Exhibit 1.6: Risk Scenario Development**

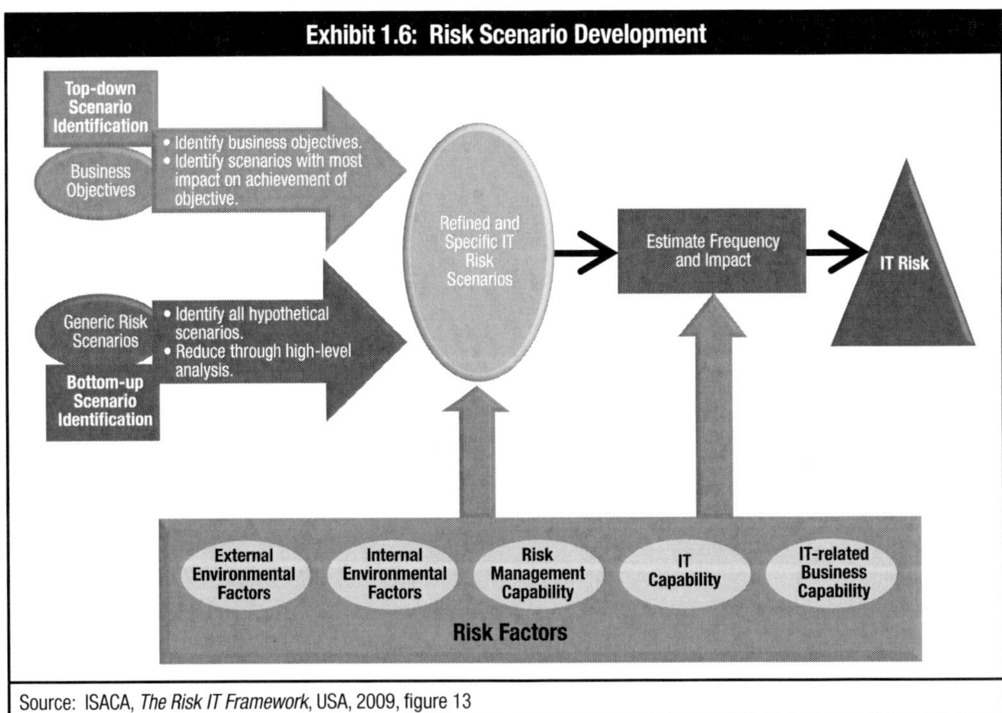


Source: ISACA, *The Risk IT Framework*, USA, 2009, figure 13

## Approaches to Risk Scenario Development

Risk scenarios can be derived via two different mechanisms:

- **Top-down approach**—From the overall business objectives, an analysis of the most relevant and probable IT risk scenarios impacting the business objectives is performed. If the impact criteria are well aligned with the real value drivers of the enterprise, relevant risk scenarios will be developed.
- **Bottom-up approach**—A list of generic scenarios is used to define a set of more concrete and customized scenarios, which are then applied to the individual enterprise situation.

**Note:** The approaches are complementary and should be used together. Risk scenarios must be realistic, but thorough. Risk scenarios should be relevant and linked to realistic business risk scenarios. On the other hand, using a wide-ranging set of generic risk scenarios helps to ensure that no risk is overlooked and provides a more comprehensive and complete view of risk. A complex enterprise may consider several hundred different risk scenarios in order to ensure that all significant risk has been considered. An enterprise should also use the lessons learned from previous incidents in risk scenario development.

## Risk Scenario Development

Business objectives drive the top-down risk scenario development approach; thus, the approach is unique to each enterprise. The approach is most beneficial in ensuring that the risk scenarios remain relevant and linked to real business risk.

In practice, using a set of generic risk scenarios is suggested since it helps ensure that no risks are overlooked and provides a comprehensive and complete view of IT risk.

The following table describes the bottom-up risk scenario development process in more detail.

| **Bottom-Up Risk Scenario Development Process Steps** | |
|---|---|
| **Step** | **Description** |
| 1 | Using a list of generic risk scenarios, define a (manageable) set of concrete risk scenarios for the enterprise.<br><br>In determining a "manageable" set of scenarios, a business may begin by considering:<br>• Commonly occurring scenarios in its industry or product area<br>• Scenarios representing threat sources that are increasing in number or severity<br>• Scenarios that involve legal and regulatory requirements applicable to the business<br><br>**Note:** Some less common situations should also be included in the scenarios. |
| 2 | Perform a validation against the business objectives of the entity. Do the selected risk scenarios address potential impacts on achievement of business objectives of the entity, in support of the overall enterprise's business objectives? |
| 3 | Refine the selected scenarios based on this validation, and detail them to a level in line with the criticality of the entity. |
| 4 | Reduce the number of scenarios to a manageable set. "Manageable" does not signify a fixed number, but should be in line with the overall importance (size) and criticality of the unit<br><br>**Note:** There is no general rule, but if scenarios are reasonably and realistically scoped, the enterprise should expect to develop at least a few dozen scenarios. |
| 5 | Keep all risk factors in a register so that they can be reevaluated in the next iteration and included for detailed analysis if they have become relevant at that time. |
| 6 | Include in the scenarios an unspecified event—how to address an incident not covered by other scenarios. |

## Risk Scenario Development *(cont.)*

**Note:** Once the set of risk scenarios is defined, it can be used for risk analysis. In risk analysis, likelihood and impact of the scenarios are assessed. Important components of this assessment are the risk factors.

**Reference:** For more information on risk factors, see section G in this chapter.

## Exhibit 1.7: Risk Scenario Components

**Exhibit 1.7** depicts the components needed to ensure that risk scenarios are complete and usable for risk analysis purposes.

**Exhibit 1.7: Risk Scenario Components**

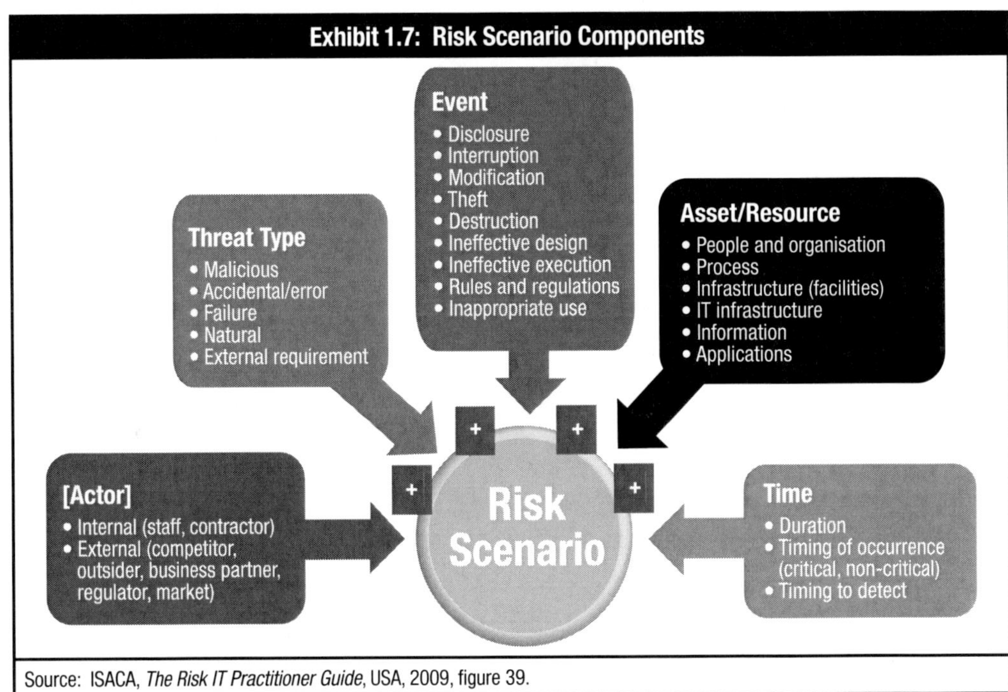


Source: ISACA, *The Risk IT Practitioner Guide*, USA, 2009, figure 39.

## Risk Scenario Component Descriptions

The following table describes the different risk scenario components in detail.

| Risk Scenario Components | |
|---|---|
| **Component** | **Description** |
| Actor | Actors generate the threat and can be internal or external, human or nonhuman:<br>• Internal actors are within the enterprise, e.g., staff, contractors.<br>• External actors include outsiders, competitors, regulators and the market.<br><br>**Note:** Not every type of threat requires an actor, e.g., failures or natural causes. |
| Threat type | The threat type is the nature of the event. Is it malicious? If not, is it accidental or is it a failure of a well-defined process? Is it a natural event (*force majeure*)? |
| Event | A scenario always has to contain an event. Is it disclosure (of confidential information), interruption (of a system or project), modification, theft, destruction, etc.? |

### Risk Scenario Component Descriptions *(cont.)*

<table>
<tr><th colspan="4">Risk Scenario Components (cont.)</th></tr>
<tr><th>Component</th><th colspan="3">Description</th></tr>
<tr><td rowspan="5">Asset</td><td colspan="3">An asset is any object of value to the enterprise that can be either tangible or intangible.</td></tr>
<tr><td></td><th>Tangible Assets</th><th>Intangible Assets</th></tr>
<tr><td>Attribute</td><td>• Has physical attributes and can be detected with the senses</td><td>• Has no physical attributes and cannot be detected with the senses</td></tr>
<tr><td>Examples</td><td>• People<br>• Infrastructure<br>• Finances</td><td>• Information<br>• Reputation<br>• Customer trust</td></tr>
<tr><td>Relevance</td><td>• Theft or misuse more easily detected than with intangible assets</td><td>• May prove difficult to quantify<br>• Focus on the tangible assets that contain, protect, affect or represent the intangible asset</td></tr>
<tr><td>Timing dimension</td><td colspan="3">The timing dimension is the relevance of the scenario to the:<br>• Time to detect, respond to or recover from the event<br>• Timing: Does the event occur at a critical moment?<br>• Duration of the event (extended outage of a service or data center)<br>• Time lag between the event and the consequence: Is there an immediate consequence, (e.g., network failure, immediate downtime) or a delayed consequence (e.g., wrong IT architecture with accumulated high costs over a time span of several years)?</td></tr>
</table>

**Note:** The risk scenario structure differentiates between:
- Loss events (events generating the negative impact)
- Vulnerabilities or vulnerability events (events contributing to the likelihood or impact of loss events occurring)
- Threat events (circumstances or events that can trigger loss events)

### Risk Scenarios: Future Circumstances

Scenario analysis should cover threats and vulnerabilities of current and possible future circumstances. Future risk could be related to emerging technologies, new regulations, demographic changes and new business initiatives.

## 2. Risk Scenario Development

### Introduction

Building a scenario requires determination of the value of an asset or a business process at risk and the potential threats and vulnerabilities that could lead to a loss event. The risk scenario should then be assessed for relevance and realism and, if found to be relevant, entered into the risk register.

**Example**

Based on a risk assessment, an enterprise has 20 key applications supported by three distinctly different technology platforms. The number of theoretically possible scenarios already approaches 100,000, which is not feasible to maintain. The number of scenarios to be developed and analyzed should be kept to a much smaller number to remain manageable because every possible combination cannot be retained.

**Changes to Risk Factors and the Enterprise Over Time**

Because risk factors and the enterprise change over time, scenarios will also change. This requires continuous risk assessment (risk monitoring).

Risk assessment should be performed at least on an annual basis, or when important internal or external changes occur.

**Creation of a Risk Register**

A risk register (or risk log) is a listing of all risks identified for the enterprise. The risk register records:
- All known risk
- Priorities of risk
- Likelihood of risk
- Potential risk impact
- Status of the risk mitigation plans
- Contingency plans
- Ownership of risk

## Exhibit 1.8: Risk Register Entry

The following table in **Exhibit 1.8** is an excerpt of a single risk register entry.

**Exhibit 1.8: Risk Register Entry**

| **Part I—Summary Data** | | | | | | |
|---|---|---|---|---|---|---|
| **Risk statement** | | | | | | |
| **Risk owner** | | | | | | |
| **Date of last risk assessment** | | | | | | |
| **Due date for update of risk assessment** | | | | | | |
| **Risk category** | ☐ Strategic (IT benefit/value enablement) | | ☐ Project Delivery (IT programme and project delivery) | | ☐ Operational (IT operations and service delivery) | |
| **Risk classification (copied from risk analysis results)** | ☐ Low | | ☐ Medium | | ☐ High | ☐ Very high |
| **Risk response** | ☐ Accept | | ☐ Transfer | | ☐ Mitigate | ☐ Avoid |
| **Part II—Risk Description** | | | | | | |
| **Title** | | | | | | |
| **High-level scenario (from list of sample high-level scenarios)** | | | | | | |
| **Detailed scenario description—scenario components** | Actor | | | | | |
| | Threat type | | | | | |
| | Event | | | | | |
| | Asset/resource | | | | | |
| | Timing | | | | | |
| **Other scenario information** | | | | | | |
| **Part III—Risk Analysis Results** | | | | | | |
| **Frequency of scenario (number of times per year)** | 0 | 1 | 2 | 3 | 4 | 5 |
| | N≤0.01 ☐ | 0.01<N≤0.1 ☐ | 0.1<N≤1 ☐ | 1<N≤10 ☐ | 10<N≤100 ☐ | 100<N ☐ |
| **Comments on frequency** | | | | | | |
| **Impact of scenario on business** | 0 | 1 | 2 | 3 | 4 | 5 |
| **1. Productivity** | **Revenue loss over one year** | | | | | |
| **Impact rating** | I≤0.1% ☐ | 0.1%<I≤1% ☐ | 1%<I≤3% ☐ | 3%<I≤5% ☐ | 5%<I≤10% ☐ | 10%<I ☐ |
| **Detailed description of impact** | | | | | | |
| **Part III—Risk Analysis Results *(cont.)*** | | | | | | |
| **2. Cost of response** | **Expenses associated with managing the loss event (US $)** | | | | | |
| **Impact rating** | I≤$10k ☐ | $10K<I≤$100K ☐ | $100K<I≤$1M ☐ | $1M<I≤$10M ☐ | $10M<I≤$100M ☐ | $100M<I ☐ |
| **Detailed description of impact** | | | | | | |
| **3. Competitive advantage** | **Drop in customer satisfaction ratings** | | | | | |
| **Impact rating** | I≤0.5 ☐ | .05≤I≤1 ☐ | 1<I≤1.5 ☐ | 1.5<I≤2 ☐ | 2<I≤2.5 ☐ | 2.5<I ☐ |
| **Detailed description of impact** | | | | | | |
| **4. Legal** | **Regulatory compliance—Fines (US $)** | | | | | |
| **Impact rating** | None ☐ | <$1M ☐ | <$10M ☐ | <$100M ☐ | <$1B ☐ | >$1B ☐ |
| **Detailed description of impact** | | | | | | |
| **Overall Impact rating (average of four impact ratings)** | | | | | | |
| **Overall rating of risk, obtained by combining frequency and impact ratings on risk map** | ☐ Low | | ☐ Medium | | ☐ High | ☐ Very high |
| **Part IV—Risk Response** | | | | | | |
| **Risk response for this risk** | ☐ Accept | | ☐ Transfer | | ☐ Mitigate | ☐ Avoid |
| **Justification** | | | | | | |
| **Detailed description of response (not in case of 'accept')** | **Response Action** | | | | **Completed** | **Action Plan** |
| | 1. | | | | ☐ | ☐ |
| | 2. | | | | ☐ | ☐ |
| | 3. | | | | ☐ | ☐ |
| | 4. | | | | ☐ | ☐ |
| | 5. | | | | ☐ | ☐ |
| | 6. | | | | ☐ | ☐ |
| **Overall status of risk action plan** | | | | | | |
| **Major issues with risk action plan** | | | | | | |
| **Overall status of completed responses** | | | | | | |
| **Major issues with completed responses** | | | | | | |
| **Part V—Risk Indicators** | | | | | | |
| **Key risk indicators for this risk** | 1.<br>2.<br>3.<br>4.<br>5.<br>6. | | | | | |

Source: ISACA, *The Risk IT Practitioner Guide*, USA, 2009, figure 36, page 48

## Managing the Number of Risk Scenarios

One technique of keeping the number of scenarios manageable is to develop:
- A set of generic scenarios throughout the enterprise
- More detailed scenarios in areas in which risk levels are higher

The assumptions made when grouping or generalizing scenarios should be well understood by all and adequately documented.

There should be a sufficient number of risk scenario scales reflecting the complexity of the enterprise and the extent of exposures to which the enterprise is subject.

The enterprise must consider risk that has not yet occurred and consider developing scenarios around unlikely, obscure or nonhistorical events.

**Example:** The term "insider threat" may not adequately explain whether the scenario addresses the threat of privileged and nonprivileged users.

## Prerequisites for Developing a Manageable Set of Risk Scenarios

Developing a manageable and relevant set of risk scenarios requires:
- Organizational buy-in or support from enterprise entities and business lines, risk management, IT, finance, compliance and other parties
- Expertise and experience to not overlook relevant scenarios and not be drawn into highly unrealistic or irrelevant scenarios
- A thorough understanding of the environment
- The involvement of all stakeholders including:
  - Senior management, which has decision-making authority
  - Business management, which has the best view of business impact
  - IT, which has the understanding of the risks associated with the underlying technology
  - Risk management, which can facilitate the risk management process

## Systemic and/or Contagious Risk

Attention should be paid to so-called "systemic" and/or "contagious" risk scenarios:
- **Systemic risk**—Something that happens with an important business partner that affects a large group of enterprises within an area or industry.

  **Example:** A nationwide air traffic control system goes down for an extended period of time (six hours), which affects air traffic on a large scale and consequently all businesses relying on air traffic.
- **Contagious risk**—Events that happen at several of the enterprise's business partners within a very short time frame

  **Example:** A financial transaction clearinghouse is prepared for an emergency by having sophisticated disaster recovery measures in place. However, when a catastrophe occurs, it finds that no transactions are sent by its providers and, consequently, with no supply chain, is temporarily out of business.

## Obscure Risk Identification

The enterprise must consider risk that has not yet occurred and consider developing scenarios around unlikely, obscure or nonhistorical events.

Developing such scenarios requires two considerations: visibility and recognition.

The enterprise must:
- Be in a position that it can observe anything going wrong
- Have the capability to recognize an observed event as something that is going wrong

# *G. Risk Factors*

## 1. Introduction to Risk Factors

### Definition of Risk Factors

Risk factors are those features that influence the likelihood and/or business impact of risk scenarios.

### Relevance

The importance of risk factors lies in the influence they have on risk. They are heavy influencers of the likelihood and impact of risk scenarios and should be taken into account during every risk analysis, when likelihood and impact are assessed.

### Threats and Risk Factors

When considering risk, the risk practitioner must be aware of the threat agents that pose a threat to the assets of the enterprise. Threats may be internal or external, intentional or accidental, skilled or amateur, motivated or curious, and natural, man-made, physical, or related to equipment or utility failure.

Thorough risk analysis will consider all types of threats and the risk that those threats pose to the enterprise. Obviously, a skilled, highly-motivated agent working for another government or enterprise will be a more serious threat than a low-skilled, curious individual casually wandering through the Internet looking for targets of opportunity.

### Exhibit 1.9: Risk Factors in Detail

**Exhibit 1.9** depicts different risk factors that are discussed in more detail in the subsequent topics in this section.

**Exhibit 1.9: Risk Factors in Detail**

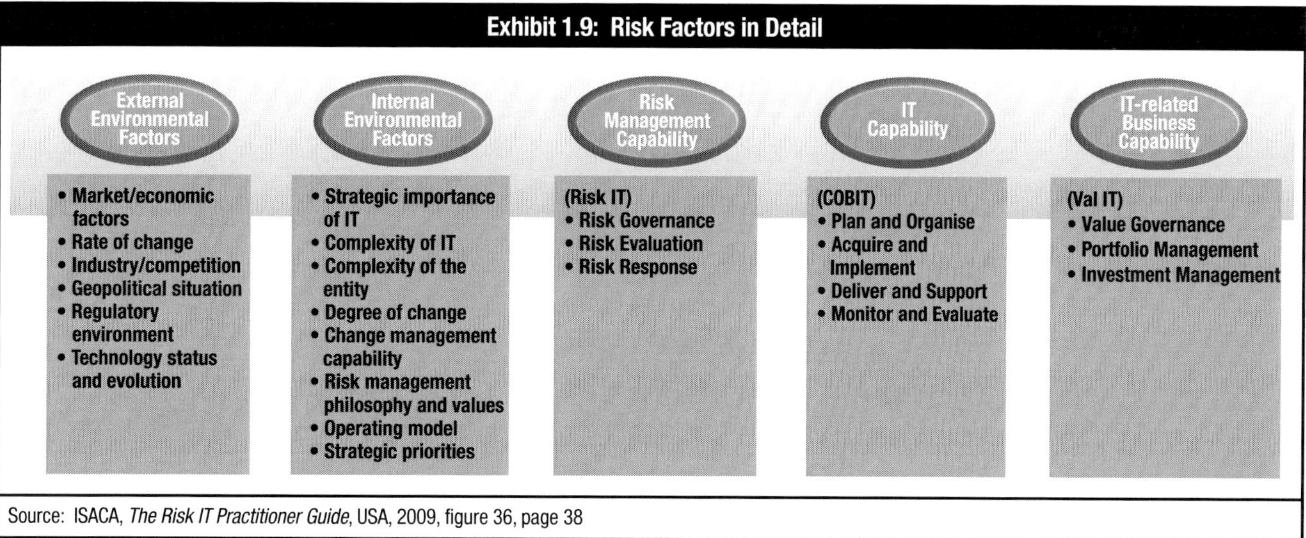


Source: ISACA, *The Risk IT Practitioner Guide*, USA, 2009, figure 36, page 38

## 2. External Risk Factors

### Introduction

Shifts in global events and the economy including financial, supply chain, regulatory and competitive market changes are key external factors.

### Importance of Understanding the External Business Environment

The external context is the external environment in which the enterprise seeks to achieve its objectives. Understanding the external context is important to ensure that external stakeholders and their objectives and concerns are considered when developing risk criteria.

The external context is based on the enterprisewide context, but with specific details of legal and regulatory requirements, stakeholder perceptions, and other aspects of risk specific to the scope of the risk management process.

The external context can include:
- The cultural, political, legal, regulatory, financial, technological, economic, natural and competitive environment, whether international, national, regional or local
- Key drivers and trends that impact the objectives of the enterprise
- Perceptions and values of external stakeholders

### External Risk Factors

External risk factors are those circumstances that can increase the likelihood or impact of an event and that are not always directly controllable by the enterprise. Because external risk lies outside the enterprise's control, the enterprise is limited in the direct actions that can be taken to handle such risk. Nonetheless, the risk can still be managed by developing strategies to prevent exposures, avoid risk and deal with an incident effectively once the risk materializes.

**Example:** Building dikes to prevent flooding, moving to an area not subject to flooding, and procuring insurance all can be used to contend with natural disasters such as floods

External risk factors are listed in the following table.

| External Risk Factors | |
|---|---|
| **Factor** | **Description** |
| Market/economy | This includes the industry sector, in which the enterprise operates, i.e., operating in the financial sector requires different IT requirements and IT capabilities than operating in a manufacturing environment.<br><br>Other economic factors also can be included, e.g., nationalization, mergers and acquisitions, and consolidations.<br><br>**Example:** Demographic trends—the aging of the population reduces demand for youth-oriented products. |
| Rate of change in the market in which the enterprise operates | Are business models changing fundamentally?<br><br>Is the product or service at the end of an important life-cycle moment? |
| Competition | This is the competitive environment in which the enterprise operates.<br><br>**Example:** Actions of competitors—the introduction of a new product that makes one of the enterprise's product lines obsolete |

## External Risk Factors *(cont.)*

| External Risk Factors *(cont.)* | |
|---|---|
| **Factor** | **Description** |
| Geopolitical situation | Is the geographic location subject to frequent natural disasters?<br><br>Does the local political and overall economic context represent an additional risk?<br><br>**Example:** Acts of nature—a sustained drought causing a dramatic drop in the output of agricultural products |
| Regulatory environment | Is the enterprise subject to new or stricter regulations?<br><br>Are there any other compliance requirements beyond regulation (e.g., industry-specific, contractual)?<br><br>**Example:** Government regulations—concern of enterprises (such as chemical companies that produce hazardous substances) that government will change environmental laws so that it becomes difficult for them to produce their products in a cost-effective manner |
| Technology innovation and evolution | Is the enterprise using state-of-the art technology, and more important, how fast are relevant technologies evolving?<br><br>Is the enterprise capable of applying new technologies at the same pace or faster than its competitors? |

**Note:** The types of risk that impact enterprises vary depending on the home country location, industry, level of globalization and many other factors. For example, financial services enterprises tend to be concerned about credit and market risk. Many enterprises are concerned with reputation and legal risk. However, one area of risk that impacts all enterprises is operational risk.

**Example:** Today, financial services enterprises, particularly banks, are addressing operational risk due to the capital adequacy accord known as Basel II. Basel II defines operational risk as the risk of losses resulting from inadequate or failed internal processes, people and systems or from external events. Although designed for banking, this definition holds true for practically all industries. The adoption of Basel II by financial services enterprises is partly dependent on the posture of the local regulatory agency (usually central bank) in mandating or promoting this risk-based standard.

## Impact of Governance Codes and Local Laws

Governance codes as well as regional and local laws can be generic or industry-specific, and they create additional demands on enterprises—normally in response to heightened expectations from society or as a result of corporate scandals that revealed a need to tighten up existing regulations. An assortment of local laws also adds to the compliance framework within which enterprises must operate. Some professions, such as law, medicine and accounting, provide various codes of conduct and specific regulations that must be adhered to by their practicing members.

The presence of diverse laws and regulations places a challenge on enterprises to either be compliant with an increasingly complex variety of conditions and restrictions or to face major sanctions and/or penalties.

## 3. Internal Environmental Risk Factors

### Introduction

The internal environment is shaped by an enterprise's history and culture and affects all components of enterprise risk management (ERM).

These things influence how strategies and objectives are established; business activities are structured; and risk is identified, assessed and acted on. They also influence the design and functioning of control activities, information and communication systems, and monitoring activities.

The internal environment of the enterprise is comprised of many elements, including:
- The enterprise's ethical values
- The competence and development of personnel
- Management's philosophy for managing risk
- How the enterprise assigns authority and responsibility

A board of directors is a critical part of the internal environment and significantly influences other internal environment elements. Although all elements are important, the extent to which each is addressed will vary with the enterprise. Nevertheless, the enterprise should have an internal environment that provides an appropriate foundation for ERM.

### Internal Risk Factors

Various internal risk factors are listed in the following table.

> **Note:** Additional internal risk factors are also discussed in this topic.

| Internal Risk Factors | |
|---|---|
| **Factor** | **Description** |
| Strategic importance of IT in the enterprise | Is IT a strategic differentiator, functional enabler or supporting function? |
| Complexity of IT | Is IT highly complex (e.g., complex architecture, recent mergers), or is it simple, standardized and streamlined? |
| Complexity of the enterprise | This includes geographic spread and value chain coverage (e.g., in a manufacturing environment).<br><br>Does the enterprise manufacture and distribute parts, and/or is IT also doing assembly activities? |
| Enterprise change | This is the degree of internal change the enterprise is experiencing, such as reorganizations and new leadership. |
| Change management capability | To what extent is the enterprise capable of organizational change? |
| Risk management philosophy | This includes the risk management philosophy of the enterprise (risk averse or risk taking) and, linked with that, the values of the enterprise. |
| Operating model | This is the degree to which the enterprise operates independently or is connected to its clients/suppliers—the degree of centralization/decentralization. |
| Enterprise priorities | These are the strategic priorities of the enterprise. |
| Technology innovation and evolution | Is the enterprise keeping abreast with technology evolution, and is it capable of supporting evolving technologies over time? |

## Importance of Integrity and Ethics of Enterprise Management

An enterprise's strategy and objectives, and the way they are implemented, are based on preferences, value judgments and management styles. Management's integrity and commitment to ethical values influence these preferences and judgments, which are translated into standards of behavior.

Because an enterprise's good reputation is so valuable, the standards of behavior must go beyond mere compliance with the law. Management values must balance the concerns of the enterprise, employees, suppliers, customers, competitors and the public. Managers of well-run enterprises increasingly have accepted the view that good ethics pays, and that ethical behavior is good business.

Management integrity is a prerequisite for ethical behavior in all aspects of an enterprise's activities. The effectiveness of ERM cannot rise above the integrity and ethical values of the people who create, administer and monitor enterprise activities. An enterprise that operates with a high degree of ethics may have a lower incidence of risk related to fraud or misappropriation. Integrity and ethical values are essential elements of an enterprise's internal environment and affect the design, administration and monitoring of other ERM components.

## Role of Enterprise Management in Determining Enterprise Culture

Top management—starting with the chief executive officer (CEO)—plays a key role in determining the corporate culture (tone at the top). As the dominant personality in an enterprise, the CEO often sets the ethical tone. Certain organizational factors also can influence the likelihood of fraudulent and questionable financial reporting practices. Those same factors are also likely to influence ethical behavior. Individuals may engage in dishonest, illegal or unethical acts simply because the enterprise gives them strong incentives or temptations to do so. Undue emphasis on results, particularly in the short term, can foster an inappropriate internal environment. Focusing solely on short-term results can hurt the enterprise, even in the short term. Concentration on the bottom line—sales or profit, at any cost—often evokes unsought actions and reactions. High-pressure sales tactics, ruthlessness in negotiations or implicit offers of kickbacks, for instance, may evoke reactions that can have immediate (and lasting) effects.

Ethical values must be:
- Communicated
- Accompanied by explicit guidance regarding what is right and wrong
- Established in formal codes of corporate conduct, which address a variety of behavioral issues such as:
  - Integrity and ethics
  - Conflicts of interest
  - Illegal or otherwise improper payments
  - Anticompetitive arrangements

## Management Determination of Competency Levels

Competence reflects the knowledge and skills needed to perform assigned tasks. Management decides how much to invest in making sure that tasks are executed properly using skilled resources, equipment and defined processes.

This requires weighing the enterprise's strategy and objectives against plans for their implementation and achievement. A trade-off often exists between competence and cost. The risk of failure is higher with untrained staff, poorly maintained or old equipment, or undefined procedures.

## Board of Directors Role in the Internal Environment

An enterprise's board of directors is a critical part of the internal environment and significantly influences its elements. The board's role in risk governance through independent oversight of management, scrutiny of activities, and appropriateness of the enterprise's risk appetite strategy all play a role. Other factors include the:

- Degree to which difficult questions are raised and pursued with management regarding strategy, plans and performance
- Interaction between the board or audit committee and internal and external auditors

An active and involved board of directors, board of trustees or comparable body should possess an appropriate degree of management, and technical and other expertise, coupled with the mindset necessary to perform its oversight responsibilities. This is critical to an effective ERM environment. Because the board must be prepared to question and scrutinize management's activities, present alternative views, and act in the face of wrongdoing, the board must also include outside directors. There must be a sufficient number of independent outside directors—not only to provide sound advice, counsel and direction, but also to serve as a necessary check and balance on management.

## Impact of the Enterprise Organizational Structure

An enterprise's organizational structure provides the framework to plan, execute, control and monitor its activities. A relevant organizational structure includes defining key areas of authority and responsibility and establishing appropriate lines of reporting.

**Example:** An enterprise that must report on its compliance with regulations will have staff that gather the necessary data and other staff that compile the reports and then submit them to senior management for approval. The reports given to senior management are often not as detailed as the reports first generated and used by local managers to evaluate the performance of their departments. For normal reporting, a scheduled reporting process is used to gather data and generate and submit the reports. In the case of a serious incident, another process may be used to escalate the issue to management and initiate recovery activities.

An enterprise develops an organizational structure suited to its needs. Some are centralized; others are decentralized. Some have direct reporting relationships, while others are more of a matrix organization. Some entities are organized by industry or product line, geographic location, business function or a particular distribution or marketing network.

Whatever the structure, an enterprise should be organized to enable effective ERM and to carry out its activities to achieve its objectives.

## Assignment of Authority and Responsibility

Assignment of authority and responsibility:

- Involves the degree to which individuals and teams are authorized (and limited by their authority) and encouraged to use initiative to address issues and solve problems
- Includes establishing reporting relationships and authorization protocols and also policies that describe:
  - Appropriate business practices
  - The knowledge and experience of key personnel
  - The resources provided for carrying out duties

Some entities push authority downward to bring decision making closer to frontline personnel—to the individuals who are closest to everyday business transactions. An enterprise may take this route to become more market-driven or quality-focused. This may involve empowerment to sell products at discount prices; negotiate long-term supply contracts, licenses or patents; or enter alliances or joint ventures. Effective delegation aligns authority with accountability to encourage individual initiative within limits.

## Impact of Delegation

A critical challenge is to delegate only to the extent required to achieve objectives. This means ensuring that:

- Decision making is based on sound practices for risk identification and assessment, including sizing risk and weighing potential losses vs. gains in determining which risk to accept and how it is to be managed.
- All personnel understand the enterprise's objectives.

It is essential that individuals know how their actions are related to one another and contribute to achievement of the objectives.

Increased delegation is sometimes the result of streamlining or flattening the organizational structure. Purposeful structural change to encourage creativity, initiative and faster response times can enhance competitiveness and customer satisfaction. Increased delegation may:

- Carry an implicit requirement for a higher level of employee competence and greater accountability
- Require effective procedures for management to monitor results so that decisions can be overruled or accepted as necessary

Along with better, market-driven decisions, delegation may increase the number of undesirable or unanticipated decisions. The internal environment is greatly influenced by the extent to which individuals recognize that they will be held accountable. This holds true all the way to the chief executive, who, with board oversight, has ultimate responsibility for all activities within an enterprise.

## Impact of Human Resources (HR) Practices

HR practices pertaining to hiring, orientation, training, evaluating, counseling, promoting, compensating and taking remedial actions send messages to employees regarding expected levels of integrity, ethical behavior and competence.

- **Hiring standards** with emphasis on educational background, prior work experience, past accomplishments and evidence of integrity and ethical behavior demonstrate an enterprise's commitment to competent and trustworthy people.
- **Transfers and promotions** driven by periodic performance appraisals demonstrate the enterprise's commitment to advancement of qualified employees.
- **Competitive compensation programs** that include bonus incentives serve to motivate and reinforce outstanding performance—although reward systems should be structured and controls should be put in place to avoid undue temptation to misrepresent reported results.
- **Performance improvement processes and disciplinary actions** send a message that underperformance and violations of expected behavior are not tolerated.
- **Mentoring and support** helps provide employees with the skills and behaviors necessary to tackle new challenges as issues and risk throughout the enterprise change and become more complex—driven in part by rapidly changing technology and increasing competition.
- **Education and training** helps personnel keep pace and deal effectively with the evolving environment. Hiring competent people and providing one-time training are not enough. The education process must be ongoing.

## 4. Risk Management Capability

### Risk Management Capability

Risk management capability is an indication of how well the enterprise is executing the core risk management processes. The better executed or more mature the processes, the more capable the risk management program. This factor is correlated with the capability of the enterprise to recognize and detect risk and adverse events and should not be neglected.

Risk management capability is a very significant element in the likelihood and impact of risk events in an enterprise because it is responsible for:
- Management's risk decisions (or lack thereof)
- The presence, absence and/or effectiveness of controls that exist within an enterprise

Risk management capability is also an important component of the overall risk profile of the enterprise.

---

## 5. IT Capability

### IT Capability

In the context of risk management, IT capabilities may be associated with the maturity level of IT processes and IT controls. Mature and well-controlled IT processes are equivalent to high IT capabilities, which can have a positive influence on reducing the:
- **Likelihood of events**—e.g., having good software development processes in place to deliver high-quality and stable software or having good security measures in place to reduce the number of security-related incidents
- **Business impact when events happen**—e.g., having good disaster recovery planning (DRP) in place when disaster strikes

The IT sourcing model is often seen as a separate risk factor. There is no doubt that the sourcing model, e.g., keeping IT in-house or outsourcing parts or complete IT departments, has an important impact on risk and how to measure it. The COBIT process model contains several processes dealing with the selection and management of sourcing models.

---

## 6. IT-related Business Capability

### IT-related Business Capability

The degree to which business management is capable of managing the direction and performance of IT is an important risk factor.

Mature IT value management processes are associated with a high capability of the business to manage IT-related affairs. The enterprise will generate more value from IT and will miss fewer opportunities ***if***:
- The business is capable of making the right IT investments.
- Correct IT partners are selected.
- Programs are well selected and managed.

> **Note:** IT-related business capability is especially a risk factor in the risk associated with IT benefit/value enablement.

# *H. Qualitative and Quantitative Risk Analysis*

### Introduction

After collecting data, risk assessment is the first process in the risk management process. Enterprises use risk assessment to determine the:
- Extent of the potential threat
- Risk associated with business processes, operations and IT systems throughout their development life cycle and use

The entire risk management process should be managed at multiple levels in the enterprise, including the operational, project and strategic levels, and should form part of the risk management practice.

IT risk should be treated in the same manner as any other business risk and should be analyzed and assessed using similar approaches. Accepted business risk assessment practices and methods should be applied consistently across different IT systems.

> **Note:** Risk management frameworks and standards such as the Committee of Sponsoring Organizations of the Treadway Commission (COSO) provide appropriate methods for risk assessment.

## 1. Key Terms and Principles

### Definition of Impact

Impact refers to the magnitude of harm that could be caused by a threat's exploitation of a vulnerability.

The level of impact is governed by the potential mission impacts and, in turn, produces a relative value for the IT assets and resources affected (e.g., the criticality and sensitivity of the IT system components and data).

### Definition of Likelihood

In nontechnical terms, "likelihood" is usually a synonym for "probability;" but in statistical usage, a clear technical distinction is made. It would be improper to switch "likelihood" and "probability" in the two following sentences:
- "If I were to flip a fair coin 100 times, what is the *probability* of it landing heads-up every time?"
- "Given that I have flipped a coin 100 times and it has landed heads-up 100 times, what is the *likelihood* that the coin is fair?"

To determine the likelihood of a future adverse event, enterprises must analyze threats to a system, potential vulnerabilities and the controls in place.

### Definition of Residual Risk

Residual risk is the remaining risk after management has implemented a risk response.

Residual risk can be used by management to determine:
- Which areas require more controls
- Whether the benefits of such controls outweigh the control costs

**US National Institute of Standards and Technology (NIST) Recommended Risk Assessment Methodology**

The following table outlines the NIST risk assessment methodology for IT-related risk at a very high level.

| NIST Risk Assessment Methodology | |
|---|---|
| **Step** | **Description** |
| 1 | System characterization |
| 2 | Threat identification |
| 3 | Vulnerability identification |
| 4 | Control analysis |
| 5 | Likelihood determination |
| 6 | Impact analysis |
| 7 | Risk determination |
| 8 | Control recommendations |
| 9 | Results documentation |

## 2. Qualitative Risk Analysis

**Overview of Qualitative Risk Analysis**

- Qualitative analysis defines risk using a scale or comparative values (i.e., defining risk factors in terms of high/medium/low or on a numeric scale from 1 to 10). It is based on judgment, intuition and experience rather than on numbers or financial values.

**Benefits of Qualitative Risk Analysis**

- The cost of qualitative analysis is generally significantly lower than the cost of quantitative analysis.
- Qualitative risk analysis usually results in a better understanding of business unit dependencies and interactions.
- Qualitative risk analysis is more consensus-based and often reflects the input of business units more accurately than quantitative analysis.
- It is often better for evaluating "soft" (intangible) risk, such as morale or reputation.

## Challenges of Qualitative Risk Analysis

When applying subjective information to analyzing risk, the following challenges must be considered:
• Subjectivity or bias in data collected
• Overemphasis on minor events
• Does not provide good data for cost-benefit analysis
• Ranking levels may not be meaningful to data providers

## Typical Qualitative Risk Analysis Methods

The following table describes typical qualitative methods used in risk assessment.

| Typical Qualitative Risk Analysis Methods | |
|---|---|
| **Method** | **Description** |
| Risk Control Self-assessment (RCSA) | RCSA:<br>• Is based on the evaluation of risk levels by line managers who are expected to be familiar with threats and incidents within their areas of responsibility<br>• Involves all levels of the enterprise in risk management and creates a risk awareness culture and increased risk ownership<br>• May be subjective and provide inaccurate results depending on the culture and openness of the enterprise<br>• Is typically a bottom-up process by business managers, but may be a top-down process by senior stakeholders, which provides a good blend—a granular view from the bottom up and an enterprise view from the top down<br><br>**Note:** It is vital for executive management to be open and active to assure RCSA participants that they will not suffer from speaking candidly. |
| Scorecards | Scorecards:<br>• Consist of generic questionnaires containing weighted, risk-based questions with multiple-choice responses<br>• Create qualitative assessments that can be:<br>– Translated into quantitative measures, such as a ranking of risk factors<br>– Used to adjust capital reserve levels<br>• Include rewards for internal control improvements<br><br>**Note:** Problems can arise due to the subjective nature of scorecards and manipulation of the process to artificially lower capital charges. |
| Key risk indicators (KRIs) | KRIs:<br>• Are used to alert the enterprise to critical changes in risk, especially early warning alerts to changes in the control environment<br>• Can be improved beyond after-the-fact loss indicators to predict KRI challenges<br>• Cannot be expected to capture all potential losses |

## Typical Qualitative Risk Analysis Methods *(cont.)*

| Typical Qualitative Risk Analysis Methods *(cont.)* | |
|---|---|
| **Method** | **Description** |
| Likelihood-impact matrix | The matrix offers a good way to categorize risk events qualitatively in terms of their probability of occurrence (likelihood) and their consequences (impact)<br><br>Example of a Risk Level Matrix From NIST SP800-30: |
| Attribute analysis | Attribute analysis:<br>• Is a creative problem-solving technique that can be employed productively when exploring possible qualitative risk impacts<br>• Enables the development of creative solutions for problems by the application of unconventional perspectives<br><br>Its main value is that it subjects something familiar to examination in different ways in a structured way. |
| Delphi forecasting | • This was developed as a forecasting tool involving subject matter experts (SMEs) who anonymously provide their responses to a questionnaire, and these responses are tabulated and portrayed graphically.<br>• Participants in the exercise may be asked to forecast such things as the cost, schedule and resource impacts associated with the occurrence of a risk event.<br>• First-round responses may be highly divergent; however, after several rounds of responses, the responses may show that a consensus has emerged on the impacts of the risk event.<br>• With the statistical summary of responses, the exercise is repeated.<br>• The objective is to see whether SMEs can achieve a consensus on an issue after they have had a chance to reflect on feedback provided by the responses of their colleagues.<br><br>**Note:** In a typical Delphi exercise, each round of questionnaire distribution leads to increased conformity in the SMEs' views on an issue. After several rounds, when it becomes clear that additional consensus will not occur, the exercise is ended. |

Example of a Risk Level Matrix (Likelihood-impact matrix):

| Typical Qualitative Risk Analysis Methods | | | |
|---|---|---|---|
| **Description** | | | |
| **Impact** | | | |
| **Threat Likelihood** | **Low (10)** | **Medium (50)** | **High (100)** |
| **High (1.0)** | Low<br>$10 \times 1.0 = 10$ | Medium<br>$50 \times 1.0 = 50$ | High<br>$100 \times 1.0 = 100$ |
| **Medium (0.5)** | Low<br>$10 \times 0.5 = 5$ | Medium<br>$50 \times 0.5 = 25$ | High<br>$100 \times 0.5 = 50$ |
| **Low (0.1)** | Low<br>$10 \times 0.1 = 1$ | Medium<br>$50 \times 0.1 = 5$ | High<br>$100 \times 1.0 = 10$ |
| *Risk scale: High (>50 to 100); Medium (>10 to 50); Low (1 to 10)* | | | |

**Typical Qualitative Risk Analysis Methods *(cont.)***

| Typical Qualitative Risk Analysis Methods *(cont.)* | |
|---|---|
| **Method** | **Description** |
| Failure Modes and Effects Analysis (FMEA) | FMEA:<br>• Evaluates the impact of risk from an immediate, near-term and long-term perspective. This methodology will assess risk from both local business unit and enterprisewide perspectives.<br>• Acknowledges that a failure in one area may result in unforeseen consequences in other areas or departments<br>• Results in the prioritization of risk—giving each risk a risk priority number (RPN) |

## 3. Quantitative Risk Analysis

**Overview of Quantitative Analysis**

Quantitative analysis is the use of numerical and statistical techniques to calculate likelihood and impact of risk. It uses financial data, percentages and ratios to provide an approximate measure of the magnitude of impact in financial terms.

This measure can be used in the cost/benefit analysis of the recommended controls. Although a computation is used to arrive at various risk aspects, the approach is still subjective to some extent. The values used to calculate likelihood, impact and asset value are subject to speculation and may be quite difficult to quantify. The results of a quantitative risk analysis must, therefore, allow for some margin of error.

**Benefits of Quantitative Analysis**

Since quantitative analysis is data driven, it:
- Allows for:
  - Data to be classified and counted
  - Statistical models to be constructed to explain what is being observed
  - Findings to be generalized to a larger population and for direct comparison to be made between two different sets of data or observations
- Produces statistically reliable results
- Allows discovery of phenomena which are likely to be genuine versus those which are merely chance occurrences

**Challenges of Quantitative Analysis**

When measuring, modeling and managing operational risk, the following challenges must be considered:
- It is not always easy to collect data on each and every process.
- Data may not be in the desired format or may not meet the needs of quantitative analysis.
- Reliable historical data are not always available for analysis to allow for quantification of process failures or the risk induced by these failures.
- Past data do not necessarily help predict future events (black swan phenomenon).
- It is difficult to apply statistical models for events that happen infrequently.
- The cost of quantitative analysis is generally significantly higher than the cost of qualitative analysis.

## Typical Quantitative Risk Analysis Methods

The following table describes typical quantitative methods used in risk assessment.

<table>
<tr><th colspan="3">Typical Quantitative Risk Analysis Methods</th></tr>
<tr><th>Method</th><th colspan="2">Description</th></tr>
<tr><td rowspan="6">Internal loss data</td><td colspan="2">Used by: Financial services enterprises—These are key to any financial services enterprise's efforts to improve operational risk management.<br><br>Issue: The biggest issue most of these enterprises face is the lack of reliable and consistent operational risk data.<br><br>The following table outlines internal loss data considerations.</td></tr>
<tr><th>Loss Type</th><th>Description</th></tr>
<tr><td>Internal loss data quality</td><td>• Many banks started accumulating internal loss data to prepare for Basel II, which requires:<br>– A minimum of three years' of data to start<br>– Five years' of data on an ongoing basis as part of the advanced measurement approach (AMA)<br>• The quality of internal loss data:<br>– Is a factor and must be available across all business lines and geographic locations<br>– Should include near-loss data</td></tr>
<tr><td>Economic losses</td><td>• It is critical to capture all economic losses, not just major or material losses, with a large impact on the bottom line.<br><br>Rationale:<br>• These data predict expected losses (ELs), even though they typically represent less than 25 percent of all losses.<br>• Classification is difficult because:<br>– Many loss events result from a variety and combination of factors.<br>– The same loss event could fall into credit, market and operational risk buckets.</td></tr>
<tr><td>Operational risk losses</td><td>• There is an issue as to the enterprise's acceptance of risk.<br>• Many enterprises are hesitant to capture operational risk losses as a negative reflection on their performance. Many enterprises view market and credit risk as an acceptable cost of doing business.</td></tr>
<tr><td colspan="2">Alternative method: Another method of validating internal loss data is to compare it with peer enterprises via externally available data and then scale the data to reflect the enterprise's environment.</td></tr>
</table>

### Typical Quantitative Risk Analysis Methods *(cont.)*

| Typical Quantitative Risk Analysis Methods *(cont.)* | |
|---|---|
| **Method** | **Description** |
| External data | **Used by:** Financial services enterprises<br><br>External data are needed because there is typically a lack of internal data—especially around unexpected losses, which represent the majority of losses in most banks.<br><br>**Issue:** The use of external data stems from their sources, which include data providers or bank consortia.<br><br>**Relevance:** External data must be mapped, scaled and adapted to each bank's business, legal, regulatory, technical, control and cultural environment. |
| Business process modeling (BPM) and simulation | Although it may appear that most operational risk is preventable with the implementation of procedures and controls, it is not an easy task to identify and control all risk.<br><br>The BPM and simulation discipline:<br>• Is an effective method of identifying and quantifying the operational risk in enterprise business processes<br>• Improves business process efficiency and effectiveness<br><br>The process simulation model aids the enterprise in:<br>• Developing insights into the operations of the business<br>• Leveraging assets and reducing costs<br>• Testing process changes before implementation (change management)<br>• Experimenting with process improvements to reduce cycle times and manage operational risk<br>• Conducting stress tests and scenario analysis |
| Statistical process control (SPC) | The typical SPC is the cumulative sum (cusum) control chart:<br>• It is very effective in detecting small process shifts, which need to be stable and should operate with minor variability.<br>• All cusum statistics incorporate all the information known about the process.<br>• The plain-vanilla cusum simply plots the cumulative sums of the deviations of the observed values and a target value.<br><br>SPC is most often used by enterprises with manufacturing environments; however, it can be very effective in general BPM. |

## 4. Methods for Discovering High-Impact Risk Types

### Introduction

This topic contains information on methods for discovering high-impact risk types.

The uncovering of uncommon, but potentially high-impact, risk needs to be recognized as an important area of focus within risk assessment. This is especially true of process-based activity and in process improvement (as in BPR).

### Relevance

Without consideration of low-probability, high-impact events, there may be serious gaps in the assessment of risk and, consequently, the completeness of coverage of risk mitigation strategies.

### Methods for Uncovering Less Obvious Risk Factors

The following table describes other methods for uncovering less obvious risk factors.

| Methods for Uncovering Less Obvious Risk Factors | |
|---|---|
| **Method** | **Description** |
| Cause-and-effect analysis | A predictive or diagnostic analytical tool used:<br>• To explore the root causes or factors that contribute to positive or negative effects or outcomes<br>• For identifying potential risk<br><br>**Note:** A typical form is the Ishikawa diagram, also known as the fishbone diagram. |
| Fault tree analysis | A technique that:<br>• Provides a systematic description of the combination of possible occurrences in a system, which can result in an undesirable outcome (top-level event)<br>• Combines hardware failures and human failures<br><br>A fault tree is constructed by:<br>• Relating the sequences of events that, individually or in combination, could lead to the top-level event<br>• Deducing the preconditions for the:<br>– Top-level event<br>– Next levels of events, until the basic causes are identified (elements of a "perfect storm" [unlikely simultaneous occurrence of multiple events that cause an extraordinary incident])<br><br>**Note:** The most serious outcome is selected as the top-level event. |
| Sensitivity analysis | A quantitative risk analysis technique that:<br>• Helps to determine which risk factors potentially have the most impact<br>• Examines the extent to which the uncertainty of each element affects the object under consideration when all other uncertain elements are held at their baseline values<br><br>**Note:** The typical display of results is in the form of a tornado diagram. |

## 5. Risk Associated With Business Continuity and Disaster Recovery Planning

### Introduction

Business continuity planning (BCP) and disaster recovery planning (DRP) complement the risk management process by ensuring preparedness at times when rare, but high-impact, events occur.

Both disciplines can (and should) leverage each other to ensure that the assets of the enterprise are protected.

Business impact analysis (BIA) is a key business continuity planning (BCP) activity that focuses on determining the impact of an event over time—how the impact increases over time and at what point in time the business is most likely no longer able to recover from the incident.

### Relevance

Since BCP and DRP are the last line of defense in the event of a catastrophic impact on the enterprise, it is imperative that these plans be robust enough to work effectively in a crisis.

The risk for many enterprises is that the business continuity (BC) and disaster recovery (DR) plans are not adequate to maintain or recover business operations in a crisis.

The risk practitioner should ensure that BC and DR plans are kept up to date, are approved by management and are tested on a regular basis.

### Definition of Business Impact Analysis (BIA)

A BIA is a specialized process to determine the impact of losing the support of any resource.

A BIA is a discovery process meant to:
- Reveal the importance of a process and the potential impact that any disruption to that process would have on the enterprise.
- Establish the escalation of loss over time.
- Answer questions about actual procedures, shortcuts, workarounds and the types of failure that may occur. It involves determining:
  - What the process does
  - Who performs the process
  - What the output is
  - The value of the process output to the enterprise
  - How the impact of the loss would escalate over time and at which point failure of the process might threaten the viability of the enterprise.

> **Note:** The use of qualitative and quantitative assessment for risk discovery and assessment has already been mentioned in the previous topic. The same qualitative and quantitative techniques can be used in the BIA.

## Discovery Exercises for Proper BIA Execution

Proper execution of the BIA process entails a series of discovery exercises, including:
- Asking questions that focus on identifying trigger events and the current sequence of the enterprise's critical processes
- Interviewing key personnel
- Reviewing existing documentation
- Collecting data by observing business processes and personnel performing actual processes (job shadowing)
- Looking for existing workarounds and alternate procedures

**Note:** Using surveys can raise issues of accuracy and consistency.

The BIA process will:
- Verify critical success factors (CSFs)
- Identify vital materials and records necessary for recovery, including:
  - Data backups
  - The vendor list
  - Inventory records
  - The customer list
  - Employee records with contact data
  - Bills of materials
  - Procedure documents
  - The business continuity (BC) plan
  - Banking information
  - Copies of all contracts and legal documents

The data gathered during the BIA can be used later to guide the formulation of the risk response strategy.

# I. IT Risk Identification and Assessment

## 1. Threats and Opportunities Inherent in Enterprise Use of IT

### Introduction

IT plays a critical role in the operation of nearly every modern business process and enterprise. It is a key part of strategic decisions, operational continuity and projects that bring about organizational change. Therefore, there are many risk factors that pertain specifically to enterprise IT systems.

The strategic importance of IT to the modern enterprise is seen through:
- High investment
- The pervasiveness of IT
- Reliance on IT's continuing operation
- The impact caused when IT does not perform as expected
- IT's critical role in realizing efficiencies
- The ways in which IT enables business to take strategic action

These possibilities carry with them significant risk—risk that is inherent from the use of IT in itself. Only by proactively recognizing and addressing this risk can business interests be safeguarded.

### Inherent Risk and Rewards of IT Use

The use of IT obviously carries a risk, just as it has potential rewards. Ignoring or focusing only on the most obvious risk is dangerous because IT has become a utility that underpins practically every business activity.

Risk is typically not easily measured, reported or monitored. This is compounded by the fact the word "risk" is often incorrectly applied to both eventualities and their likelihood with statements such as: "This risk is low-risk."

Monitoring and reporting on risk is made even more difficult because there is no shared language between those estimating the risk and those making risk response decisions. Clarity in defining the business impact (both positive and negative) of IT-related risk is, therefore, critical for understanding where there are threats and vulnerabilities and where there are opportunities.

**Relevance:** Being able to clearly differentiate between the terms "risk," "threat" and "vulnerability" is crucial:
- **Risk**—A derived value that refers to the likelihood (or frequency) and magnitude of loss that exists from a combination of asset(s), threat(s) and control conditions
- **Threat**—An action or actor that/who may act in a manner that can result in loss or harm
- **Vulnerability**—A weakness in design, implementation, operation or internal control

## Measuring the Adverse Impact of IT Risk

The adverse impact of a risk event can be described in terms of loss or degradation of any, or a combination of any, of the following three basic IT risk goals:
- Integrity
- Availability
- Confidentiality

The following table provides a brief description of each business information requirement and the consequence (or impact) of the goal's not being met.

| Business Information Requirements and Related Impacts | |
|---|---|
| **Requirement** | **Description/Impact of Unmet Goals** |
| Integrity | Relates to the accuracy and completeness of information and its validity in accordance with business values and expectations.<br><br>System and data integrity refers to the requirement that information be protected from improper modification.<br><br>**Impact:** Loss of integrity—If unauthorized changes are made to the data or information system by either intentional or accidental acts and the loss of system or data integrity is not corrected:<br>• Continued use of the contaminated system or corrupted data may result in inaccuracy, fraud or erroneous decisions.<br>• Violation of integrity may be the first step in a successful attack against system availability or confidentiality. |
| Availability | Relates to the information being accessible, when required by the business process, and also concerns the safeguarding of necessary resources and associated capabilities.<br><br>**Impact:** Loss of availability—If a mission-critical IT system is unavailable to its end users, the enterprise's objectives may be affected. Loss of system functionality and operational effectiveness, for example, may result in loss of productive time, thus impeding the end users' performance of their functions in supporting the enterprise's objectives. |
| Confidentiality | Relates to the protection of information from unauthorized disclosure<br><br>**Note:** Data must be protected from improper disclosure depending on the sensitivity of the data and associated legal requirements.<br><br>**Impact:** Loss of confidentiality—Unauthorized disclosure of confidential or sensitive information can range from jeopardizing national security to the disclosure of data covered under the local privacy law. Unauthorized, unanticipated or unintentional disclosure of such information can result in loss of public confidence loss of competitive advantage, embarrassment or legal action against the enterprise. |

## Measuring the Business Impact of IT-related Risk

Some tangible impacts of IT risk can be measured quantitatively as in:
- Lost revenue
- The cost of repairing the system
- The level of effort required to correct problems caused by a successful threat action

Other impacts are difficult to measure in specific units, but can be qualified or described in terms of high, medium and low impacts. Such impacts include:
- Loss of public confidence
- Loss of credibility
- Damage to an enterprise's interest
- Impact on morale in the enterprise

## Business-related IT Risk Types

The business-related IT risk types listed in the following table can be used as a guide.

| Business-related IT Risk Types | |
|---|---|
| **Type** | **Description** |
| Investment or expense risk | The risk that the IT investment fails to provide value for money or is otherwise excessive or wasteful<br><br>This includes consideration of the overall IT investment portfolio. |
| Access or security risk | The risk that confidential or otherwise sensitive information may be divulged or made available to those without appropriate authority<br><br>An aspect of this risk is noncompliance with local, national and international laws related to privacy and protection of personal information. |
| Integrity risk | The risk that data cannot be relied on because they are unauthorized, incomplete or inaccurate |
| Relevance risk | The risk associated with not getting the right information to the right people (or process or systems) at the right time to allow the right action to be taken |
| Availability risk | The risk of loss of service or the risk that data are not available when needed |
| Infrastructure risk | The risk that an enterprise does not have an IT infrastructure and systems that can effectively support the current and future needs of the business in an efficient, cost-effective and well-controlled fashion (includes hardware, networks, software, people and processes) |
| Project ownership risk | The risk of IT projects failing to meet objectives through lack of accountability and commitment |

## Threats and Vulnerabilities Associated with SDLC

Most enterprises follow some form of a system development life cycle (SDLC) for IT project management. This structure is intended to ensure that the IT project is properly managed and will meet business requirements.

An important function for the risk practitioner during any IT project—or business process reengineering project—is to ensure that the system is designed and built with adequate levels of protection so that the end product will be built to mitigate any risk that the new system may pose to the enterprise.

This requires that risk assessment be built into each phase of the SDLC—starting with the functional requirements definition phase and continuing on to the implementation and operation of the system.

## Functional Requirements Definition Phase

The key task related to risk assessment during the functional requirements definition phase is to ensure that the risk associated with the system is identified. The level of risk associated with the system will depend on the criticality of the system (how important the system is to support business operations) as well as the sensitivity (privacy) needs and criticality needs of the data being processed on the system.

Once the risk requirements are known, they must be mitigated through the design of appropriate controls. Some of the controls will be to build redundancy and access controls (to prevent improper modification or disclosure) into the systems design.

## Risk Associated With Outsourcing

Many enterprises are choosing an outsourcing strategy to provide IT, data management, web hosting and other supporting services. This presents a new risk to the enterprise since the enterprise now becomes reliant on another enterprise for its supporting infrastructure. There is also risk related to privacy laws that may affect where data are stored and the need to protect sensitive data in transit, storage and processing from inadvertent or unauthorized disclosure.

One risk that the enterprise must be aware of is that in many jurisdictions the responsibility for protection of data remains with the original enterprise, not the firm providing the outsourcing services, so the original enterprise must ensure that the security requirements and the right to audit are included in contracts.

## IT Project-related Risk Factors

IT projects—the implementation of new technology, business process reengineering, or upgrades and modification to existing systems—are subject to risk of failure due to factors such as:
- Unrealistic delivery schedules or budget
- Lack of skilled resources
- Unclear or changing business requirements
- Challenges with technology
- Poor project management
- Resistance from users

## IT Project-related Risk

The risk associated with an IT project that fails to deliver the expected results on time or on schedule may have a serious impact on business operations. Such impact may be related to:
- Loss of opportunity or market share
- Inability to meet customer or regulatory demand
- Lost revenue
- Other tangible or intangible consequences

Risk associated with changing an existing process must be identified. The ISACA IT Audit and Assurance Standards, Guidelines, and Tools and Techniques identify several risk areas to consider when planning a BPR project. The risk factors can be broken down into the three broad areas, as listed in the following table.

## IT Project-related Risk *(cont.)*

<table>
<tr><th colspan="3">IT Project-related Risk</th></tr>
<tr><th>Risk Area</th><th colspan="2">Description</th></tr>
<tr><td rowspan="6">Design risk</td><td colspan="2">A good design can improve profitability while satisfying customers. Conversely, a design failure would spell doom to any BPR project. It would be reckless to undertake new projects without dedicated resources capable of committing the time and attention necessary to develop a quality solution. Often, this type of detailed planning may consume more money and time than is available from key personnel. Recognition should be given to the types of risk that may occur in the BPR design listed in the following table.</td></tr>
<tr><th>Risk Type</th><th>Description</th></tr>
<tr><td>Sponsorship risk</td><td>• C-level management is not supportive of the effort.<br>• There is insufficient commitment from the top or inappropriate project leadership.<br>• Poor communication is a major problem.</td></tr>
<tr><td>Scope risk</td><td>• The BPR project must be related to the vision and the specifications of the strategic plan.<br>• Serious problems will arise if the scope is improperly defined.<br>• It is a design failure if politically sensitive processes and existing jobs are excluded from the scope of change.</td></tr>
<tr><td>Skill risk</td><td>• Absence of radical, out-of-the-box thinking will create a failure by dismissing new ideas that should have been explored.<br>• "Thinking big" is the most effective way to achieve the highest return on investment (ROI).<br>• Participants without broad skills will experience serious difficulty because the project vision is beyond their ability to define an effective action plan.</td></tr>
<tr><td>Political risk</td><td>• Sabotage or passive resistance is always possible from people fearing a loss of power or resistance to change.<br>• Uncontrolled rumors lead to fear and subversion of the concept.<br>• People will resist change unless the benefits are well understood and accepted.</td></tr>
</table>

## IT Project-related Risk *(cont.)*

<table>
<tr><th colspan="2">IT Project-related Risk (cont.)</th></tr>
<tr><th>Risk Area</th><th>Description</th></tr>
<tr><td>Implementation risk</td><td>The implementation risk factors represent another source of potential failures that could occur during the BPR project. The most common implementation risk factors include those listed in the following table.
<table>
<tr><th>Risk Type</th><th>Description</th></tr>
<tr><td>Leadership risk</td><td>• Leadership failures include disputes over ownership and project scope.<br>• Management changes during the BPR project may signal wavering needs that may cause the loss of momentum.<br>• Strong sponsors will provide money, time and resources while serving as project champions with their political support.</td></tr>
<tr><td>Technical risk</td><td>• Technical complexity may exceed the initial project scope.<br>• The required capability may be beyond that of prepackaged software.<br>• Custom functions and design may exceed IT's creative capability or available time.<br>• Delays in implementation could signal that the complexity of scope was underestimated.<br>• If the key issues are not fully identified, disputes will arise about the definitions of deliverables, which leads to scope changes during implementation.</td></tr>
<tr><td>Transition risk</td><td>• The loss of key personnel may create a loss of focus during implementation.</td></tr>
<tr><td>Personnel risk</td><td>• Personnel may feel burned out because of workload or their perception that the project is not worth the effort.<br>• Reward and recognition are necessary during transition to prevent the project from losing momentum.</td></tr>
<tr><td>Scope risk</td><td>• Improperly defined project scope will produce excessive costs with schedule overruns (variance from schedule).<br>• Poor planning may neglect the human resources (HR) requirements, which will lead team members to feel that the magnitude of effort is overwhelming.<br>• The reaction will cause a narrowing of the scope during implementation, which usually leads to a failure of the original BPR objectives.</td></tr>
</table></td></tr>
</table>

## IT Project-related Risk *(cont.)*

<table>
<tr><th colspan="3">IT Project-related Risk (cont.)</th></tr>
<tr><th>Risk Area</th><th colspan="2">Description</th></tr>
<tr><td rowspan="5">Operation or rollout risk</td><td colspan="2">It is still possible for the BPR project to fail after careful planning. Common failures during production implementation include negative attitudes and technical flaws. These problems manifest in the form of management risk, technical risk and cultural risk, as described in the following table.</td></tr>
<tr><th>Risk Type</th><th>Description</th></tr>
<tr><td>Management risk</td><td>• Strong, respected leadership is required to resolve power struggles over ownership.<br>• Communication problems must be cured to prevent resistance and sabotage.<br>• Executive sponsors need to provide sufficient training to prevent an unsuccessful implementation.</td></tr>
<tr><td>Technical risk</td><td>• Insufficient support is the most obvious cause of failure in a rollout.<br>• Inadequate testing leads to operational problems caused by software problems.<br>• Data integrity problems represent a root problem capable of escalating into user dissatisfaction.<br>• Perceptions of a flawed system will undermine everyone's confidence.</td></tr>
<tr><td>Cultural risk</td><td>• Resistance in the enterprise is a result of failing to achieve user buy-in.<br>• Resistance will increase to erode the benefits.<br>• Effective training is often successful in solving user problems.<br>• Dysfunctional behavior will increase unless the new benefits are well understood and achieved.</td></tr>
</table>

## Risk Components

In assessing risk, each of the risk components listed in the following table can be considered to determine the levels of component risk within each category

<table>
<tr><th colspan="3">Risk Components</th></tr>
<tr><th>Component</th><th colspan="2">Description</th></tr>
<tr><td rowspan="4">Inherent risk</td><td colspan="2">The risk level/exposure without taking controls or other management actions into account<br><br>Example:<br>• The inherent risk associated with operating system (OS) security is ordinarily high because changes to, or even disclosure of, data or programs through OS security weaknesses could result in system failure, security breach or regulatory penalties.<br><br>Inherent risk for most areas of IT is ordinarily high because the potential effect of errors ordinarily spans several business systems and many users. In assessing the inherent risk, there should be consideration for pervasive and detailed IT controls, as outlined in the following table.</td></tr>
<tr><th>Control Type</th><th>Investigation Area</th></tr>
<tr><td>Pervasive IT controls</td><td>• Integrity of IT management and IT management experience and knowledge<br>• Pressures on IT management that may predispose it to conceal or misstate information (e.g., large business-critical project overruns, hacker activity)<br>• Nature of the enterprise's business and systems (e.g., plans for electronic commerce [e-commerce], complexity of the systems, lack of integrated systems)<br>• Factors affecting the enterprise's industry as a whole (e.g., changes in technology, IS staff availability)</td></tr>
<tr><td>Specific IT controls</td><td>• Complexity of the systems involved<br>• Level of manual intervention required<br>• Susceptibility to loss or misappropriation of the assets controlled by the system (e.g., inventory, payroll)<br>• Likelihood of activity peaks at certain times in the period of investigation<br>• Poor change control procedures<br>• Integrity, experience and skills of the management and staff involved in applying the IT controls</td></tr>
</table>

**Risk Components *(cont.)***

| Risk Components *(cont.)* | |
|---|---|
| **Component** | **Description** |
| Residual risk | The risk that remains after management has implemented a risk response |
| Control risk | The risk of a failure of the internal control systems to prevent, detect or correct an incident in a timely manner.<br><br>**Example:**<br>• The control risk associated with manual reviews of computer logs can be high because activities requiring investigation are often easily missed due to the volume of logged information.<br><br>The control risk associated with computerized data validation procedures is ordinarily low because the processes are applied consistently. |
| Detection risk | The risk that the prescribed controls, substantive testing procedures, or monitoring will not detect an error that could be material, individually or in combination with other errors.<br><br>**Example:**<br>An intrusion detection system (IDS), an antivirus system or firewall is unable to detect or notice an adverse condition and trigger an adequate response (sometimes called a false negative—an indication that everything is fine when there actually is a problem). |

**Information Systems Architecture**

The calculation of risk for IT systems is directly affected by the type of architecture that the enterprise is using. Some types of architecture are much more robust or secure than others. The risk associated with a ring topology or a centralized system are different and may be less than the risk associated with a bus, star or tree topology. The risk practitioner must consider many IT factors when determining IT risk including the age of equipment, maintenance schedules, location and users.

Risk management also considers the risk at all levels of the IT infrastructure. The risk at the application, database, network, operating system, utility and hardware levels all must be considered.

## 2. Types of Business Risk and Threats That Can Be Addressed Using IT Resources

**Introduction**

Whether for compliance, effectiveness or efficiency, IT enablement of business has dramatically increased in recent years. As complexities of business evolve, the integral role of IT is extended to that of assisting business in handling risk that is an inevitable part of business strategies, processes and operations.

Therefore, included as part of business enablement, IT has the task of assisting in the management of business risk.

### Key Role of IT in the Enterprise Control Environment

When IT is used in the execution of business strategies and operations, it is inevitably drawn into the arena of business risk. A majority of large enterprises have implemented stringent internal controls and empowered internal auditors to conduct more intensive auditing of internal business processes and supporting IT processes and systems.

IT is intimately associated with a range of business activities that are sources of risk and, as such, has a key part to play in the enterprise's control environment. IT risk managers, teaming with personnel managing enterprise risk from other perspectives, can ensure that IT risk is given the right priority, and that opportunities for IT systems and services to assist in managing risk factors of different types are leveraged. Just as IT can be applied to yield results that were not previously possible in many fields, IT can also prove itself in the field of risk management.

IT can facilitate the wiring up, locking down and constant surveillance of the business; specifically in the domain of risk management information systems (RMISs), IT is relied on for advanced risk analytics and reporting.

## 3. Enterprise Risk Management

### Enterprise Risk Management (ERM) Model Control Objectives for Risk

An ERM model must address the enterprise's objectives with control objectives for risk in the following categories of business:
- **Planning**—High-level planning, resource allocation and budgeting
- **Operational**—Day-to-day activities
- **Financial reporting**—Presentation of financial results
- **Compliance**—Adherence to statutory requirements of all jurisdictions within which the enterprise does business

The internal controls in each area ensure that:
- The business is being run in accordance with the overall plan.
- Financial statements and management reporting present an accurate view of the operations.
- All activities (including reporting) that are covered by statutory regulations are being carried out within the constraints of those regulations.

### Segregation of Duties (SoD) as a Key Component of a Strong Internal Control Environment

Separation of duties (also called segregation of duties) is a key component to maintaining a strong internal control environment because it reduces the risk of errors and fraudulent transactions. When duties for a business process or transaction are segregated so that it requires the involvement of more than one person to accomplish a task, it becomes more difficult for fraudulent activity to occur because it would require collusion among several employees.

There are a wide variety of automated (i.e., IT-based) compliance solutions that address the issue of SoD, as seen in the next section. Prior to these tools being available, enterprises typically addressed SoD through a combination of controls, such as:
- Defining transaction authorizations
- Assigning custody of assets
- Granting access to data
- Reviewing or approving authorization forms
- Creating user authorization tables
- Creating manual SoD tables

## Automated Tools to Assist in SoD

Automated tools are typically used to address SoD and also to provide the enterprise with reporting functionality on SoD violations (i.e., detective controls) and to implement preventive controls.

In general, automated control systems contain the following three elements:
- **Access controls**—Restrict access to the underlying business systems and data to ensure that only authorized individuals have access, and that each user is only granted the minimal level of access require to perform their job function
- **Process controls**—Restrict the activities performed by authorized users. This employs techniques such as dual control (requiring two people to take action simultaneously to perform a task) or mutual exclusivity (if one person has executed one task, they are prohibited from executing subsequent tasks).
- **Continuous monitoring**—Employs automation to detect system transactions, setup or data changes that contravene corporate policy. These systems may be used to block certain activities from unauthorized users or to limit the ability of a user to make a change without higher-level approval.

**Example:** Each of these elements may be subject to access control protection to ensure that only authorized individuals can view or change the access rules. Similarly, process controls ensure that only correct actions are permitted and monitoring controls will track any invalid operations after the fact.

## Difficulties in Maintaining Manual Controls

Automated compliance solutions aim to provide enterprises with timely and efficient internal controls that do not disrupt their normal business process.

Many systems can now be updated automatically—such as antivirus signatures, software patches and access controls.

As enterprises grow and the reliance on IT for both internal and external users increases, it becomes impossible to manage access rights and privileges in a manual fashion. Resources, systems, employees and business partners are added to the infrastructure and employee job functions are changed to mirror the ongoing changes within an enterprise. This causes manual access privileges to become quickly outdated.

Without automating identity management and user access controls, there is the potential of impacting the ability of employees and customers to access enterprise systems and data in a timely manner. Manual authorizations are often time consuming and require significant administrator time. It may not even be possible to maintain a manual system and the time taken for system administration is time taken away from other important tasks.

## Contribution Areas for IT in Managing Business Risk

The following table lists the two specific contribution areas for IT in managing business risk.

| IT Contributions to Manage Business Risk | |
|---|---|
| **Contribution Area** | **Description** |
| Locked-down operating | IT can be used to build in business process controls. Historically, for business processes with low levels of automation, there was an *ad hoc* option—at the discretion of the operators and done on an *ad hoc* basis. However, automation with IT requires precision.<br><br>As a consequence of automation, the routine aspects of business are increasingly locked down to a repeatable and predictable pattern. Where variation and variance is to be avoided as much as possible, predictable translates as low risk. Automation with IT is the preferred route toward a six-sigma (one defect per million cycles) or zero-defect goal.<br><br>Applications enforce business rules, such as:<br>• Mandatory fields are required before a record can be saved.<br>• Lookup fields can be used to ensure that valid codes are entered.<br>• Approvals above a certain value can be routed via work flow for management approval.<br>• Automated teller machines (ATMs) will not discharge money without a valid account and personal identification number (PIN) combination.<br><br>This is an essential part of controlling normal business operations. It also allows HR to be channeled to doing other things, as long as the IT systems reliably perform the handling and also the checking and balancing. In the creative realm, in which predictability rapidly leads to commoditization and loss of competitiveness, IT tools are available to knowledge workers as enablers. Even here, prescribed forms are typically common, suggesting or encouraging through IT rather than explicit enforcement.<br><br>**Example:** It is much easier to create a letter according to an enterprise's template than to start one from scratch. Also, in using the template, it is far more likely to achieve a compliant result.<br><br>Constant surveillance IT can maintain a watchful eye on the enterprise information and maintain records needed for the provision of evidence in litigation or with which to prosecute. |

### Contribution Areas for IT in Managing Business Risk *(cont.)*

| IT Contributions to Manage Business Risk *(cont.)* | |
|---|---|
| **Contribution Area** | **Description** |
| Decision support, risk analytics and reporting | Advanced risk/return decision making requires advanced IT support. It is not feasible to manually calculate the risk/opportunity related to today's credit portfolio. Data volumes are huge, and the sophisticated models require precise calibration and consistent fine-tuning. Quantitative analysis will inevitably turn to IT for the large-scale analysis of risk factors because the use of IT is most feasible when a large number of inputs and mathematical complexity are involved.<br><br>The objective of all management information systems is to enable faster and better decision making. In the case of risk management information, the decision making relates to known and potential risk.<br><br>The goal of risk management for information systems is to achieve compatible and efficient IT monitoring and reporting processes for capturing; analyzing; and, ultimately, reporting risk factors of all types across the entire enterprise. The consumers of output from the reporting systems are both internal—across all layers of management—and external. Automating risk information management can assist in the embedding of required practices into the enterprise by making "business as usual" risk management activities efficient rather than onerous. |

### Threat and Vulnerabilities Related to IT Operations Management

A system must be deployed in a secure manner that has adequate controls to mitigate risk and allow the system to be implemented without causing an undue level of risk to the enterprise as a whole, to other systems or networks or to individuals or departments.

However, once installed, the system must continue to operate in a secure manner. This requires the use of operational controls to ensure that the risk management controls built into the system continue to operate correctly. An enterprise that has poor IT management or change control procedures may be exposed to the risk that the controls will not work properly or that they may be bypassed by future development projects.

## 4. Methods/Frameworks for Describing IT Risk in Business Terms

### Describing the Business Impact of IT Risk

Meaningful IT risk assessments and risk-based decisions require IT risk to be expressed in unambiguous and clear, business-relevant terms. Effective risk management requires mutual understanding between IT and the business over which risk needs to be managed and why.

All stakeholders must have the ability to understand and express how adverse events may affect business objectives. This means that:
- IT personnel should understand how IT-related failures or events can impact enterprise objectives and cause direct or indirect loss to the enterprise.
- Business personnel should understand how IT-related failures or events can affect key services and processes.

> **IMPORTANT:** The link between IT risk and the ultimate business impact needs to be established to understand the effects of adverse events.

## Exhibit 1.10: Expressing IT Risk in Business Terms

Several techniques and options exist that can help the enterprise describe IT risk in business terms. **Exhibit 1.10** and the following table depict and describe some available methods.

**Exhibit 1.10: Expressing IT Risk in Business Terms**

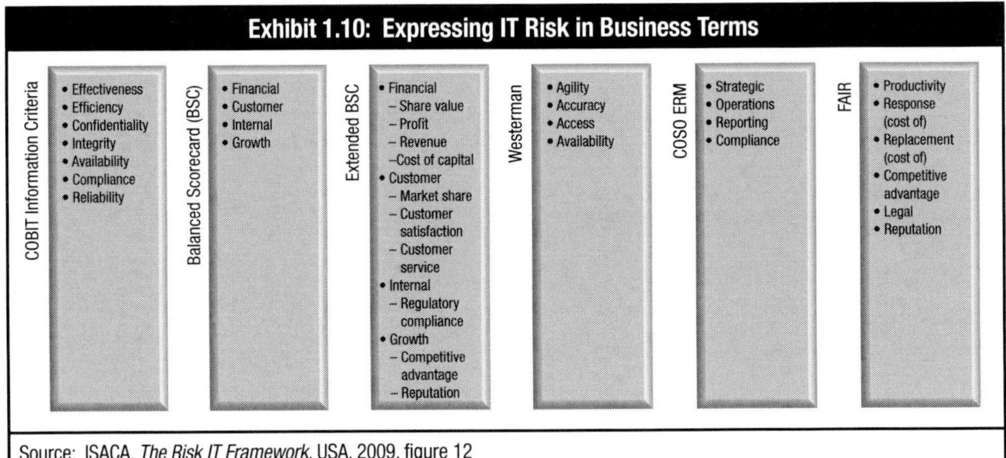


Source: ISACA, *The Risk IT Framework*, USA, 2009, figure 12

**Techniques for Expressing IT Risk in Business Terms**

| Technique | Description |
|---|---|
| COBIT Information Criteria (business requirements for information) | Allow for the expression of business aspects related to the use of IT<br><br>They express a condition to which information (in the widest sense), as provided through IT, must conform for it to be beneficial to the enterprise. The business impact of any IT-related event lies in the consequence of not achieving the information criteria. By describing impact in these terms, this remains a sort of intermediate technique that does not fully describe business impact, e.g., impact on customers or in financial terms. |
| COBIT Business Goals and Balanced Scorecard (BSC) | Based on the "business goals" concept introduced in COBIT<br><br>Business risk lies in any combination of those business goals not being achieved. The COBIT business goals are structured to align with the four classic BSC perspectives: financial, customer, internal and growth. |
| Extended BSC Criteria | A variant of the approach described in COBIT business goals and balanced scorecard<br><br>The extended BSC criteria go one step further, linking the BSC dimensions to a limited set of more tangible criteria. |
| Westerman 4As | Based on the 4A framework, which defines IT risk as the potential for an unplanned event involving IT to threaten any of the following four interrelated enterprise objectives:<br>• **Agility**—Possess the capability to change with managed cost and speed<br>• **Accuracy**—Provide correct, timely and complete information that meets the requirements of management, staff, customers, suppliers and regulators<br>• **Access**—Ensure appropriate access to data and systems so that the right people have the access that they need and the wrong people do not have access<br>• **Availability**—Keep the systems (and their business processes) running and recover from interruptions |

**Exhibit 1.10: Expressing IT Risk in Business Terms *(cont.)***

| Techniques for Expressing IT Risk in Business Terms *(cont.)* | |
|---|---|
| **Technique** | **Description** |
| The Committee of Sponsoring Organizations of the Treadway Commission (COSO)Enterprise Risk Management—Integrated Framework | Lists the following criteria that pertain to:<br>• **Strategic**—High-level goals, aligned with and supporting the enterprise objectives. Strategic objectives reflect management's choice as to how the enterprise will seek to create value for its stakeholders.<br>• **Operations**—The effectiveness and efficiency of the enterprise's operations, including performance and profitability goals, and safeguarding resources against loss<br>• **Reporting**—The reliability of reporting, including internal and external reporting and may involve financial and nonfinancial information<br>• **Compliance**—Adherence to relevant laws and regulations |
| Factor Analysis of Information Risk (FAIR) | While security-oriented in origin, impact criteria (productivity, cost of response, cost of replacement, competitive advantage, legal and reputational) apply to all IT-related risk factors. |

**Note:** The challenge of describing IT risk in business terms requires, among other things:
- Identifying various risk factors
- Developing risk scenarios to make risk more specific and allow for proper risk analysis and assessment

## 5. Risk Awareness and Communication

### Defining Risk Awareness

Risk awareness acknowledges that risk is an integral part of the business. This does not imply that all risk is to be avoided or eliminated, but rather that:
- Risk is well understood and known.
- IT risk issues are identifiable.
- The enterprise recognizes and uses the means to manage risk.

### Importance of Risk Communication

Risk communication is a critical part of the risk management process. People are naturally uncomfortable talking about risk and tend to put off admitting that risk is involved—let alone communicating about issues, incidents or crises.

If risk is to be managed and mitigated, it must first be discussed and effectively communicated at an appropriate level to the various stakeholders and personnel throughout the enterprise.

### Benefits of Effective Risk Communication

The benefits of open communication on risk include:
- Assistance in executive management's understanding of the actual exposure to IT risk, enabling the definition of appropriate and informed risk responses
- Awareness among all internal stakeholders of the importance of integrating risk management into their daily duties
- Transparency to external stakeholders regarding the actual level of risk and risk management processes in use

## Consequences of Poor Risk Communication

The consequences of poor communication of risk include:

- A false sense of confidence at all levels of the enterprise and a higher risk of a breach or incident that could have been prevented. Risk ignorance is an unacceptable risk management strategy.
- Lack of direction or strategic planning to mandate risk management efforts.
- Unbalanced communication to the external world on risk, especially in cases of high, but managed, risk, which may lead to an incorrect perception on actual risk by third parties such as:
  - Clients
  - Investors
  - Regulators
- The perception that the enterprise is trying to cover up known risk from stakeholders

## Exhibit 1.11: IT Risk Communication Components

**Exhibit 1.11** and the following table depict and describe the broad array of information flows and the major types of IT risk information that should be communicated.

**Exhibit 1.11: Risk Communication Components**

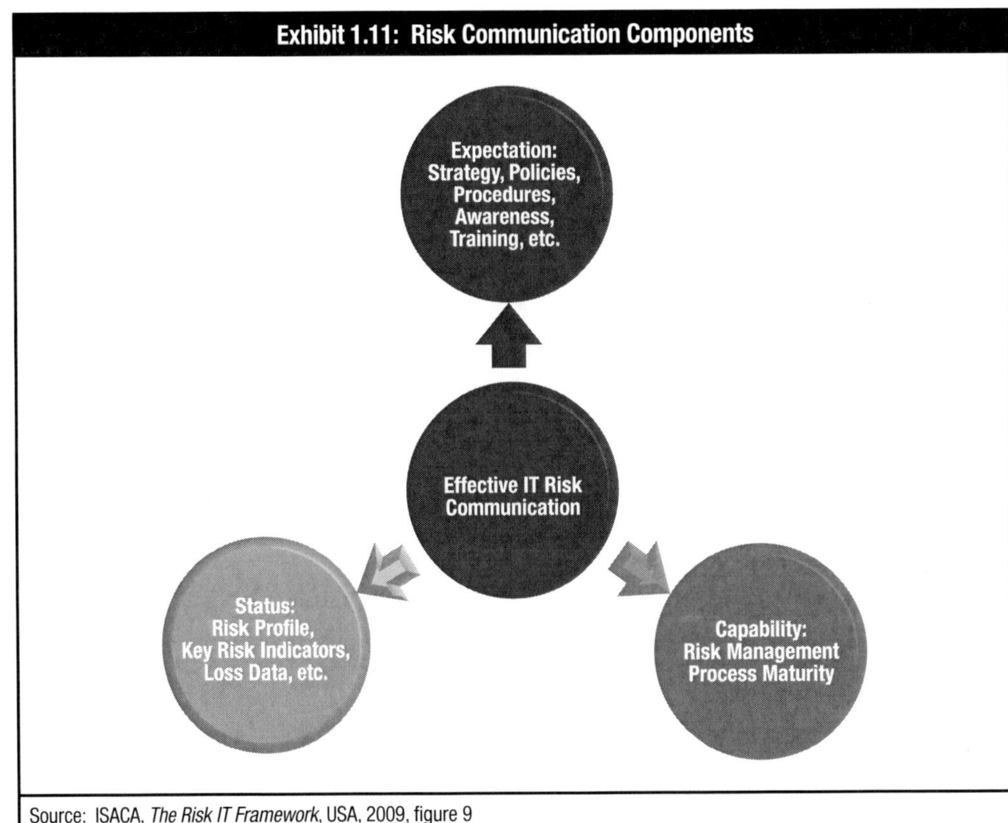


Source: ISACA, *The Risk IT Framework*, USA, 2009, figure 9

**Exhibit 1.11: IT Risk Communication Components *(cont.)***

| Description—Risk Components to Be Communicated | |
|---|---|
| Expectations from risk management | The risk components that must be communicated throughout the enterprise include risk strategy, policies, procedures, awareness training, continuous reinforcement of principles, etc. This is essential communication regarding the enterprise's overall strategy toward IT risk and:<br>• Drives all subsequent efforts on risk management<br>• Sets the overall expectations about the risk management program |
| Current risk management capability | This information:<br>• Allows for monitoring of the state of the "risk management engine" in the enterprise<br>• Is a key indicator for good risk management<br>• Has predictive value for how well the enterprise is managing risk and reducing exposure |
| Status | This includes the actual status with regard to IT risk, including information such as:<br>• The risk profile of the enterprise, i.e., the overall portfolio of (identified) risk to which the enterprise is exposed<br>• Key risk indicators (KRIs) to support management reporting on risk<br>• Event/loss data<br>• The root cause of loss events<br>• Options to mitigate risk (including cost and benefits) |

## Effective Communication

The following table lists the required elements for effective communication.

| Effective Communication | |
|---|---|
| **Communication Element** | **Description** |
| Clear | Risk information must be known and understood by all stakeholders. |
| Concise | Information or communication should not inundate the recipients. All ground rules of good communication apply to communication on risk. This includes the avoidance of jargon and technical terms regarding risk that may not be understood by the intended audience. |
| Useful | Any communication on risk must be relevant to the audience. Technical information that is too detailed and/or is sent to inappropriate parties will hinder, rather than enable, a clear view of risk. |
| Timely | The timing of when to communicate about risk is important for the effectiveness of the communication. Communication at the wrong time may be too late to be effective, whereas communication that is too early may be ignored as being irrelevant.<br><br>**Examples:**<br>• A risk that is not addressed when an IT system is being set up may result in an unacceptable level of risk associated with system operations and expensive rework to implement controls postimplementation.<br>• Failure to anticipate risk during project planning may result in project failure; the business consequence would be delayed business initiatives.<br><br>Communication is timely when it allows action to be taken at the appropriate moments to identify and treat the risk. It serves no useful purpose to communicate about a project delay a week before the deadline. |

## Effective Communication *(cont.)*

| Effective Communication *(cont.)* | |
|---|---|
| **Communication Element** | **Description** |
| Aimed at the correct target audience | Information must:<br>• Be communicated at the right level of detail<br>• Be adapted for the audience<br>• Enable informed decisions<br><br>In this process, aggregation of data must not hide root causes of risk.<br><br>**Example:** A security officer needs technical IT data on intrusions and viruses to deploy solutions. An IT steering committee may not need this level of detail, but it does need aggregated information to decide on policy changes or additional budgets to treat the same risk. |
| Available on a need-to-know basis | Information related to risk should be known and communicated to all parties with a genuine need. A risk register with all documented risk is not public information and should be properly protected against internal and external parties with no need of access. Communication does not always need to be formal, through written reports or messages. Timely face-to-face meetings between stakeholders are an important means of communication for information related to business or IT-related risk. |

**Reference:**

For more information on:
- Risk scenarios, see section F in this chapter
- Risk factors, see section G in this chapter

## J. Suggested Resources for Further Study

### Suggested Resources for Further Study

In addition to the resources cited throughout this manual, the following resources are suggested for further study of risk management; risk governance; and related frameworks, standards and leading practices:

- ISACA:
  - COBIT 4.1, 2007
    **Note:** The COBIT 4.1 framework is available at no charge from ISACA and can be downloaded at *www.isaca.org/cobit*. The new COBIT 5 framework will be available in 2012.
  - *The Risk IT Framework*, 2009
  - *The Risk IT Practitioner Guide*, 2009
  - *Enterprise Value: Governance of IT Investments, The Val IT Framework 2.0*, 2008
  - *Implementing and Continually Improving IT Governance*, 2010
- Committee of Sponsoring Organizations of the Treadway Commission (COSO), *Enterprise Risk Management—Integrated Framework*, USA, 2004
- International Organization for Standardization (ISO):
  - ISO/IEC 27001:2005, *Information technology—Security techniques—Information security management systems—Requirements*, Switzerland, 2005
  - ISO/IEC 27002:2005, *Information technology—Security techniques—Code of practice for information security management*, Switzerland, 2005
  - ISO/IEC 27004:2009, *Information technology—Security techniques—Information security management—Measurement*, Switzerland, 2009
  - ISO/IEC 27005:2011, *Information technology—Security techniques—Information security risk management,* Switzerland, 2011
- ISO/IEC 27005:2009
- Jones, J.; *An Introduction to Factor Analysis of Information Risk (FAIR)*, Risk Management Insight LLC, November 2006, USA
- Kovacich, Gerald L.; Edward Halibozek; *The Manager's Handbook for Corporate Security: Establishing and Managing a Successful Assets Protection Program*, Butterworth-Heinemann, USA, 2003
- National Institute of Technology and Standards (NIST), *Risk Management Guide for Information Technology Systems*, Special Publication (SP) 800-30, *www.csrc.nist.gov*
- Peltier, Thomas A.; *Information Security Risk Analysis, 3rd Edition*, Auerbach Publications, USA, 2010
- Westerman, George; Richard Hunter; *IT Risk—Turning Business Threats Into Competitive Advantage*, Harvard Business School Press, USA, 2007

**Page intentionally left blank**

# Domain 2–Risk Response

## A. Chapter Overview

**Introduction**

This chapter provides an introduction to the principles of risk response.

The focus of this chapter is to evaluate the various risk response recommendations provided to the analyst and to select the best mitigation strategy—considering all of the factors that will influence the risk response decision.

The risk practitioner may assist in the investigation and evaluation of the risk response options—cost, time, skill, effectiveness, etc.—so that the enterprise can put in place a risk response strategy that is ideal for the enterprise. The best response will be based on several factors such as risk levels, urgency and impact, the cost of the risk response strategy chosen, and the enterprise's risk appetite and risk response capability.

**Inputs (Tie-back)**

Risk response activities rely directly on the input from the previous chapter—risk identification, assessment and evaluation—where the risk that the enterprise is facing is identified and the risk response strategies are recommended for the enterprise's consideration.

**Relevance**

This domain is an important function of the risk practitioner through the determination of the cost-benefit analysis and recommendation of the:
- Prioritization of risk responses, including IS control implementation efforts
- Appropriate risk response
- Necessary level of control

**Outputs (Tie-forward)**

The risk response decision process results in the selection of the best risk response strategy for a specific risk, for a group of risk and for the risk profile of the enterprise as a whole.

Risk mitigation is addressed in more detail in Domain 4—Information Systems Control Design and Implementation. Concepts related to IS control design and implementation can, in most cases, be transferred to non-IS controls for a wider applicability.

**Learning Objectives**

As a result of completing this chapter, the CRISC candidate should be able to:
- List various parameters for risk response selection.
- List the different risk response options.
- Describe risk responses that may be most suitable for a high-level risk scenario.
- Describe how exception management relates to risk management.
- Describe how residual risk relates to inherent risk and risk appetite.
- Describe the need for performing a cost-benefit analysis when determining a risk response.
- Describe the attributes of a business case to support project management.
- Identify standards, frameworks and leading practices related to risk response.

## Contents

This chapter contains the following sections:

## B. Task and Knowledge Statements

### Introduction

This section describes the task and knowledge statements for Domain 2, which focuses on the development and implementation of risk responses to ensure that risk issues, opportunities, and events are addressed in a cost-effective manner and in line with business objectives.

### Domain 2 Task Statements

The following table describes the task statements for Domain 2.

| No. | Task Statement (TS) |
|---|---|
| TS2.1 | Identify and evaluate risk response options and provide management with information to enable risk response decisions. |
| TS2.2 | Review risk responses with the relevant stakeholders for validation of efficiency, effectiveness and economy. |
| TS2.3 | Apply risk criteria to assist in the development of the risk profile for management approval. |
| TS2.4 | Assist in the development of risk response action plans to address risk factors identified in the organizational risk profile. |
| TS2.5 | Assist in the development of business cases supporting the investment plan to ensure risk responses are aligned with the identified business objectives. |

### Domain 2 Knowledge Statements

The following table describes the knowledge statements for Domain 2.

| | Knowledge Statement (KS) |
|---|---|
| No. | Knowledge of: |
| KS2.1 | Standards, frameworks and leading practices related to risk response |
| KS2.2 | Risk response options |
| KS2.3 | Cost/benefit analysis and return on investment (ROI) |
| KS2.4 | Risk appetite and tolerance |
| KS2.5 | Organizational risk management policies |
| KS2.6 | Parameters for risk response selection |
| KS2.7 | Project management tools and techniques |
| KS2.8 | Portfolio, investment and value management |
| KS2.9 | Exception management |
| KS2.10 | Residual risk |

# C. The Risk Response Process

## Introduction

This section provides knowledge on developing and selecting an appropriate response to a given risk based on cost/benefit analysis, business objectives and risk culture.

## Importance of Defining a Risk Response

The purpose of defining a risk response is to ensure that the residual risk is within the limits of the risk appetite and tolerance of the enterprise.

Risk response is based on selecting the correct, prioritized response to risk, based on the level of risk, the enterprise's risk tolerance and the cost-benefit advantages of the selected risk response option.

The risk response process integrates with the other risk management processes (identification, assessment, evaluation and monitoring) to ensure that management is provided accurate reports on:
- The level of risk faced by the enterprise
- The types of incidents that have occurred
- Any change to the enterprise's risk profile based on changes in the (internal and external) risk environment

> **Note:** Risk should always be reported based on the risk to the business, the ability of the business to meet its objectives and the risk to IT systems.

## 1. Overview of the Risk Response Process

### Description of the Risk Response Process

The risk response process is triggered when a risk exceeds the enterprise's risk tolerance level.

The prioritization of the risk responses and development of the risk response plan is influenced by several parameters:
- Cost of the response to reduce risk to within tolerance levels
- Importance of the risk
- Capability to implement the response
- Effectiveness of the response
- Efficiency of the response

Since not all risk can be addressed at the same time and remediation may take considerable investment in time and resources, a risk prioritization strategy is used to create a risk response plan and implementation schedule. Risk with a greater likelihood and impact on the enterprise will—in most cases—be prioritized above other risk that is considered less likely or less damaging.

## Exhibit 2.1: Risk Response Process

**Exhibit 2.1** illustrates how the risk response process is driven by input from the risk assessment process.

**Exhibit 2.1: Risk Response Process**

Risk Analysis
Risk Scenarios
Risk
Risk Tolerance
Risk Analysis
Phase 1
Phase 2
Risks Exceeding Risk Tolerance Level
Select Risk Response Options
Parameters for Risk Response Selection
Cost of Response to Reduce Risk Within Tolerance Levels
Importance of Risk
Capability to Implement Response
Effectiveness of Response
Efficiency of the Response
Risk Response Options
1. Avoid
2. Reduce/Mitigate
3. Share/Transfer
4. Accept
Risk Responses
Prioritise Risk Response Options
Risk Response Prioritisation
Business Case
Quick Wins
Defer
Business Case
Current Risk Level
Effectiveness/Cost Ratio
Prioritised Risk Responses
Risk Action Plan
Risk Response

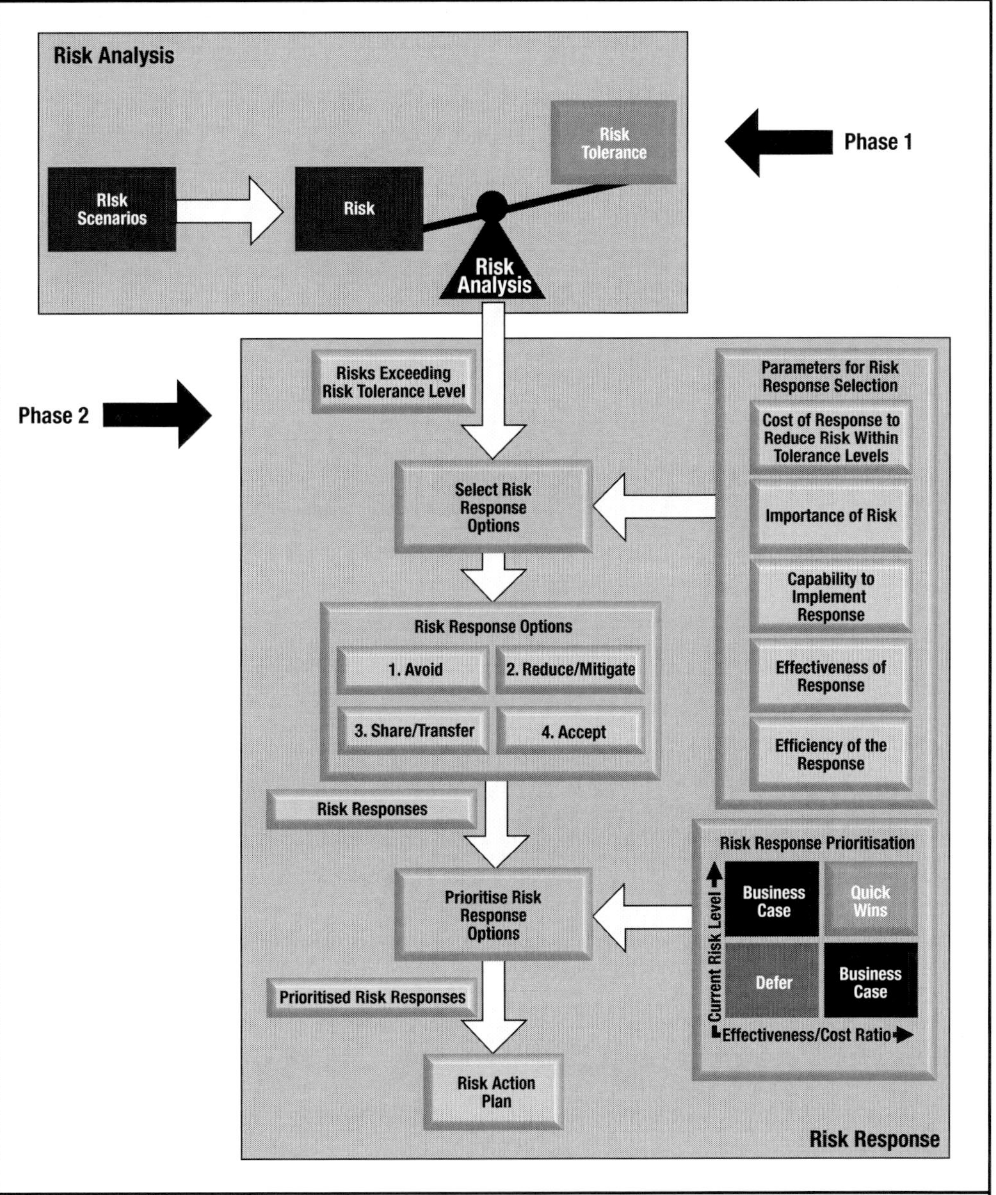

Source: ISACA, *The Risk IT Practitioner Guide*, USA, 2009, figure 46

## High-level Risk Response Process

The following table describes the risk response process at a high level.

| High-level Risk Response Process Phases | | |
|---|---|---|
| **Phase** | **Description** | **Notes** |
| 1 | Review results of the risk analysis.<br><br>Determine if the risk level exceeds the risk tolerance level:<br>• If **yes**, go to phase 2.<br>• If **no**, no action is required. | Risk analysis identifies and prioritizes risk levels according to risk scenarios and makes assessments based on the likelihood and magnitude of the risk, taking into account potential business impact.<br><br>The risk assessment should also provide recommendations for risk response. |
| 2 | Select risk response options. | In instances where the risk analysis shows that risk is not within the defined risk tolerance levels, weigh projected risk vs. the potential cost of implementing and maintaining controls and select the most appropriate response. |
| 3 | Prioritize the risk response option. | Select a risk response implementation strategy according to the priorities for addressing risk, and develop the risk action plan. |
| 4 | Implement the risk action plan. | Implement the selected risk response according to the risk action plan. This is covered in more detail in domain 4, with a focus on IS control design and implementation. |

## 2. Risk Response Options

### Introduction

This topic describes the four key risk response options utilized to ensure that the enterprise follows the best possible risk response strategy and manages its risk environment to gain the positive benefits of risk while minimizing the negative effects.

### Risk Response Options

Enterprises must choose between the following risk response options:
- Risk avoidance (avoid)
- Risk mitigation (reduce/mitigate)
- Risk sharing (share/transfer)
- Risk acceptance (accept)

### Risk Avoidance (Avoid)

Risk avoidance means that activities or conditions that give rise to risk are discontinued.

Risk avoidance applies when the level of risk, even after the selection of controls, would be greater than the risk tolerance level of the enterprise. This is the case when:
- There is no other cost-effective response that can succeed in reducing the likelihood and magnitude below the defined thresholds for risk appetite.
- The risk cannot be shared or transferred.
- The risk is deemed unacceptable by management.

**Examples:**
- Not engaging in electronic commerce (e-commerce) to avoid the risk associated with that line of business
- Not engaging in a very large project when the business case shows a significant risk of failure
- Not operating in some countries or regions due to safety concerns

### Risk Mitigation (Reduce/Mitigate)

Risk mitigation means that actions are taken to reduce:
- The likelihood and/or
- The impact of risk.

Risk mitigation can utilize various forms of control. A complete risk mitigation portfolio will consist of all types of controls carefully integrated together. The main control types to be considered are:
- Managerial (e.g., policies)
- Technical (e.g., tools such as firewalls and intrusion detection systems [IDSs])
- Operational (e.g., procedures, separation of duties)
- Preparedness activities

**Examples:**
- Strengthening overall risk management practices, such as implementing sufficiently mature risk management processes
- Deploying new technical, management or operational controls that reduce either the likelihood or the impact of an adverse event
- Installing a new access control system
- Implementing policies or operational procedures
- Developing an effective incident response and business continuity plan

### Risk Sharing (Share/Transfer)

Risk sharing means that risk impact is reduced by transferring or otherwise sharing a portion of the risk with an external enterprise or another internal entity.

> **IMPORTANT:** In both a physical and legal sense, these techniques **do not relieve an enterprise of a risk**, but can involve the skills of another party in managing the risk and can reduce the financial consequence if an adverse event occurs.

**Examples:**
- Taking out insurance coverage for disasters or incidents
- Outsourcing unique business processes
- Sharing project risk with other organizations through fixed price arrangements or shared investment arrangements

### Risk Acceptance (Accept)

Risk acceptance means that no action is taken relative to a particular risk; loss is accepted when/if it occurs.

> **Note:** This is different from being ignorant of risk. Accepting risk assumes that the risk is known; that is, an informed decision has been made by management to accept it as such.

If an enterprise adopts a risk acceptance stance, it should carefully consider who can accept the risk. Risk should be accepted only by senior business management in collaboration with senior management and the board.

**Examples:**
- Choosing not to implement costly controls to comply with regulatory requirements and paying the penalty for noncompliance, as applicable
- Selecting a start-up as a software supplier, which is not necessarily a "going concern," because the potential opportunity is promising
- Opting to conduct prosperous business in a politically volatile country

## 3. Risk Response Selection and Prioritization

**Introduction**

This topic contains information on the selection of an appropriate response and the prioritization of risk responses.

**Risk Response Considerations**

Consider the goals and objectives of an enterprise when selecting any of the risk mitigation options. It may not be practical to address all identified risk, so priority should be given to the threat and vulnerability pairs that have the highest potential to cause significant impact or harm to business objectives.

Also, in safeguarding an enterprise's objectives and assets, the option used to mitigate the risk, and the methods used to implement controls, may vary because of each enterprise's unique environment and objectives.

The "best in class" approach is to use appropriate technologies from among the various vendor solutions, along with the appropriate risk mitigation options and nontechnical, administrative measures.

**Exhibit 2.2: Risk Response Options and Parameters**

**Exhibit 2.2** illustrates different high-level risk response options and the parameters that influence the selection of these options.

Exhibit 2.2: Risk Response Options and Parameters

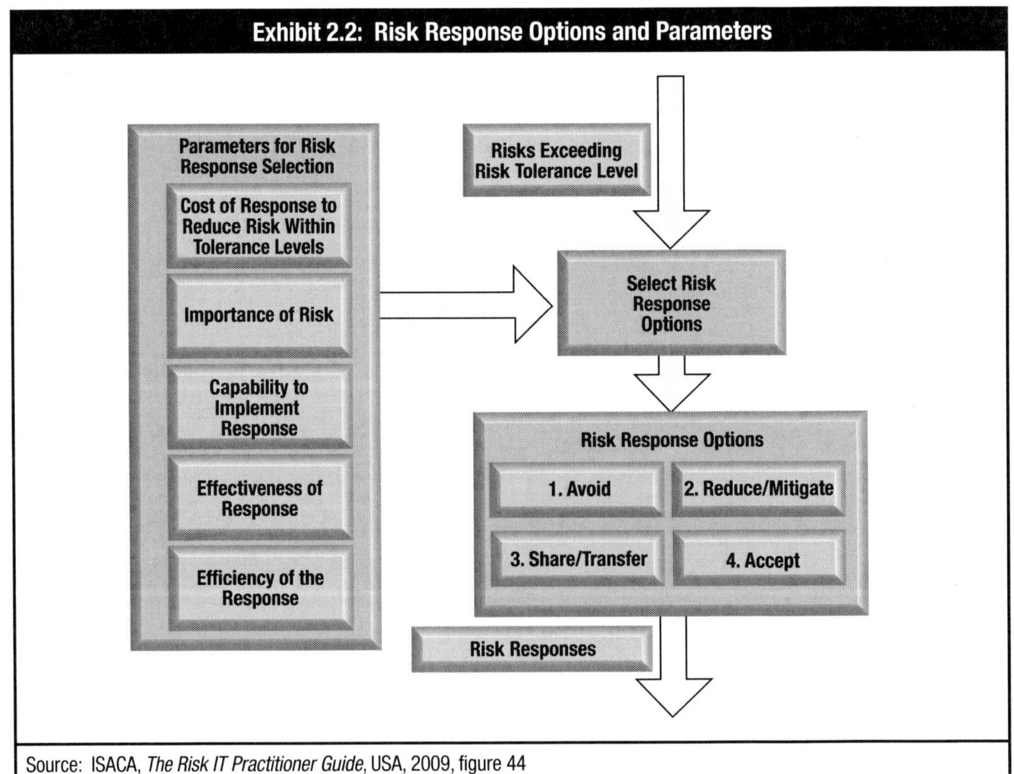


Source: ISACA, *The Risk IT Practitioner Guide*, USA, 2009, figure 44

## Parameters for Risk Response Selection

The following table lists the parameters to be considered when selecting a specific risk response option, as illustrated in **exhibit 2.3**.

<table>
<tr><th colspan="3">Risk Response Selection Parameters</th></tr>
<tr><th>Parameter</th><th colspan="2">Description</th></tr>
<tr><td rowspan="4">Cost of response</td><td colspan="2">The cost of the response to reduce risk to within tolerance levels</td></tr>
<tr><th>In the case of ...</th><th>The cost of response is the cost ...</th></tr>
<tr><td>Risk transfer,</td><td>Of the insurance premium.</td></tr>
<tr><td>Risk mitigation,</td><td>To implement and maintain control measures.<br><br>Examples: Capital expense, salaries and consulting, licensing, maintenance, training</td></tr>
<tr><td>Importance of risk</td><td colspan="2">The importance of the risk is reflected by:<br>• The combination of likelihood and magnitude (impact) levels (both quantitative and qualitative impact measures)<br>• Its position on the risk map compared to other risk</td></tr>
<tr><td rowspan="4">Capability to implement response</td><td colspan="2">The enterprise's capability to implement the response</td></tr>
<tr><th>When the risk management process is ...</th><th>The appropriate risk response may be...</th></tr>
<tr><td>Mature,</td><td>More sophisticated.</td></tr>
<tr><td>Immature,</td><td>Very basic.</td></tr>
<tr><td>Effectiveness of response</td><td colspan="2">The extent to which the response will reduce the likelihood and/or the impact of the risk</td></tr>
<tr><td>Efficiency of response</td><td colspan="2">The relative benefits promised compared to those listed in the "Risk Response Prioritization Option Descriptions" table (following)<br><br>Example: One type of risk response control may effectively address several risk factors while another may not.</td></tr>
</table>

## Need for Risk Response Prioritization

It is likely that the aggregated required effort for the mitigation responses, i.e., the collection of controls that need to be implemented or strengthened, will exceed available resources.

In this case, prioritization of the risk response is required.

## Exhibit 2.3: Risk Response Prioritization Options

**Exhibit 2.3** and the following table depict and describe the prioritization of risk responses based on the outcomes that they offer by placing the probable outcomes in a quadrant.

**Exhibit 2.3: Risk Response Prioritization Options**

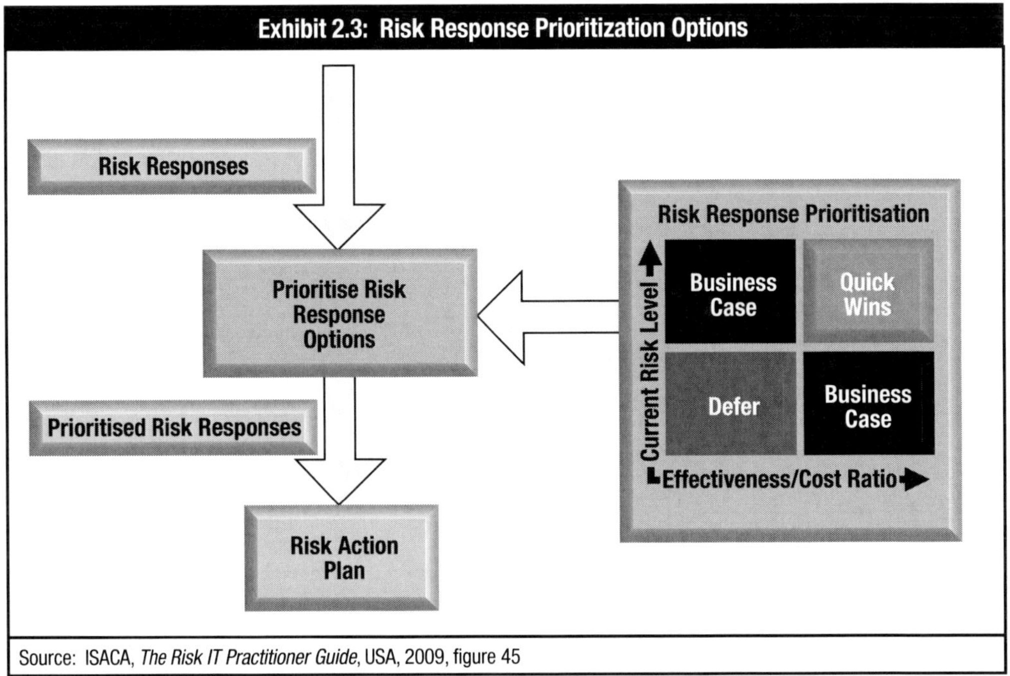


Source: ISACA, *The Risk IT Practitioner Guide*, USA, 2009, figure 45

| Risk Response Prioritization Options | |
|---|---|
| **Option** | **Description** |
| Quick win | Very effective and efficient response that addresses medium to high risk |
| Business case to be made | Requires careful analysis and management decisions on investments:<br>• More expensive or difficult risk responses to medium to high risk<br>• Efficient, effective responses to low to medium risk |
| Deferral | Costly risk response to a low risk |

## IT-related Risk Response Prioritization Examples

**Example 1:** A risk has been identified that the enterprise's IT system and application landscape is so complex that, within a few years, extending capacity will become difficult and maintaining software will become very expensive.

- The response alternatives consist of:
  - Major rearchitecture and redesign of the existing system
  - Purchase of a new, integrated system
- This is categorized as a "business case to be made" because of the project cost.

**Example 2:** A risk of noncompliance with regulations is identified because a number of relatively simple procedures are missing.

- The response consists of creating the missing procedures and implementing them.
- This is categorized as a "quick win" because the allocation of existing resources or a minor resource investment provides measurable benefits.

**Note:** Risk response should be selected with careful consideration of the impact of any controls on the business. Controls should not be so restrictive that they impede the ability of the business to be productive or profitable. Controls should also be reviewed to ensure that they are legal, implemented fairly, and accepted by management and employees.

## 4. Risk Response Implementation and Reporting

**Risk Response Plan**

Once the risk response strategy has been decided, the risk practitioner will create a risk response plan. This plan will outline the steps, timelines, budgets, people and tools needed to implement the risk response strategy.

**Risk Response Selection and Prioritization Guidelines**

Risk response prioritization considerations should take into account factors such as:
- Stakeholder interests
- Acceptance of change
- Balance of technical and nontechnical solutions
- Cost
- Impact on productivity
- Ownership of controls
- Ability to audit and monitor risk
- Regulations
- Changing market conditions

**Risk Response Tracking**

As part of risk response, the ongoing status of risk mitigation processes must be tracked. This tracking is often done using a risk register. This is important to ensure that the risk response strategy remains active and that proposed controls are implemented according to schedule. When an enterprise is aware of a risk, but does not have a justifiable risk response strategy, or is not following its strategy, the liability of the enterprise to adverse publicity or even civil or criminal penalties increases.

**Risk Response Integration**

The risk practitioner should always look for opportunities to achieve greater efficiency by integrating risk response options to address more than one risk. The use of techniques that are versatile and enterprisewide, rather than individual solutions, provides better justification for risk response strategies and related costs.

**Example:** Deploying an access control system that supports more than one system

**Risk Response Implementation**

The implementation of IS controls should consider the following:
- Controls are tested prior to implementation whenever possible.
- People are trained in the use of the tools.
- A control owner is clearly identified and responsible for the control.
- The control is measureable.
- The control is monitored to ensure that it remains effective over time (see Domain 5—Information Systems Control Monitoring and Maintenance).

> **Note:** The implementation of IS controls is addressed in detail in Domain 4—Information Systems Control Design and Implementation.

# D. Risk Response Process Details

## Risk Response Process

The overview below outlines the phases, tasks and steps in the risk response process.

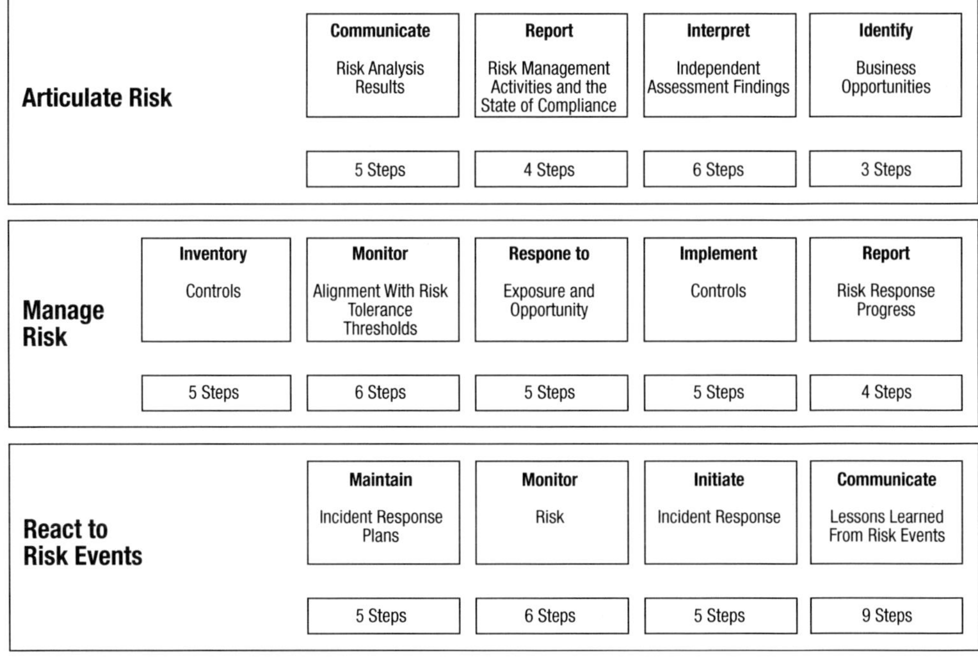


## Phase 1—Articulate Risk

### Introduction

Phase 1 of the risk response process requires articulating (documenting and reporting) risk to ensure that information on the true state of exposures and opportunities is made available:
- In a timely manner
- To the right people to enable the appropriate response

### Tasks Associated With Phase 1

The following table lists the tasks to articulate risk.

| Tasks to Articulate Risk | |
|---|---|
| **Task** | **Name/Description** |
| 1 | Communicate risk analysis results. |
| 2 | Report risk management activities and the state of compliance. |
| 3 | Interpret independent risk assessment findings. |
| 4 | Identify business opportunities. |

## Steps Associated With Task 1

The following table describes the steps to communicate risk analysis results.

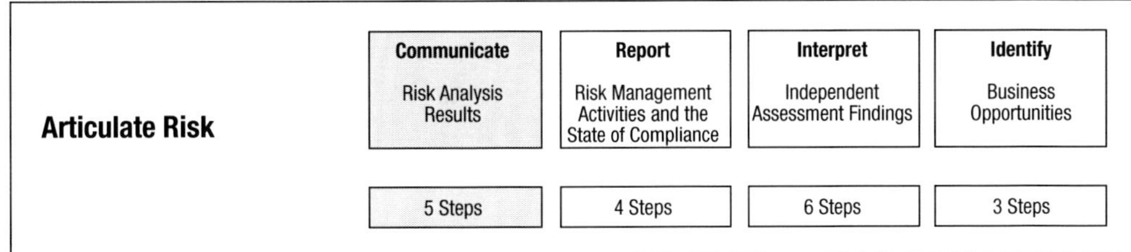


| Steps to Communicate Risk Analysis Results | |
|---|---|
| **Step** | **Action** |
| 1 | Report the results of risk analysis in terms and formats useful to support business and risk management decisions. |
| 2 | Coordinate additional risk analysis activity as required by decision makers.<br><br>**Example:** Report rejection and scope adjustment |
| 3 | Clearly communicate the risk-return context, including, wherever possible, probabilities of loss and/or gain, ranges, and confidence levels that enable management to balance risk-return ratios. |
| 4 | Identify the:<br>• Negative impacts of events/scenarios that drive response decisions<br>• Positive impacts of events/scenarios that represent opportunities that management should channel back into the strategy- and objective-setting process. |
| 5 | Provide decision makers with an understanding of:<br>• Worst-case and most probable scenarios<br>• Due diligence exposures<br>• Significant reputation, legal or regulatory considerations including:<br>– Key components of risk (e.g., likelihood, magnitude, impact) and key risk factors and their estimated effects<br>– Estimated probable loss magnitude or probable future gain<br>– Estimated high-end loss/gain potential and the most probable loss/gain scenario(s) (e.g., a probable loss likelihood of between three and five times per year and a probable loss magnitude of between US $50,000 and US $100,000 with 90 percent confidence)<br>– Additional relevant information to support the conclusions and recommendations of the analysis |

## Steps Associated With Task 2

The following overview and table describe the steps to report risk management activities and the state of compliance.

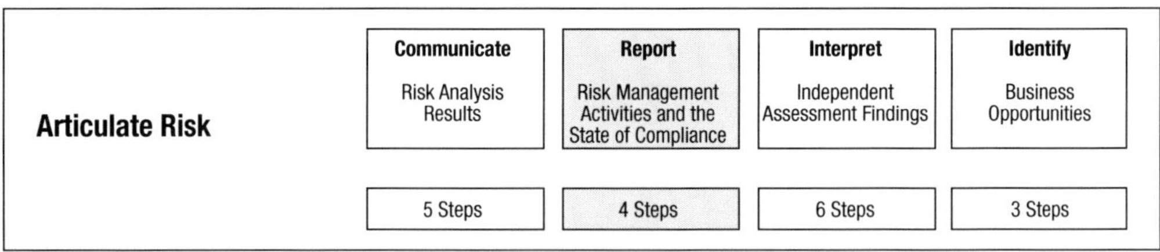


| Steps to Report Risk Management Activities and the State of Compliance | |
|---|---|
| **Step** | **Action** |
| 1 | Meet the risk reporting needs of various stakeholders.<br><br>**Example:** Board, risk committee, risk control functions, business unit management |
| 2 | Apply the principles of relevance, efficiency, timeliness and accuracy to ensure strategic and efficient reporting on risk issues and status. |
| 3 | When reporting, include the following:<br>• Control effectiveness and performance<br>• Issues and gaps<br>• Remediation status<br>• Events and incidents<br>• Impacts of events and incidents on the risk profile<br>• Performance of risk management processes |
| 4 | Provide inputs to integrated enterprise reporting. |

## Steps Associated With Task 3

The following overview and table describe the steps to interpret independent risk assessment findings.

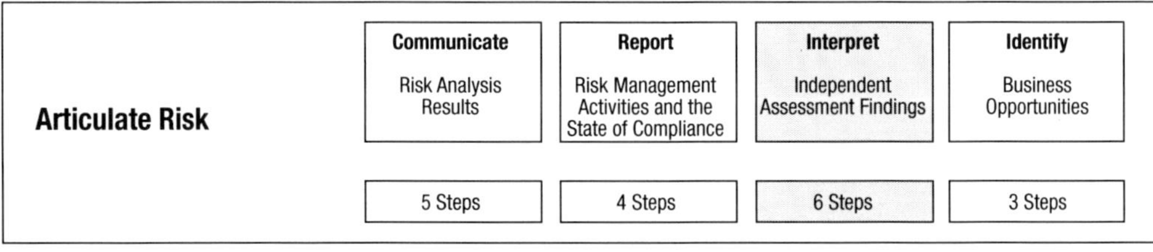


| Steps to Interpret Independent Risk Assessment Findings | |
|---|---|
| **Step** | **Action** |
| 1 | Review the results and specific findings of objective third parties, internal audit, quality assurance, self-assessment activities, etc. |
| 2 | Map the results/findings to the risk profile and the risk and control baseline. |
| 3 | Consider the established risk tolerance. |
| 4 | Take gaps and exposures to the business for its decision on disposition or the need for risk analysis. |
| 5 | Help the business understand how corrective action plans will affect the overall risk profile. |
| 6 | Identify opportunities for integration with:<br>• Other remediation efforts<br>• Ongoing risk management activities |

## Steps Associated With Task 4

The following overview and table describe the steps to identify business opportunities.

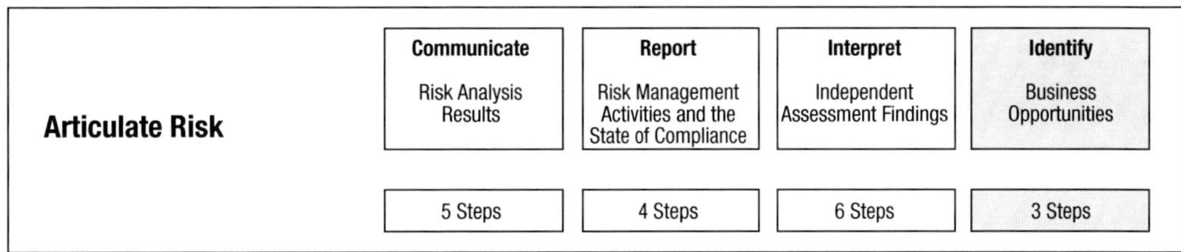


| Steps to Identify Business Opportunities | |
|---|---|
| **Step** | **Action** |
| 1 | On a recurrent basis, consider the relative levels of risk to risk management capacity for specific business processes, business units, products, etc. |
| 2 | For areas with relatively high-risk capacity capability (i.e., indicating an ability to take on more risk), identify opportunities that could:<br>• Enable the area to accept greater risk<br>• Enhance growth and return |
| 3 | Look for opportunities in which resources can be leveraged to:<br>• Create competitive advantage (e.g., use existing information in new ways, better leverage human and business resources)Reduce enterprise coordination costs<br>• Exploit scale and scope economies in certain key strategic resources common to several lines of business<br>• Coordinate activities among business units or in the value chain |

## Phase 2—Manage Risk

### Introduction

Phase 2 of the risk response process requires managing risk to ensure that measures for seizing strategic opportunities and reducing risk to an acceptable level are managed as a portfolio.

### Tasks Associated With Phase 2

The following table lists the tasks to manage risk.

> **Note:** It must be understood that the tasks are not sequential in nature and that monitoring is a continuous process.

| Tasks to Manage Risk | |
|---|---|
| **Task** | **Name/Description** |
| 1 | Inventory controls. |
| 2 | Monitor operational alignment with risk tolerance thresholds. |
| 3 | Respond to discovered risk exposure and opportunity. |
| 4 | Implement controls. |
| 5 | Report IT risk response plan progress. |

## Steps Associated With Task 1

The following overview and table describe the steps to inventory controls.

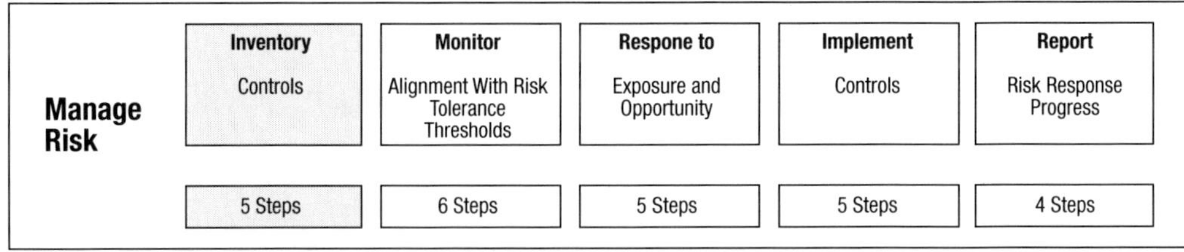


| Steps to Inventory Controls | |
|---|---|
| **Step** | **Action** |
| 1 | Inventory the controls in place across the risk focus areas.<br><br>**Rationale:** Manages and enables risk to be taken in line with risk appetite and tolerance |
| 2 | Classify controls, and map them to specific IT risk statements and aggregations of IT risk.<br><br>**Examples:** Predictive, preventive, detective and corrective controls |
| 3 | Develop tests for control design and control operating effectiveness. |
| 4 | Identify procedures and technology used to monitor the operation of controls.<br><br>**Examples:** Monitoring of controls when IT is involved or the automation of enterprise monitoring processes |
| 5 | Partition operational controls into the following categories:<br>• Controls deployed in line with expectations with no known operating deficiencies<br>• Controls deployed in line with expectations with known operating deficiencies<br>• Controls deployed beyond expectations with no known operating deficiencies<br><br>**Note:** This third category of controls may not be justified and could indicate opportunity for cost reduction while maintaining the same level of risk. |

## Steps Associated With Task 2

The following overview and table list the steps to monitor operational alignment with risk tolerance thresholds.

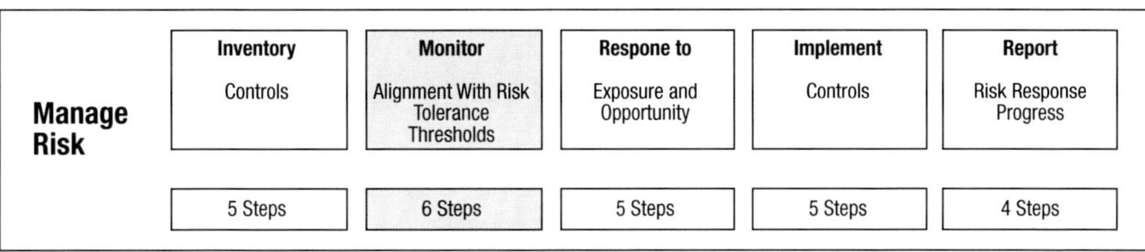


| Steps to Monitor Operational Alignment With Risk Tolerance Thresholds | |
|---|---|
| **Step** | **Action** |
| 1 | Ensure that each business line accepts accountability for:<br>• Operating within its individual and portfolio tolerance levels<br>• Embedding monitoring tools into key operating processes<br>• Monitoring control performance<br>• Measuring variance from thresholds against objectives |
| 2 | Periodically test control design and operating effectiveness for key risk issues. |
| 3 | Obtain buy-in from management on indicators that will function as key risk indicators (KRIs). |
| 4 | When implementing KRIs:<br>• Set thresholds and checkpoints (e.g., weekly, daily, continuously).<br>• Configure where to send notifications (e.g., line management, senior management, internal audit) so that the recipients can respond or adjust their plans. |
| 5 | Integrate KRI data into ongoing performance indicator reporting. |
| 6 | Ensure that there is a detailed examination of areas of residual risk outside of tolerance thresholds (e.g., request risk analysis). |

## Steps Associated With Task 3

The following overview and table describe the steps to respond to discovered risk exposure and opportunity.

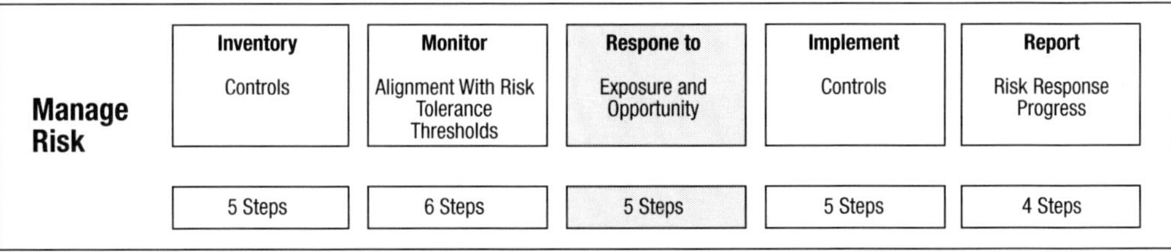


| Steps to Respond to Discovered Risk Exposure and Opportunity | |
|---|---|
| **Step** | **Action** |
| 1 | Emphasize projects that are expected to reduce the potential likelihood and magnitude of adverse events/losses, and balance them with projects that enable the seizing of strategic business opportunities. |
| 2 | Hold cost-benefit discussions regarding the contribution of new or existing controls toward operating within risk tolerance. |
| 3 | Select candidate controls based on:<br>• Specific threats<br>• The degree of risk exposure<br>• Probable loss<br>• Mandatory requirements specified in internal and/or external standards |
| 4 | Monitor changes to the underlying business operational risk profiles. |
| 5 | Adjust the rankings of risk response projects. |

## Steps Associated With Task 4

The following overview and table describe the steps to implement controls.

**Note:** The design and implementation of information systems controls is addressed in more detail in Domain 4—Information Systems Control Design and Implementation.

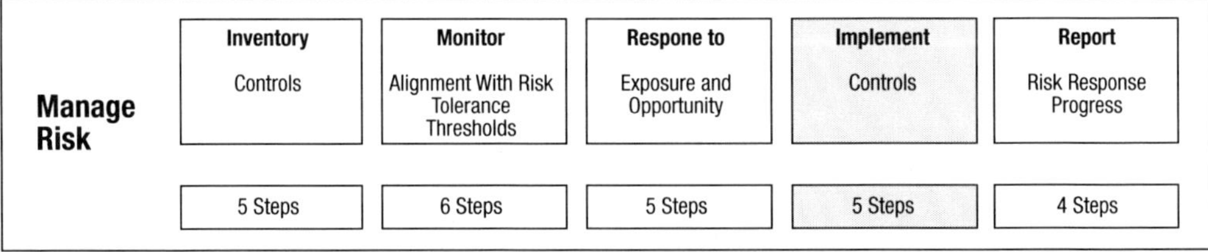


| Steps to Implement Controls | |
|---|---|
| **Step** | **Action** |
| 1 | Take appropriate steps to ensure the effective deployment of new controls and adjustments to existing controls. |
| 2 | Communicate with key stakeholders early in the process. |
| 3 | Before relying on the control:<br>• Conduct pilot testing.<br>• Review performance data to verify operation against design. |
| 4 | Map new and updated operational controls to monitoring mechanisms that will:<br>• Measure control performance over time<br>• Prompt management corrective action when needed |
| 5 | Identify and train staff on new procedures as they are deployed. |

## Steps Associated With Task 5

The following overview and table describe the steps to report IT risk response plan progress.

**Note:** Monitoring of risk responses, particularly information systems controls, is covered in more detail in Domain 5—Information System Monitoring and Maintenance.

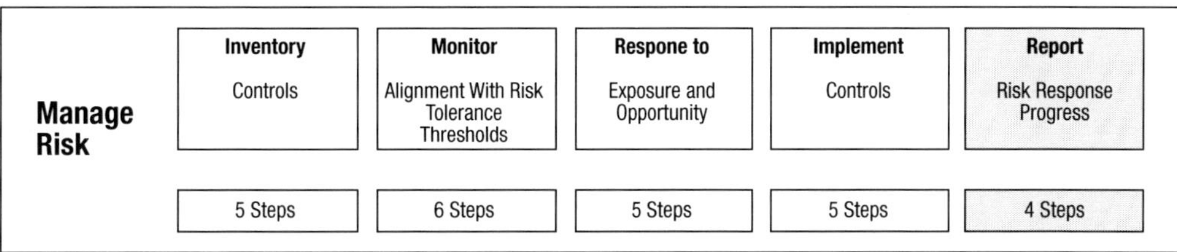


| Steps to Report Risk Response Plan Progress | |
|---|---|
| **Step** | **Action** |
| 1 | Monitor risk response plans at all levels. |
| 2 | Ensure the effectiveness of required risk responses. |
| 3 | Determine whether acceptance of residual risk has been obtained. |
| 4 | Ensure that committed risk responses are owned by the affected process owner(s) and that deviations are reported to senior management. |

## Phase 3—React to Risk Events

### Introduction

Phase 3 of the risk response process requires reacting to risk events to ensure that measures for seizing immediate opportunities or limiting the magnitude of loss from events are activated in a timely manner and are effective.

### Tasks Associated With Phase 3

The following table lists the tasks to react to risk events.

| Tasks to React to Risk Events | |
|---|---|
| **Task** | **Name/Description** |
| 1 | Maintain incident response plans. |
| 2 | Monitor risk. |
| 3 | Initiate incident response. |
| 4 | Communicate lessons learned from risk events. |

## Steps Associated With Task 1

The following overview and table describe the steps to maintain incident response plans.

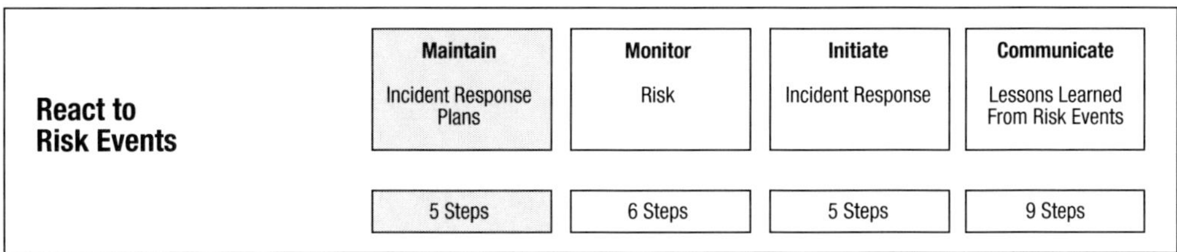


| Steps to Maintain Incident Response Plans | |
|---|---|
| **Step** | **Action** |
| 1 | Prepare for the materialization of threats through plans that document the specific steps to take when a risk event may cause an operational, developmental and/or strategic business impact (i.e., IT-related incident) or has already caused a business impact. |
| 2 | Maintain open communication about risk acceptance, risk management activities, analysis techniques and results available to assist with plan preparation. |
| 3 | When developing action plans, consider how long the enterprise may be exposed and the time it may take to recover. |
| 4 | Based on the potential or known impact, define pathways of escalation across the enterprise, from line management to executive committees. |
| 5 | Verify that incident response plans for highly critical processes are adequate. |

## Steps Associated With Task 2

The following overview and table describe the steps to monitor risk.

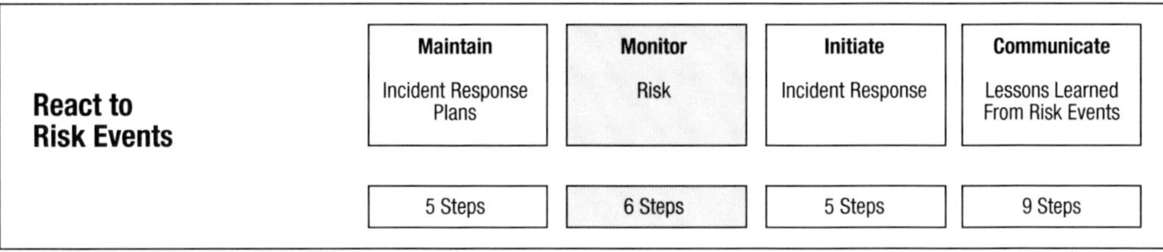


| Steps to Monitor Risk | |
|---|---|
| **Step** | **Action** |
| 1 | Monitor the environment. |
| 2 | When a control limit has been breached, either escalate to the next step or confirm that the measure is back within limits. |
| 3 | Categorize incidents (e.g., loss of business, policy violation, system failure, fraud, lawsuit), and compare actual exposures against acceptable thresholds. |
| 4 | Communicate business impacts to decision makers. |
| 5 | Continue to take action and drive desired outcomes. |
| 6 | Ensure that policy is followed and that there is clear accountability for follow-up actions. |

## Steps Associated With Task 3

The following overview and table describe the steps to initiate incident response.

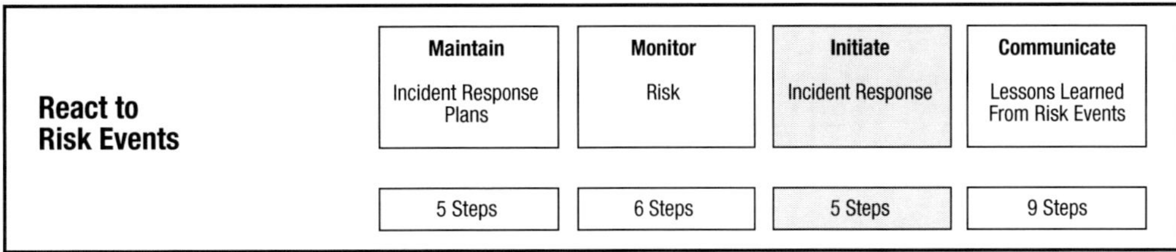


| Steps to Initiate Incident Response | |
|---|---|
| **Step** | **Action** |
| 1 | Take action to minimize the impact of an incident in progress. |
| 2 | Identify the category of the incident, and follow the steps in the response plan. |
| 3 | Inform all stakeholders and affected parties that an incident is occurring. |
| 4 | Identify the amount of time required to carry out the plan, and make adjustments, as necessary, for the situation at hand. |
| 5 | Ensure that the correct action is taken. |

## Steps Associated With Task 4

The following overview and table describe the steps to communicate lessons learned from risk events.

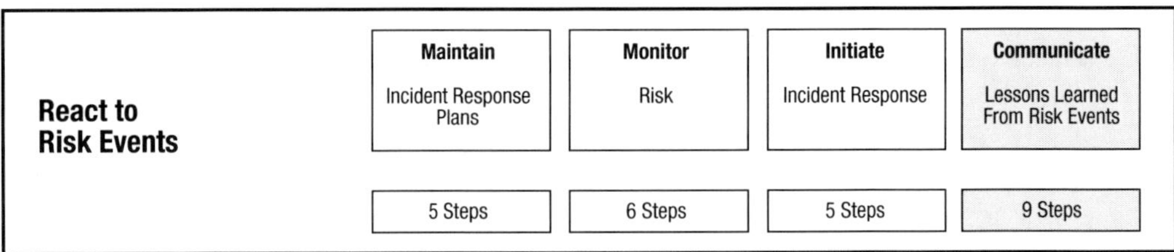


| Steps to Communicate Lessons Learned From Risk Events | |
|---|---|
| **Step** | **Action** |
| 1 | Examine past adverse events/losses and missed opportunities. |
| 2 | Determine whether there was a failure stemming from lack of awareness, capability or motivation. |
| 3 | Research the root cause of similar risk events and the relative effectiveness of actions taken then and now. |
| 4 | For behavioral incidents, determine the extent of any underlying problems.<br><br>**Example:** A serious systemic problem vs. an isolated case that could be managed through staff training or greater documentation of procedures |
| 5 | Identify tactical corrections; potential investments in projects; or adjustments to overall risk governance, evaluation and/or response processes. |
| 6 | To identify and correct the underlying root causes of operations and service delivery incidents and service levels (e.g., defects, rework), integrate with the:<br>• Service center<br>• Incident response process<br>• Problem management process |
| 7 | Identity the root cause of incidents through open communication across business and IT functions. |
| 8 | Request additional risk analysis as needed. |
| 9 | Communicate root cause, additional risk response requirements and process improvements to risk governance processes and appropriate decision makers. |

## Summary

This domain is an important function of the risk management process through the determination of a cost-benefit analysis and recommendation of:
- Prioritization of risk responses, including IS control implementation efforts
- Appropriate risk response
  - Necessary level of control

The response strategies selected throughout this domain will be implemented during the next domains and will also be expanded in more detail in Domain 4—Information Systems Control Design and Implementation.

## E. Suggested Resources for Further Study

### Suggested Resources for Further Study

In addition to the resources cited throughout this manual, the following resources are suggested for further study:

- ISACA:
  - COBIT 4.1, 2007
    **Note:** The COBIT 4.1 framework is available at no charge from ISACA and can be downloaded at *www.isaca.org/cobit*. The new COBIT 5 framework will be available in 2012.
  - *Enterprise Value: Governance of IT Investments, The Val IT Framework 2.0*, 2008
  - *The Risk IT Framework*, 2009
  - *The Risk IT Practitioner Guide*, 2009
- International Organization for Standardization (ISO), ISO 27005:2011, *Information technology—Security techniques—Information security risk management*, Switzerland, 2011
- Project Management Institute, *A Guide to the Project Management Body of Knowledge (PMBOK), 4th Edition*, USA, 2008

**Page intentionally left blank**

# Domain 3–Risk Monitoring

## A. Chapter Overview

### Introduction

This chapter provides information on monitoring risk and communicating information to the relevant stakeholders to ensure the continued effectiveness of the enterprise's risk management strategy.

### Inputs From Other Domains

In the previous domains—*Risk Identification, Assessment and Evaluation; and Risk Response*—risk was identified and prioritized and risk response activities were chosen and implemented to ensure that risk was within acceptable limits.

### Relevance

This phase of risk management concerns itself with monitoring and reporting on risk, taking into consideration risk events, threat events and vulnerability events. Risk monitoring ensures that current and emerging risk is within the risk tolerance levels of the enterprise.

This requires the determination of key risk indicators (KRIs) and risk reporting schedules to help ensure that:

- Senior management, operational managers, auditors, regulators, business continuity planners and security staff are aware of current risk and that
- Risk responses effectively mitigate the risk facing the enterprise.

### Outputs to Other Domains

Risk monitoring provides input into the risk response and information systems control design and monitoring processes.

Information systems control monitoring can be seen as a distinct subset of risk monitoring.

### Learning Objectives

As a result of completing this chapter, the CRISC candidate should be able to:

- Explain the principles of risk ownership
- List common risk and compliance reporting requirements, tools and techniques.
- Describe various risk assessment methodologies.
- Differentiate between key performance indicators (KPIs) and key risk indicators (KRIs).
- Describe data extraction, aggregation, and analysis tools and techniques at a high level.
- Differentiate between various types of processes to review the enterprise's risk monitoring process.
- List various standards, frameworks and practices related to risk monitoring.

## Contents

This chapter contains the following sections:

# B. Task and Knowledge Statements

## Introduction

This section describes the task and knowledge statements for Domain 3, which focuses on developing and implementing risk responses to ensure that risk factors and events are addressed in a cost-effective manner and in line with business objectives.

## Domain 3 Task Statements

The following table describes the task statements for Domain 3.

| No. | Task Statement (TS) |
|---|---|
| TS3.1 | Collect and validate data that measure key risk indicators (KRIs) to monitor and communicate their status to relevant stakeholders. |
| TS3.2 | Monitor and communicate key risk indicators (KRIs) and management activities to assist relevant stakeholders in their decision-making process. |
| TS3.3 | Facilitate independent risk assessments and risk management process reviews to ensure that they are performed efficiently and effectively. |
| TS3.4 | Identify and report on risk, including compliance, to initiate corrective action and meet business and regulatory requirements. |

## Domain 3 Knowledge Statements

The following table describes the knowledge statements for Domain 3.

| No. | Knowledge Statement (KS)<br>Knowledge of: |
|---|---|
| KS3.1 | Standards, frameworks and leading practices related to risk monitoring |
| KS3.2 | Principles of risk ownership |
| KS3.3 | Risk and compliance reporting requirements, tools and techniques |
| KS3.4 | Key performance indicators (KPIs) and key risk indicators (KRIs) |
| KS3.5 | Risk assessment methodologies |
| KS3.6 | Data extraction, validation, aggregation and analysis tools and techniques |
| KS3.7 | Various types of reviews of the organization's risk monitoring process (e.g., internal and external audits, peer reviews, regulatory reviews, quality reviews) |

# C. Essentials of Risk Monitoring

## 1. Key Terms and Principles

### Introduction

This topic provides information on:
- Risk indicators
- Key risk indicators (KRIs)
- Key performance indicators (KPIs)

### Definition of Risk Indicators

Risk indicators are metrics capable of showing that the enterprise is subject to, or has a high probability of being subject to, a risk that exceeds the defined risk appetite.

They are used to measure levels of risk in comparison to defined risk thresholds and alert the enterprise when a risk level approaches a high or unacceptable level of risk. The purpose of a risk indicator is to set in place tracking and reporting mechanisms that alert staff to a developing or potential risk.

Risk indicators are specific to each enterprise, and their selection depends on a number of parameters in the internal and external environment, including, but not limited to the:
- Strategic focus of the enterprise
- Size and complexity of the enterprise
- Type of market in which the enterprise operates (e.g., highly regulated)

### Definition of Key Risk Indicator (KRI)

KRIs are those risk indicators that:
- are deemed highly relevant
  and
- possess a high probability of predicting or indicating important risk

KRIs are the prime risk monitoring indicators for the enterprise.

### Definition of Key Performance Indicator (KPI)

KPIs are measures that determine how well a process is performing in enabling the goal to be reached

A KPI is a lead indicator and provides an activity goal. It can be used to measure the success of a particular activity or as an indicator of the capabilities, practices and skills related to such an activity.

**Example:** A KPI may indicate that an error rate of five percent is acceptable. An error rate higher than two percent would be unacceptable and require escalation and some form of response.

## 2. Key Risk Indicator Selection

### Introduction

A common mistake when implementing KRIs—other than selecting too many KRIs—includes choosing KRIs that:
- Are not linked to specific risk
- Are incomplete or inaccurate due to unclear specifications
- Are difficult to measure
- Are difficult to aggregate, compare and interpret
- Provide results that cannot be compared over time

The selection and maintenance of appropriate KRIs is critical to the ongoing success of the risk monitoring process.

This topic provides an overview of:
- KRI benefits
- Factors influencing the selection of KRIs
- Criteria for KRI effectiveness
- Example of KRI reliability vs. sensitivity
- KRI maintenance

### KRI Benefits

The selection of the right set of KRIs benefits the enterprise by:
- Providing an early warning (forward-looking) signal that a high risk is emerging to enable management to take proactive action (before the risk actually becomes a loss)
- Providing a backward-looking view on risk events that have occurred, enabling risk responses and management to be improved
- Enabling the documentation and analysis of trends
- Providing an indication of the enterprise's risk appetite and tolerance through metric setting (that is, KRI thresholds)
- Increasing the likelihood of achieving the enterprise's strategic objectives
- Assisting in continually optimizing the risk governance and management environment

## Factors Influencing the Selection of Key Risk Indicators

The following table describes some factors to be considered when selecting key risk indicators.

| Factors Influencing the Selection of Key Risk Indicators | |
|---|---|
| **Factor** | **Description** |
| Stakeholders | Select risk indicators with the involvement of relevant stakeholders to ensure greater buy-in and ownership.<br><br>Risk indicators *should* be identified for all stakeholders and *should not* focus solely on IT, but should include the operational and strategic risk. |
| Balance | Make a balanced selection of risk indicators, covering:<br>• Lag indicators (indicating risk after events have occurred)<br>• Lead indicators (indicating which controls are in place to prevent events from occurring)<br>• Trends (analyzing indicators over time or correlating indicators to gain insights) |
| Root cause | Ensure that selected indicators drill down to the root cause of events and not just the symptoms.<br><br>The relation of a risk indicator to a root cause is not necessarily a one-to-one relationship. Therefore, it is important to map the unique root cause to a single or predefined set of indicators and vice versa to avoid false conclusions about the most appropriate risk response. |

## Criteria for KRI Effectiveness

The following table describes the criteria to be considered when selecting KRIs.

| Criteria for KRI Effectiveness | |
|---|---|
| **Criterion** | **Description** |
| Impact | Indicators of risk with high business impact are more likely to be KRIs. |
| Effort | For different indicators that are equivalent in sensitivity, the one that is easier to measure and maintain is preferred. |
| Reliability | The indicator must possess a high correlation with the risk and be a good predictor or outcome measure. |
| Sensitivity | The indicator must be representative of risk and capable of accurately indicating risk variances. |

**Note:** The complete set of KRIs should also balance indicators for risk, root causes and business impact.

## KRI Optimization

To ensure accurate and meaningful reporting, KRIs will need to be optimized to ensure that: 1) the right data are being collected and reported on, and 2) that the KRI thresholds are set correctly. KRIs that are reporting on the data points that cannot be controlled by the enterprise, or are not alerting management at the correct time to an adverse condition, must be adjusted (optimized) to be more precise, more relevant or more accurate.

The following table describes a few examples in which KRIs may need to be optimized.

| Examples in Which KRIs Should Be Optimized | |
|---|---|
| **Metric Criterion** | **Description** |
| Sensitivity | Management has implemented an automated tool to analyze and report on access control logs based on severity; the tool generates excessively large amounts of results. Management performs a risk assessment and decides to configure the monitoring tool to report only on alerts marked "critical." |
| Timing | Management has implemented strong segregation of duties (SoD) within the enterprise resource planning (ERP) system. One monitoring process tracks system transactions that violate the defined SoD rules before month-end processing is completed so that suspicious transactions can be investigated before reconciliation reports are generated. |
| Frequency | Management has implemented a key control that is performed multiple times a day. Based on a risk assessment, management decides that the monitoring activity can be performed weekly because this will capture a control failure in sufficient time for remediation. |
| Corrective action | Automated monitoring of controls is especially conducive to being integrated into the remediation process. This can often be achieved by using existing problem management tools, which help prioritize existing gaps, assign problem owners and track remediation efforts. |

## Example: Reliability vs. Sensitivity

**Example:** A smoke detector can be used to illustrate the difference between reliability and sensitivity:
- **Reliability**—The smoke detector will sound an alarm every time there is smoke.
- **Sensitivity**—The smoke detector will sound an alarm when a specified threshold of smoke density (number of particles per cubic foot) is reached.

## KRI Maintenance

Since the enterprise's internal and external environments are constantly changing, the risk environment is also highly dynamic and the set of KRIs needs to be changed over time.

Each KRI is related to the risk appetite and tolerance levels of the enterprise. KRI trigger levels should be defined at a point that enables stakeholders to take appropriate action in a timely manner.

## 3. Data Extraction, Validation, Aggregation and Analysis

### Introduction

Because KRIs often rely on information from diverse sources, it is important that the risk practitioner understands the basic concepts related to data extraction, validation, aggregation and analysis.

This section provides an overview of the tasks related to data extraction, validation, aggregation and analysis:
- Requirements gathering
- Data access
- Data validation
- Data analysis
- Reporting and corrective action

**Note:** Variations of these phases exist, but the steps within them and the principles used are generally the same.

### Requirements Gathering

It is important to first understand the risk to be monitored, prepare a detailed plan and define the project's scope. In the case of a monitoring project, this step should involve process owners, data owners, system custodians and other process stakeholders.

### Data Access

As part of the data access step, management identifies which data are available and how they be acquired in a format that can be used for analysis. There are two options for data extraction:
- Extracting data directly from the source system(s) after system owner approval
- Receiving data extracts from the system custodian (IT) after system owner approval

The recommended course of action is direct extract, especially since this risk monitoring generally involves management monitoring its own controls rather than auditors/third parties monitoring management's controls. If it is not feasible to get direct access, a data access request form should be submitted to the data owner(s) that details the appropriate data fields to be extracted. The request should specify the method of delivery for the file (i.e., posting on a dedicated server, via e-mail, or on CD/DVD). Most of the data analysis tools can handle any delimited text file, fixed length text file or spreadsheet.

### Data Validation

Data validation ensures that extracted data are ready (in the correct format) and accurate enough for analysis. One objective is to perform data quality tests to ensure data are valid, complete and free of errors or duplication. This may also involve reformatting data that have been gathered from different sources to make the data suitable for comparative analysis.

There is a risk that the data could become corrupted; either through data manipulation, in storage, or during data transfer. As most of the data used for risk monitoring is the result of a download, report, transfer or other form of operation, it is imperative to validate data before they are analyzed. Omitting the step of data validation may compromise the results of the analysis. Whereas specific data validation techniques may vary depending on the tool, one should consider the concepts in the following table every time data are extracted.

## Data Validation Practices

The following table provides an overview of common data validation concepts and practices.

| Concepts to Consider When Validating Data | |
|---|---|
| **Checking for ...** | **Helps ...** |
| Validity | Ensure that data match definitions in the table layout. |
| Control totals | Ensure that the data are complete. |
| Ranges | Ensure that extracted data contain only the data requested, e.g., if data are requested for the first quarter of the year (January, February and March), one should not be provided data for the entire year. |
| Missing items | Identify missing data, such as gaps in sequence or blank fields/records. |
| Duplicates | Identify and confirm the validity of duplicates. For example, many alarms may report a problem repeatedly. The data analyst may want to eliminate duplicate alarm reporting or set thresholds that will only report on certain alarm conditions if the alarm occurs more than ten times in a short time period. Such reporting thresholds may be called clipping levels. |
| Reliability | Determine the confidence that an analyst may have in the integrity of the data—Were the data extracted directly from a source? Were the data computed from other values? |
| Reasonableness | Make certain assumptions about the information; e.g., if the average number of transactions per month is 2,400, and the report from one month is substantially different from normal volumes, this may be an indication that a data source file is missing or duplicated. A data value that is substantially different from normal values (a statistical anomaly, for example) may need further validation.<br><br>Doing statistical analysis of data to determine data norms, standards of deviation and outliers, may help the analyst identify exceptional or erroneous data.<br><br>**Note:** This test requires familiarity with the data used for monitoring. |
| Relationships (sequencing) | Identify table fields that relate to each other. For example, in an invoice table, the due date should always be later than the invoice date. To test relationships, use a filter to compare one field to another. |
| Orphan records | Record, in a transaction or detail table, that the record has no match in a master table.<br><br>**Example:** There are two purchase order (PO) tables—one for header records and the other for line items. Every PO header record should match to at least one line; one or more line item records should match to one header record. Orphan records on either side could mean that orders are not properly completed and processed. |

**Note:** Statistical analysis helps assess data validity through a variety of techniques.

### Data Analysis

Analysis can involve a simple set of steps or can be a complex combination of commands and other functionality. Data analysis must be designed to achieve the stated objectives from the project plan. Although this may be applicable to any monitoring activity, it is beneficial to keep transferability and scalability in mind. This may include robust documentation, use of software development standards and naming conventions.

After the data extracts have been validated (this process can be automated), the enterprise should develop the logic to be used for data analysis. The logic should then be executed and reviewed for errors.

At this stage, it may be necessary to troubleshoot testing issues and refine the logic to ensure that the outcome is valid and accurate. This step also includes formatting the output for reporting purposes. The output should be reviewed again to ensure that it is providing the correct results. If necessary, the logic and output must be further refined and the previous steps repeated. The monitoring process can then be set up to be run on a repeatable basis at the appropriate levels of sensitivity, range, time and frequency.

**Note:** The final step in data analysis, which involves making logical conclusions about the data, remains an important task for the CRISC.

### Reporting and Corrective Action

Reporting structure and distribution depends on the requirements of the business and on regulatory demands. The format of the report may be determined by the monitoring objectives and the technology being used. Reporting procedures include the manner in which reports are distributed, and who should get the reports so that they are directed to the right people and in the right format, (e.g., dashboards, workflows). Similar to the data analysis stage, reporting may also identify areas in which changes to the sensitivity of the reporting parameters or the timing and frequency of the monitoring activity may be required.

### Risk Monitoring Capabilities

Risk monitoring tasks can range from *ad hoc* queries performed on demand, to scheduled, repeatable monitoring processes, or even to solutions that integrate risk monitoring processes into strategic, risk management and performance management processes. Each approach has different challenges and benefits.

## 4. Risk Monitoring

### Introduction

Risk monitoring provides timely information on the actual status of the enterprise with regard to risk. This includes information such as:
- The risk profile of the enterprise; i.e., the overall portfolio of (identified) risk to which the enterprise is exposed
- KRIs to support management reporting on risk
- Event/loss data
- The root cause of loss events
- Options to mitigate risk (cost and benefit calculations)

This topic provides information on:
- Internal and external risk monitoring sources

## Risk Monitoring Sources

A key factor in the value of the risk monitoring process is to ensure that risk reporting is accurate, timely and complete. If the data used to generate the reports are biased or incomplete, management will have an incorrect understanding of the true risk levels and the appropriate risk responses may not be implemented where required.

To ensure that the reports are correct and complete, it is important to gather input from all available sources.

The following sources of risk monitoring information should not be considered all-inclusive. In determining the propriety of any specific source risk practitioners should apply their professional judgment:

- Suppliers or manufacturers of hardware, software or applications
- Antivirus/antispam/content filters
- Devices placed in the demilitarized zone (DMZ)
- Mail servers
- Communication software
- Trade unions
- Computer emergency response team (CERT) alerts/blogs/newsgroups
- Governmental advice (US Department of Homeland Security)
- Outcomes from conferences (including DEF CON®)
- Newspapers
- Online news
- Amnesty International
- ArcSight, Archer
- Bloomberg
- *Citicus.com*, *www.limesurvey.org*
- Greenpeace
- iDefense Labs Vulnerability Contributor Program (VCP)
- *Nimsoft.com*, *www.curasoftware.com/Pages/default.asp*
- Red Cross
- Reuters
- Risk & Opportunity Management, Software Engineering Institute (SEI), Carnegie Mellon University (CMU), *www.sei.cmu.edu/risk/research/index.cfm*
- RSA FraudAction
- SANS: @Risk Critical Vulnerability List
- Symantec™ DeepSight™
- The SysAdmin, Audit, Network, Security [SANS™] Institute
- TippingPoint DVLabs
- United Nations
- US Computer Emergency Readiness Team (US-CERT), *www.us-cert.gov*
- US-CERT National Cyber Alert System, *www.us-cert.gov/cas/alldocs.html*
- US National Institute of Standards and Technology (NIST) National Vulnerability Database
- The World Factbook, US Central Intelligence Agency (CIA)
- World Health Organization (WHO)
- Swiss Confederation, Reporting and Analysis Centre for Information Assurance (MELANI), *www.melani.admin.ch*

**Specific, nongeneric sources are listed in alphabetical order.**

## 5. Process Capability Models

### Value of Using Process Capability Models

Process capability models are an excellent tool to measure the maturity or development of the processes and operational procedures of the enterprise. A capability model allows the enterprise to measure and track its progress to developing, implementing and following reliable, consistent and reportable procedures.

Capability models enable the enterprise to rate itself from the least mature level (having nonexistent or unstructured processes) to the most mature (having adopted and optimized the use of good practices).

Using a capability model, management can identify:
- **The actual performance of the enterprise**—Where is the enterprise today?
- **A benchmark of what other organizations are doing**—What are the industry leading practices or international standards as represented within the model?
- **The enterprise's target for improvement**—Where does the enterprise want to be at a point in the future?

### Relevance of Process Capability Models

Use of a process capability model helps determine the current risk management process capability level and allows management to determine whether it is in alignment with the desired state.

The model helps determine how to close the gap between actual and desired state and tracks process performance over time.

### Management Responsibility for Risk Management Process Capability

Boards and executive management need to consider how effective their enterprises are at managing risk and should be able to answer the following related questions:
- What is the enterprise's current risk management capability level?
- What is the enterprise's desired risk management capability level?
- How does the enterprise ensure that risk prioritization reflects an enterprisewide view and is followed consistently across the enterprise?
- How does the enterprise identify which activities are necessary to reach the desired risk management process capability level?

### Process Capability Levels

The levels within a process capability model are designed as profiles that allow an enterprise to identify symptoms or descriptions of its current and possible future states. In general, the purpose is to:
- Identify where enterprises are in relation to the consistent and reliable performance of certain activities or practices.
- Suggest how to identify areas for, and set priorities for, improvements.

## Exhibit 3.1: Process Capability Levels

Process capability levels and related performance attributes can be summarized as shown in **exhibit 3.1:**

**Exhibit 3.1: Process Capability Levels and Related Performance Attributes**

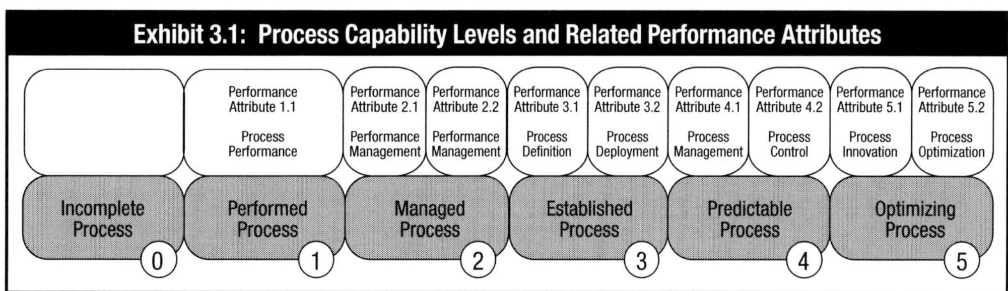


## Process Capability Levels

The following table describes the levels in the process capability levels:

| COBIT 5 ISO/ IEC 155-4-based Capability Levels | Meaning of the COBIT 5 ISO/IEC 155-4-based Capability Levels | Context |
|---|---|---|
| 5 Optimized | The previously described predictable process is continuously improved to meet relevant current and projected business goals. | Enterprise view/Corporate knowledge |
| 4 Predictable | The previously described established process now operates within defined limits to achieve its process outcomes. | |
| 3 Established | The previously described managed process is now implemented using a defined process that is capable of achieving its process outcomes. | |
| 2 Managed | The previously described performed process is now implemented in a managed fashion (planned, monitored and adjusted) and its work products are appropriately established, controlled and maintained. | Instance view/Individual knowledge |
| 1 Performed | The implemented process achieves its process purpose. | |
| 0 Incomplete | The process is not implemented or fails to achieve its process purpose. At this level, there is little or no evidence of any systematic achievement of the process purpose. | |

**Note:** The COBIT 4.1 framework is available at no charge from ISACA and can be downloaded at *www.isaca.org/cobit*. The new COBIT 5 framework will be available in 2012.

## Exhibit 3.2: Current and Desired Process Capability

To make the results of a process capability model assessment easily usable in management briefings, where they should be presented as a means to support the case for future plans to improve risk governance and management efforts, the following graphic presentation method, **exhibit 3.2**, may be useful.

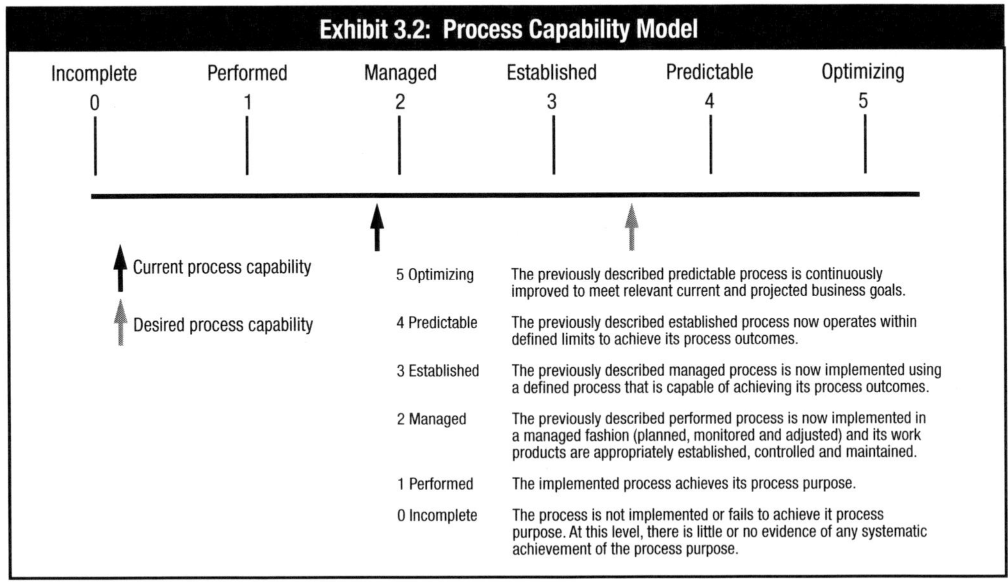


## 6. Threat Analysis

### Importance of Periodic Threat Analysis

Technical and behavioral threats to an enterprise evolve as a result of several internal and external factors, including:
- Implementation of new technologies
- Broader network and application access to partners and customers
- The ever-growing capabilities of attackers
- Lack of staff education or attentiveness

The risk practitioner must always be aware of changing threat levels or new and emerging threat vectors. The risk monitoring activity is a key tool for the risk practitioner to notice new threats and measure the effectiveness of the risk management program. All of these changes to threats warrant periodic reassessment of the threat landscape that an enterprise faces.

### CRISC Threat Analysis Responsibilities

Internal factors such as new business units, new or upgraded technologies, changes to products and services, and changes in roles and responsibilities all represent areas in which new threats may emerge.

This process of analyzing and communicating the impact on the enterprise's risk posture is critical to ensuring that stakeholders:
- Are aware of potential business impact
- May take actions to mitigate risk accordingly

**Note:** As new threats are identified and prioritized in terms of impact, the risk practitioner must help evaluate the ability of existing controls to mitigate risk associated with new threats and, in some cases, facilitate the:
- Modification of the technical architecture
- Deployment of a threat-specific countermeasure
- Implementation of a compensating mechanism or process until mitigating controls are developed
- Education of staff or business partners

### Timing of Threat Analysis—Annual and Incremental Approaches

The enterprise should perform a threat analysis at least annually by evaluating changes in the technical and operating environments of the enterprise, particularly where external entities are granted access to organizational resources.

**Note:** The enterprise may also choose to take an incremental approach, analyzing portions of the enterprise monthly or quarterly. This is often required for regulatory compliance or compliance with industry standards such as the Payment Card Industry Data Security Standard (PCI-DSS).

### Incremental Risk Assessment—Changing Asset Values and Risk Characteristics

The risk practitioner must recognize that asset values and risk characteristics can change, requiring reanalysis of risk posture.

**Example:** A company can grow increasingly revenue-dependent on an application that was initially not considered to be critical to the enterprise.

Asset value can increase or decrease over time in terms of real monetary value or strategic value to the enterprise. In addition, the risk associated with an asset can grow.

**Example:** A small database may initially contain only a few dozen personal information records; five years later, the same database may contain 10,000, representing a much higher impact if compromised.

## 7. Risk Reporting

### Introduction

Risk reporting, based on effective risk monitoring, helps highlight issues and enables management to make educated decisions on risk response activities. Much of the risk reporting will focus on the effectiveness of the controls and countermeasures implemented in the risk response phase. The risk practitioner must be able to report on whether the controls that were implemented are working effectively to mitigate risk and whether the residual risk to the enterprise is acceptable. The objective is to initiate corrective action and meet regulatory requirements.

## Reporting Content

Risk reporting covers a broad array of information flows and may include the major types of risk communication as shown in the following table.

| Types of Risk Communication | |
|---|---|
| **Reporting Content** | **Function** |
| Expectation | This is essential communication about the overall strategy the enterprise takes toward risk and drives all subsequent efforts on risk management. It sets the overall expectations from risk management.<br><br>**Example:** Risk strategy, policies, procedures, awareness training, continuous reinforcement of principles, etc. |
| Status | This includes information on the actual status with regard to risk, such as:<br>• The risk profile of the enterprise, i.e., the overall portfolio of (identified) risk to which the enterprise is exposed<br>• The root cause of loss events<br>• Thresholds for risk<br>• Options to mitigate (cost and benefits) risk<br>• Event/loss data<br>• KRIs to support management reporting on risk |
| Risk management capability | This information allows monitoring of the state of the "risk management engine" in the enterprise and is a key indicator for good risk management. It has predictive value for how well the enterprise is managing risk and reducing exposure. |
| Actionable items | When actionable items and roles and responsibilities of the risk owners are included in the risk reporting matrix, the risk report overlaps with the risk response plan. |

## Risk Reporting Criteria

To be effective and enable decision making, risk reporting has to be:
- Clear
- Concise
- Useful
- Timely
- Designed for the correct target audience
- Available on a need-to-know basis

The following table depicts the key focus areas for risk reporting.

| Focus Areas for Risk Reporting Criteria | |
|---|---|
| **Communication must be ...** | **To ...** |
| Clear | Enable understanding by all stakeholders. |
| Concise | Focus the reader on the key points.<br><br>Concise information is well structured and complete and avoids peripheral information, jargon and technical terms, unless necessary. |

## Risk Reporting Criteria *(cont.)*

| Focus Areas for Risk Reporting Criteria *(cont.)* | |
|---|---|
| **Communication must be ...** | **To ...** |
| Useful | Enable decision making.<br><br>Useful information is relevant and presented at the appropriate level of detail. Usefulness includes consideration of the target audience because information that may be useful to one party may not be useful to another. |
| Timely | Allow action at the appropriate moment to identify and treat the risk.<br><br>For each risk, critical moments exist between its origination and its potential business consequence; a delay in reporting may increase the level of impact.<br><br>**Example:** Communicating a project delay a week before the deadline serves no useful purpose. |
| Designed for the correct target audience | Enable informed decisions. Information must be communicated at the right level of aggregation and adapted for the audience.<br><br>Aggregation must not hide root causes of risk.<br><br>**Example:** A security officer may need technical data on intrusions and viruses to deploy solutions. An IT steering committee may not need this level of detail, but it does need aggregated information to decide on policy changes or additional budgets to treat the same risk. |
| Available on a need-to-know basis | Ensure that information related to IT risk is known and communicated to only those parties with a genuine need.<br><br>A risk register with all documented risk is not public information and should be properly protected against internal and external parties with no need for it. |

## Risk Reporting Channels

As risk management is an enterprisewide effort, communication flows to and from the CRISC.

The following table provides a quick overview of the most important communication channels for effective and efficient risk management including, but not limited to, those from the risk practitioner to other stakeholders. The list does not include the source and destination of the information nor the actions that should be taken on it.

| Risk Reporting Channels | | |
|---|---|---|
| **Input** | **Stakeholder** | **Output** |
| • Executive summary risk reports<br>• Current risk exposure/profile<br>• KRIs | Executive management and board | • Enterprise appetite for IT risk<br>• Key performance objectives<br>• IT risk Responsible, Accountable, Consulted and/or Informed (RACI) charts<br>• IT risk policies that express management's IT risk tolerance<br>• Risk awareness expectations<br>• Risk culture<br>• Risk analysis request |

## Risk Reporting Channels *(cont.)*

| Risk Reporting Channels *(cont.)* | | |
|---|---|---|
| **Input** | **Stakeholder** | **Output** |
| • Risk management scope and plan<br>• Risk register<br>• Risk analysis results<br>• Executive summary risk reports<br>• Integrated/aggregated risk report<br>• KRIs<br>• Risk analysis request | Chief risk officer (CRO) and enterprise risk committee | • Enterprise appetite for IT risk<br>• Residual IT risk exposures<br>• IT risk action plan |
| • Enterprise appetite for risk<br>• Risk management scope and plan<br>• Key performance objectives<br>• Risk RACI charts<br>• Risk framework and scoring methodology<br>• Risk register | Chief information officer (CIO) | • Residual risk exposures<br>• Operational risk information<br>• Business impact of the risk and impacted business units<br>• Ongoing changes to risk factors |
| • Key performance objectives | Chief financial officer (CFO) | • Financial information with regard to programs and projects (budget, actual, trends, etc.) |
| • Risk management scope<br>• Plans for ongoing business and risk communication<br>• Risk culture<br>• Business impact of the IT-related business risk and impacted business units<br>• Ongoing changes to IT risk factors | Business management and business process owners | • Control and compliance monitoring<br>• Risk analysis request |
| • Key performance objectives<br>• Risk management plan<br>• Risk framework and scoring methodology<br>• Risk register<br>• Risk culture | IT management (including security, service management) | • Residual IT risk exposures |
| • Key performance objectives<br>• Risk responsible, accountable, consulted, informed (RACI) charts<br>• Risk management plan<br>• Control and compliance monitoring | Compliance and audit | • Audit findings |

## Risk Reporting Channels *(cont.)*

| Risk Reporting Channels *(cont.)* | | |
|---|---|---|
| **Input** | **Stakeholder** | **Output** |
| • Key performance objectives<br>• Risk management plan<br>• Risk framework and scoring methodology<br>• Risk register<br>• Audit findings | Risk control functions | • Residual risk exposures<br>• Risk reports |
| • Risk awareness expectations<br>• Risk culture | Human resources (HR) | • Potential risk<br>• Support on risk awareness initiatives |

## Reporting Results of Periodic Risk Assessment

Periodic risk assessment results should be provided to the steering committee and/or senior management for use in guiding risk management priorities and activities.

## Risk Reporting Tools and Techniques

Risk reporting can range from a face-to-face meeting with stakeholders to structured e-mails and reports on risk in specific focus areas to integrated governance, risk and compliance (GRC) solutions with automated workflows and system-generated heat maps and dashboards. The tools and techniques will differ significantly between enterprises and the intended target audience.

## Link to Other Domains

In the previous domains—*Risk Identification, Assessment and Evaluation and Risk Response*—risk was identified and prioritized and risk response activities were chosen and implemented to ensure that risk was within acceptable limits. In this domain, we set in place the mechanisms to monitor and report on the operation and effectiveness of the controls selected. In the next domain, we will examine the specific design and selection of the information systems related controls. In Domain 5, we will examine how to monitor and maintain those IS controls.

## D. Suggested Resources for Further Study

### Suggested Resource for Further Study

In addition to the resources cited throughout this manual, the following resource is suggested for further study:

- ISACA:
  - *The Risk IT Practitioner Guide*, 2009

# Domain 4–Information Systems Control Design and Implementation

## A. Chapter Overview

### Introduction

This chapter provides information on the system development life cycle (SDLC) and project management with particular focus on IS control design and implementation.

### Inputs From Other Domains

The IS control design and implementation process is a distinct subset of *Domain 2—Risk Response*, particularly risk mitigation/reduction. The enterprise must determine the appropriate response to the risk to the enterprise, whether the risk has an internal or external source. This chapter focuses on the specific risk response required for handling risk related to information systems.

Risk responses are directly driven by the risk (scenarios) that the enterprise identified and prioritized in *Domain 1—Risk Identification, Assessment and Evaluation.*

### Relevance

IS control design and implementation focuses on the evaluation and selection of the information system (IS) controls that need to be designed and implemented into information systems to meet organizational risk management requirements. The selection of IS controls is based on the enterprise's risk profile, its risk appetite, risk tolerance and risk culture.

Consideration for implementing specific IS controls are cost-benefit, available solutions, ease of maintenance, existing IT architecture and strategy, and the ability to adapt to a changing risk environment.

This CRISC must be knowledgeable in how to design and implement IS controls that both mitigate risk and still align with business objectives and are in compliance with the enterprise's risk appetite and risk tolerance levels.

### Outputs to Other Domains

The IS control design and implementation process provides documentation of IS controls and related metrics and key performance indicators (KPIs) used in *Domain 5—Information Systems Control Monitoring and Maintenance* to enable control monitoring and maintenance.

### Learning Objectives

As a result of completing this chapter, the CRISC candidate should be able to:
- List different control categories and their effects.
- Judge control strength.
- Explain the importance of balancing control cost and benefit.
- Leverage understanding of the SDLC process to implement IS controls efficiently and effectively.
- Differentiate between the four high-level stages of the SDLC.
- Relate each SDLC phase to specific tasks and objectives.
- Apply core project management tools and techniques to the implementation of IS controls.

## Contents

This chapter contains three primary topic areas:
- IS controls
- Design and implementation of controls through the phases of the systems development life cycle (SDLC)
- Project risk management

The chapter is formatted into the following sections.

**Contents *(cont.)***

Process-specific controls for the following processes are addressed in Part II of this manual:

1. Determining the IT Strategy
2. Project and Program Management
3. Change Management
4. Third-party Service Management
5. Continuous Service Assurance
6. Information Security Management
7. Configuration Management
8. Problem Management
9. Data Management
10. Physical Environment Management
11. IT Operations

## *B. Task and Knowledge Statements*

### Introduction

This section describes the task and knowledge statements for Domain 4, which focus on the design and implementation of IS controls that mitigate risk and are in alignment with the enterprise's risk appetite and tolerance levels to support business objectives.

### Domain 4 Task Statements

The following table describes the task statements for Domain 4.

| No. | Task Statement (TS) |
|---|---|
| TS4.1 | Interview process owners and review process design documentation to gain an understanding of the business process objectives. |
| TS4.2 | Analyze and document business process objectives and design to identify required information systems controls. |
| TS4.3 | Design information systems controls in consultation with the process owners to ensure alignment with business needs and objectives. |
| TS4.4 | Facilitate the identification of resources (e.g., people, infrastructure, information, architecture) required to implement and operate information systems controls at an optimal level. |
| TS4.5 | Monitor the information systems control design and implementation process to ensure that it is implemented effectively and within time, budget and scope. |
| TS4.6 | Provide progress reports on the implementation of information systems controls to inform stakeholders and to ensure that deviations are promptly addressed. |
| TS4.7 | Test information systems controls to verify effectiveness and efficiency prior to implementation. |
| TS4.8 | Implement information systems controls to mitigate risk. |
| TS4.9 | Facilitate the identification of metrics and key performance indicators (KPIs) to enable the measurement of information systems control performance in meeting business objectives. |
| TS4.10 | Assess and recommend tools to automate information systems control processes. |
| TS4.11 | Provide documentation and training to ensure that information systems controls are effectively performed. |
| TS4.12 | Ensure that all controls are assigned control owners to establish accountability. |
| TS4.13 | Establish control criteria to enable control life cycle management. |

## Domain 4 Knowledge Statements

The following table describes the knowledge statements for Domain 4.

| No. | Knowledge Statement (KS)<br>Knowledge of: |
|---|---|
| KS4.1 | Standards, frameworks and leading practices related to information systems control design and implementation |
| KS4.2 | Business process review tools and techniques |
| KS4.3 | Testing methodologies and practices related to information systems control design and implementation |
| KS4.4 | Control practices related to business processes and initiatives |
| KS4.5 | The information systems architecture (e.g., platforms, networks, applications, databases and operating systems) |
| KS4.6 | Controls related to information security |
| KS4.7 | Controls related to third-party management |
| KS4.8 | Controls related to data management |
| KS4.9 | Controls related to the system development life cycle |
| KS4.10 | Controls related to project and program management |
| KS4.11 | Controls related to business continuity and disaster recovery management |
| KS4.12 | Controls related to management of IT operations |
| KS4.13 | Software and hardware certification and accreditation practices |
| KS4.14 | The concept of control objectives |
| KS4.15 | Governance, risk and compliance (GRC) tools |
| KS4.16 | Tools and techniques to educate and train users |

**Note:** Knowledge statements 4.4 and 4.6-4.12 are process specific and addressed in Part II of this manual.

# C. IS Controls

## Introduction

This section provides an overview of IS controls because a thorough understanding of controls in general, and IS controls in particular, is crucial for the effective design and implementation of IS controls.

Project risk management and design and implementation of IS controls throughout the SDLC are addressed in separate chapter sections.

Process-specific controls and a practitioner level of IS control design and implementation are addressed in Part II of this manual.

## 1. Key Terms and Principles

### Introduction

This section provides an overview of key terms and principles related to controls in general and IS controls in particular.

The relative mix and importance of process, application and general controls will be unique to each enterprise.

### Definition of Controls

Controls are the policies, procedures, practices and guidelines designed to provide reasonable assurance that:
- Business objectives are achieved.
- Undesired events are prevented or detected and corrected.

### Definition of Defense-in-Depth

Defense-in-depth—also called layered defense—is the concept of deploying a combination of controls to ensure that if one control was compromised or malfunctioning, other controls would prevent total compromise of the system and restrict access to protected assets.

Defense-in-depth can be deployed along horizontal or vertical vectors.

**Examples:**
- Horizontal defense-in-depth—controls are placed along a network path—a firewall, then a network–based intrusion detection system (IDS), an intrusion prevention system (IPS), compartmentalization or segmentation of the network, and host-based controls.
- Vertical defense-in-depth—controls are placed along various layers of the system—on the hardware, operating systems, database, application and user levels.

## Control Categories

The following table describes control categories.

| Control Categories | |
|---|---|
| **Category** | **Description** |
| Compensating controls | Are an alternate form of control that corrects a deficiency or weakness in the control structure of the enterprise<br><br>*Compensating controls may be considered when an entity cannot meet a requirement explicitly, as stated, due to legitimate technical or business constraints, but has sufficiently mitigated the risk associated with the requirement through implementation of other controls.*<br><br>**Example:** Adding a challenge response component to weak access controls can compensate for the deficiency in the access control mechanism. |
| Corrective controls | Remediate errors, omissions and unauthorized uses and intrusions, once they are detected<br><br>**Example:** Backup restore procedures enable a system to be recovered if harm is so extensive that processing cannot continue without recourse to corrective measures. |
| Detective controls | Warn of violations or attempted violations of security policy and include such controls as audit trails, intrusion detection methods and checksums |
| Deterrent controls | Provide warnings that can deter potential compromise such as warning banners on login screens or offering rewards for the arrest of hackers |
| Directive controls | Directive controls mandate the behavior of an entity by specifying what actions are, or are not, permitted.<br><br>**Example:** A policy is an example of a directive control. |
| Preventive controls | Inhibit attempts to violate security policy and include such controls as access control enforcement, encryption and authentication |

## Exhibit 4.1: Control Category Interdependencies

**Exhibit 4.1** describes the interdependencies of different control categories.

**Exhibit 4.1: Control Category Interdependencies**

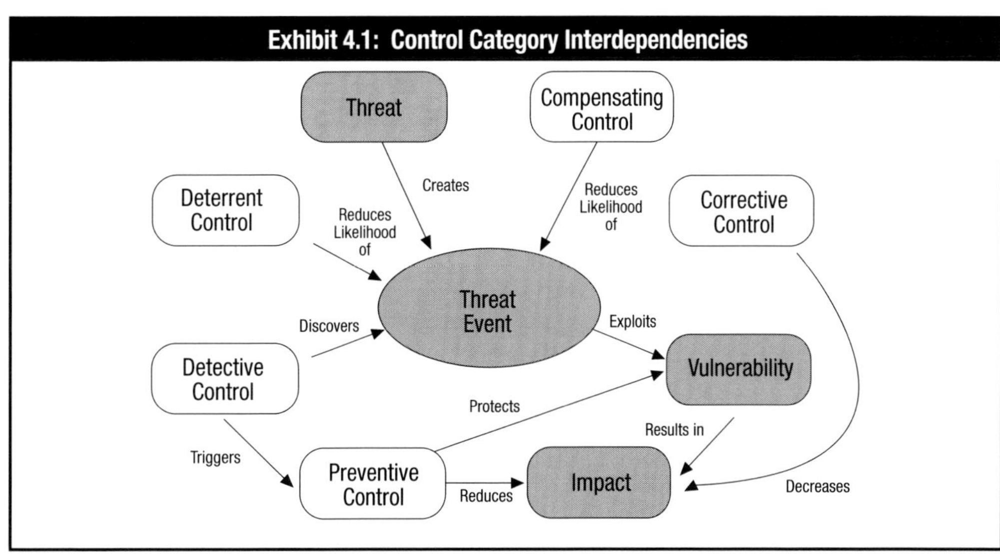

### Technical and Nontechnical Control Methods

Technical controls (e.g., a firewall) must be supported by nontechnical controls such as operational controls (i.e., a configuration management process for rule changes to the firewall configuration) and managerial controls (ownership and oversight for the control functionality).

### Examples of Technical and Nontechnical Control Methods

An effective security management system will consist of a variety of control methods:
- Technical controls
- Nontechnical controls
  - Managerial controls
  - Operational controls

The following table describes technical and nontechnical control methods.

| Technical and Nontechnical Control Methods | | |
|---|---|---|
| **Method** | **Description** | **Examples** |
| Technical controls | Safeguards and countermeasures that are incorporated into computer hardware, software or firmware | • Access control mechanisms<br>• Identification and authentication mechanisms<br>• Encryption methods<br>• Intrusion detection software |
| Nontechnical controls | Management and operational controls | • Security policies<br>• Operational procedures<br>• Personnel, physical and environmental security |

> **Note:** Controls, such as two-factor authentication required for high-security situations, can include both automated and manual processes; for example, smart cards requiring a personal identification number (PIN).

## 2. The Control Life Cycle

### Introduction

This section provides an overview of the control life cycle as well as the related concepts and principles.

The relative mix and importance of process, application and general controls will be unique to each enterprise.

## Exhibit 4.2: The Control Life Cycle

The control life cycle maps the various phases in the life of a control from the initial selection/design through development, implementation, maintenance and disposal of the system.

The CRISC must ensure that the controls are properly managed throughout each phase of the life cycle.

**Exhibit 4.2** shows the phases of the control life cycle.

**Exhibit 4.2: The Control Life Cycle**

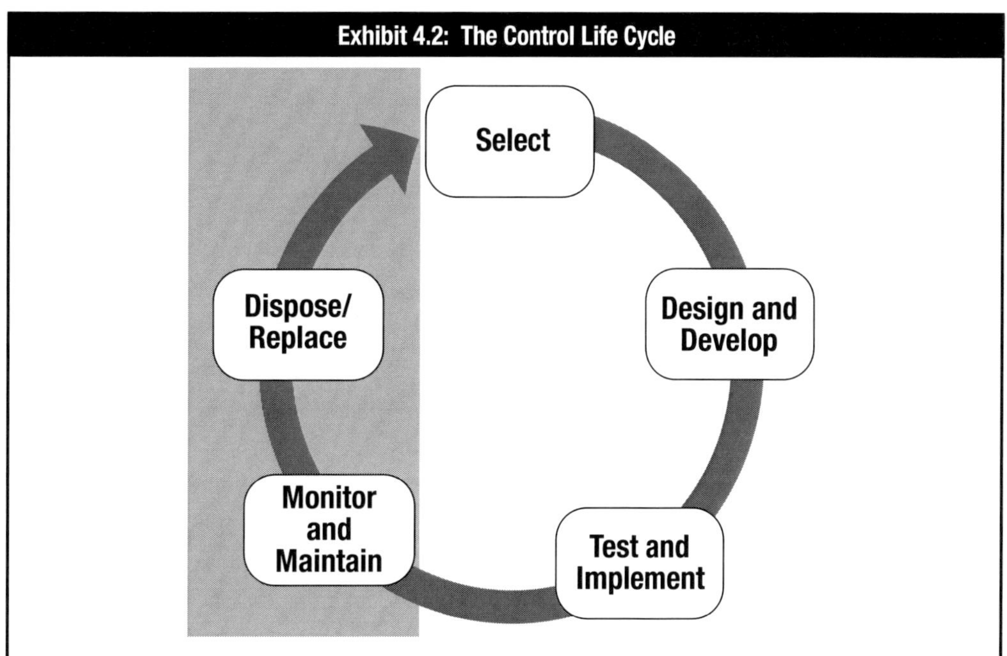


The shaded areas indicate those steps of the control life cycle that are addressed in *Domain 5—Information Systems Control Monitoring and Maintenance*.

## 3. Control Selection

### Introduction

Control selection addresses the high-level or control architecture considerations.

The selection of the appropriate level of control is based on various critical factors. Controls place limitations on the way that the business can operate and may result in an impact on system performance, operational cost and productivity.

The selection of IS controls must be made according to the risk mitigating control options available, cost, time constraints, availability of skilled personnel and business priorities.

The risk practitioner should ensure that the business requirements and input of all relevant stakeholders is considered in the selection of controls. Different stakeholders, from different parts of the enterprise, may have substantially different priorities and requirements. Selection of an ideal control for one group may be unacceptable for other areas of the enterprise.

## Control Costs and Benefits

When controls or countermeasures are planned, an enterprise should consider the costs and benefits.

Cost-benefit analysis helps:
- Provide a monetary impact view of risk
- Determine the cost of protecting what is important
- Make smart choices based on potential:
  - Risk mitigation costs
  - Losses (risk exposure)

> **Note:** If the costs of specific controls or countermeasures (control overhead) exceed the benefits of mitigating a given risk, the enterprise may choose to accept the risk rather than incur the cost of mitigation. This acceptance of risk follows the general principle that the cost of a control should never exceed the expected benefit.

## Calculating the Benefits of a Control

Calculating the benefits to be realized from the implementation of a control can be a challenge since benefits will be both tangible and intangible. The benefits are measured against the resulting reduction in quantitative and qualitative risk and the expected exposure or impact from an incident.

The calculation of risk and residual risk was covered earlier in this book; however, these are some of the factors that will affect the benefit portion of the cost-benefit calculation for IS controls:
- Reduction in financial cost:
  - Less impact on productivity, overtime and staff costs
  - Lower insurance premiums
  - Avoidance of penalties for non-compliance
  - Fewer calls to help desk or production support
- Impact on employee morale (less staff turnover and training)
- More time to work on quality improvement of systems instead of repair
- Financial savings
  - Less system downtime
  - Increase in productivity
  - Reduced cost of incidents
- Easier management of controls and consistency
- Accurate compliance reporting

## Frameworks for Control Selection

The selection of IS controls may be influenced by international standards, guidelines and leading practices. Some of the tools that may be used to select and/or justify specific controls are:
- International Organization for Standardization (ISO)—Standard 27001
- Sherwood Applied Business Security Architecture Framework (SABSA)
- ISACA—COBIT 4.1 Framework
  **Note:** The COBIT 4.1 framework is available at no charge from ISACA and can be downloaded at *www.isaca.org/cobit*. The new COBIT 5 framework will be available in 2012.
- ISACA—Risk IT Framework
- Committee of Sponsoring Organizations of the Treadway Commission (COSO)—Enterprise Risk Management—Integrated Framework
- Benchmarking activities

### Total Cost of Ownership (TCO) for Controls

When considering costs, the TCO must be considered for the full life cycle of the control or countermeasure. This can include such elements as:
- Acquisition and licensing costs
- Deployment and implementation costs
- Recurring maintenance costs
- Testing and assessment costs
- Compliance monitoring and enforcement
- Inconvenience to users
- Reduced throughput of controlled processes
- Training in new procedures or technologies as applicable
- End of life decommissioning

## 4. Control Design and Development

### Introduction

Control design and implementation is a crucial part of the control life cycle. The complexity of information systems has a distinct influence on IS controls. Thus, the design has to consider "breadth" and "depth," where breadth represents the flow of information across multiple applications and depth represents the different layers on which controls can function and how those may cross-influence each other.

The IS control design process includes the determination of which tools and control types are available to meet policy and business requirements, selection of the best controls, and the design and implementation of the selected controls.

### Developing Controls in Depth

IS controls may be implemented at many levels in an information system. Some of the control locations may be:
- Network-based controls
- Application-level controls
- Database controls
- Operating system controls
- Platform-specific controls
- Physical controls

It is important to **design** and implement controls at the correct place within the overall enterprise, system or network. A control that is placed incorrectly, or not configured properly, will provide little benefit and may in fact create a sense of false security.

Controls must be selected that will be interoperable, and yet independent so that a breach of one control does not cause the failure of other controls. The controls must provide a complete framework of protection from risk so that no gaps remain that could be exploited by an attacker, or bypassed through unintentional misuse by an internal resource.

### Establishing Key Performance Indicators (KPIs)

As controls are selected for implementation, criteria should also be established to determine the operational level and effectiveness of the controls. These criteria will often be based on KPIs that indicate whether a control is functioning correctly.

Another measurement criterion may be based on key risk indicators (KRIs), which are used to alert monitoring personnel about trends or alarm thresholds that may indicate a potentially hazardous or marginal condition.

### Control Ownership

The selection of a control requires the identification of a control owner. The control owner is responsible for the management of the control and for ensuring that the control is operated and maintained in the correct manner. Effective control operation includes reporting performance metrics and abnormal conditions to appropriate personnel.

### Designing for Easy Control Maintenance

Design of the controls implemented must include maintenance considerations, including measurability. Effectiveness of controls cannot be evaluated unless they can be tested and measured. Further, confidence levels and sampling sizes for testing the effectiveness of these controls should closely mirror audit and regulatory compliance objectives.

> **Note:** The risk practitioner must ensure that the enterprise is in adequate compliance with relevant legal and regulatory requirements.

### Example of Designing for Easy Control Maintenance

The design of a control includes designing the supporting processes and procedures for monitoring and reporting on the effectiveness of the control. This is covered in more detail in Domain 5, but will include monitoring actions:

- That show daily affirmative reviews (review notes, sign-offs, approvals, etc.)
- That have a follow-up process for anomalous/suspicious events divined
- With appropriate sampling sizes
- For which information security controls are tested accordingly and compliance is built in

## 5. Control Testing and Implementation

### Introduction

Design and implementation of the selected IS controls requires a defined process for the secure configuration, deployment, change control and approval, as well as designing the ability to monitor and report on the effectiveness of the controls, once implemented. The actual monitoring and maintenance of the controls will be covered in Domain 5.

Control testing and implementation includes:

- Testing control effectiveness
- Documentation
- User and operator training

## Testing Control Effectiveness

Control effectiveness cannot be determined by simply identifying the control category (preventive, detective, manual, automated, etc.).

Control effectiveness can be assessed by its quantitative and qualitative compliance testing results.

Control effectiveness must be assessed within context: While the initial test results may indicate that an automated control is highly reliable, a more comprehensive test may reveal that it is habitually circumvented.

Controls can be effectively assessed only by determining how well they achieve the control objective within the environment in which they are operating. Specific considerations for designing meaningful controls may include the control's:
- Design effectiveness
- Operating effectiveness
- Alignment with its operating environment (organization, people, processes and technology)

## Documentation

Policy—The first control that must be implemented is policy. Policy is an administrative or managerial control that specifies the behaviors and actions that are/are not permitted. Policy proclaims the intent of management, states compliance with regulations, mandates security procedures and sets the tone for the enterprise. Without policy there is no authority for the implementation and supervision of security controls.

Some examples of documentation include:
- **Risk register**—documentation of risks and the status of risk response efforts
- **Risk assessment report**—generated by a risk identification and evaluation effort that lists the identified risk, the priority of risk, and recommended controls.
- **Plan of action and milestones (POAM)**—listing of ongoing risk response projects and tracking of the status of the projects

## Training

**Awareness and Training**—A very important control. A strong and consistent awareness program heightens the ability of staff to detect, prevent and respond to many types of security incidents. Awareness should be a part of the annual review of all users. Staff must also be trained in the proper use of the tools that they will utilize. An IS control will not be effective if it is not configured, monitored and maintained correctly, but no staff can be expected to properly maintain a tool for which they have not be trained in its proper use.

## Establishing KPIs

As controls are selected for implementation, criteria should also be established to determine the operational level and effectiveness of the controls. These criteria will often be based on KPIs that indicate whether a control is functioning correctly.

Other measurement criteria may be based on KRIs, which are used to alert monitoring personnel about trends or alarm thresholds that may indicate a potentially hazardous or marginal condition.

## Control Ownership

The selection of a control requires the identification of the owner of the control. The owner will be responsible for the management of the control and ensuring that the control is operated and maintained in the correct manner. This will include the reporting of abnormal conditions or reporting statistics to appropriate personnel.

## Control-specific Training

Users, administrators, managers, auditors and security staff may require training on the operation, maintenance, monitoring and reporting on the new or enhanced controls selected for the enterprise.

The training should be practical, relevant and tailored to the needs of the audience.

## Summary

At this point, the CRISC candidate should be able to evaluate the various controls available to mitigate risk, and to select the best control for the enterprise based on cost-benefit analysis, risk levels and acceptance levels, stakeholder requirements, organizational factors and priorities.

The CRISC candidate should ensure that adequate IS controls have been selected to mitigate risk, that the owner has been determined for each control, that the supporting evaluation criteria have been set (KPIs) and that the training needs are addressed.

## The Life Cycle of Control Design

IS controls require continuous attention to risk levels, changes in the operational or threat environment and availability of new controls or control strategies. As IS systems go through the process of development and implementation, the risk manager must reevaluate and assess the risk levels to ensure that the controls are being designed, developed, tested, implemented and maintained properly. This requires the integration of risk management and IS control selection into each phase of the systems development life cycle (SDLC).

# D. Building Control Design Into the SDLC

## Introduction

The design and deployment of IS controls will often be undertaken as a systems development project. While there are several project management techniques that can be used to manage system development projects, they can be described by the generic term "system development life cycle (SDLC)." The CRISC candidate should be familiar with the steps necessary to select, design, develop, test and deploy and maintain IS controls throughout their life cycle.

## Relevance

The risk practitioner should advise on the design of appropriate IS controls to treat identified risk and oversee the implementation of IS controls, either as part of a system implementation or via the operational change management process.

> **Note:** The CRISC candidate should become involved as early as possible during the SDLC process and maintain that involvement throughout the remainder of the control life cycle. It is important to implement controls properly and ensure that they continue to operate securely throughout their operational life. The risk practitioner can add value during the initiation phases of the SDLC or even prior to the start of a project, when project ideas are generated, developed and communicated.

## 1. Introduction to the SDLC

### Introduction

Companies often commit significant resources (e.g., people, applications, facilities and technology) to develop, acquire, integrate and maintain application systems that are critical to the effective functioning of key business processes.

The SDLC process governs the phases deployed in the development or acquisition of a software system and, depending on the methodology, may even include the controlled retirement of the system.

> **Note:** Typical phases of SDLC include the feasibility study, requirements study, requirements definition, detailed design, programming, testing, installation and postimplementation review. Today's SDLC also tends to include the maintenance and operational phases of the control's life cycle, up to and possibly including, its date of termination.

### Objectives of the SDLC Process

The SDLC includes:
- IT processes for managing and controlling project activity
- An objective for each phase of the life cycle that is typically described with key deliverables, a description of recommended tasks and a summary of related control objectives for effective management
- Incremental steps or deliverables that lay the foundation for the next phase

### Relevance

This section contains information on the SDLC process and its relationship to the:
- Achievement of business objectives
- High-level SDLC process flow and architecture

> **Note:** High-level knowledge of this process is relevant to the risk practitioner's ability to design and implement IS controls.

## 2. Risk Associated With Software Development

### Introduction

This topic contains information on risk associated with software design and development. The concepts can be applied to risk management, where IS controls are designed and implemented to reduce risk.

### Business Risk vs. Project Risk

The following table separates and compares the risk related to designing and developing software systems along two major categories: business risk (or benefit risk) and project risk (or delivery risk).

| Risk Related to Software Systems Design and Development | |
|---|---|
| **Category** | **Description** |
| Business Risk (Benefit Risk) | This relates to the likelihood that the new system may not meet the user business needs, requirements and expectations.<br><br>**Example:** The business requirements that were to be addressed by the new system are still unfulfilled, and the process has been a waste of resources. Even if the system is implemented, it will most likely be underutilized and not maintained, making it obsolete in a short period of time. |
| Project Risk (Delivery Risk) | The project activities to design and develop the system exceed the limits of the financial resources set aside for the project. As a result, the project may be completed late, if ever. Project-related risk is addressed in more detail in section F, Managing Project Risks |

### Root Causes of Project Delivery Risk

The foremost root cause of project risk is:
- A lack of discipline in managing the software development process
- Selection of a project methodology that is unsuitable to the system being developed\

In such instances:
- Enterprises are not providing the infrastructure and support necessary to help projects avoid these problems.
- If successful projects occur, they are not repeatable and SDLC activities are not defined or followed adequately (i.e., insufficient maturity).
- With effective management, SDLC management activities can be controlled, measured and improved.

### CRISC Responsibilities Related to Project Risk

Merely following an SDLC management methodology does not ensure successful completion of a development project. The CRISC needs to enforce management discipline over a project to ensure that:
- The project addresses the specific risk management requirements.
- The project follows a defined process.
- Project planning is performed, including effective estimates of resources, budget and time.
- Scope creep is managed proactively.
- Management tracks software design and development activities.
- Senior management provides support to the software project's design and development efforts.
- Periodic review and risk analysis are performed in each project phase.

# E. System Development Life Cycle (SDLC) Phases

## 1. Project Initiation

### Introduction

This section contains information on projects initiated by sponsors who gather the information required to gain approval for the project to be created.

Information often compiled into the terms of a project charter includes the:
- Objective of the project
- Business case and/or problem statement
- Stakeholders in the system to be produced
- Project manager and sponsor

> **Note:** Approval of a project initiation document (PID) or a project request document (PRD) is the authorization for a project to begin.

### Tasks Associated With Phase 1

The following table lists the tasks to initiate the project.

| Tasks to Initiate the Project | | |
|---|---|---|
| **Task** | **Name/Description** | **Starting Page** |
| 1 | Conduct a feasibility study. | 127 |
| 2 | Define requirements. | 128 |

## *Task 1—Conduct a Feasibility Study*

### Introduction

This topic contains information on conducting a feasibility study, which:
- Begins once initial approval has been given to move forward with a project
- Includes an analysis to clearly define the need and to identify alternatives for addressing the need

### Description of a Feasibility Study

A feasibility study:
- Involves analyzing the benefits and solutions for the identified problem area
- Includes development of a business case, which:
  - States the strategic benefits of implementing the system either in productivity gains or in future cost avoidance
  - Identifies and quantifies the cost savings of the new system
  - Estimates a payback schedule for the cost incurred in implementing the system or shows the projected return on investment (ROI)

> **Note:** Intangible benefits such as improved customer relations may also be identified; however, quantify the benefits wherever possible. Nonfunctional requirements must also be known to avoid surprises in later project phases.

### Main Components of a Feasibility Study

Within the feasibility study, the following are typically addressed:
- Definition of a time frame for the implementation of the required solution
- Determination of an optimum alternative risk-based solution for meeting business needs and general information resource requirements (e.g., whether to develop or acquire a system)

## Factors to Consider During the Feasibility Study

The following table describes factors to consider during the feasibility study to assist both in determining whether the project should be undertaken and in determining whether to develop or acquire a system or whether to update or replace an existing system.

| Feasibility Study Consideration Factors | |
|---|---|
| **Factor** | **Description** |
| Date | The date by which the system needs to be functional, based on the business requirements |
| Cost | The level of effort (LOE), total cost of ownership (TCO) and anticipated ROI estimates to develop the system internally as opposed to a vendor's statement of work |
| Resources | Consist of the staff (availability and skill sets), consultants, licenses and hardware required to develop and implement the solution |
| Vendor system | The skill set of the support personnel; solvency of the vendor; system's reputation in the marketplace and with its clients; platform and compliance with security policies and regulations to which the enterprise is subject; packaging and customization design costs; consulting service costs; training costs; license characteristics (e.g., yearly renewal, perpetual) and maintenance costs |
| Interfaces | The other systems that will interface with the new/enhanced system through a pull, push or bidirectional relationship |
| Compatibility | Compatible with:<br>• Strategic business plans<br>• Other applications and systems in the environment<br>• Alignment with the enterprise's:<br>– Information security policies<br>– Regulatory and legal requirements<br>– Existing IT infrastructure |
| Future requirements | The system's ability to grow as requirements change or are enhanced to meet the enterprise's strategic plan (five to seven years) |

## Feasibility Study Results

The completed feasibility study results should include a cost-benefit analysis report that:

- Provides the results of criteria analyzed (e.g., costs, benefits, risk, resources required and organizational impact)
- Recommends one of the alternatives/solutions and a course of action

# Task 2—Define Requirements

## Introduction

This topic contains information on defining requirements, which is concerned with identifying and specifying the requirements for the solution.

## Description of Requirements

Requirements include:

- Business requirements containing descriptions of what a system should do
- Functional requirements and the use of case models describing how users will interact with a system
- Technical requirements and design specifications and coding specifications describing how the system will interact, conditions under which the system will operate and the information criteria that the system should meet

## Framework for Defining Requirements

The COBIT framework defines information criteria that should be incorporated into the requirements to address issues associated with effectiveness, efficiency, confidentiality, integrity, availability, compliance and reliability.

The requirements definition task of the project initiation phase also deals with issues that are sometimes called nonfunctional requirements (e.g., training or business continuity).

## Defining the Requirements

The users in this process specify:
- IS control needs: nonautomated and automated
- How they wish to have those needs addressed by the system (e.g., access controls, regulatory requirements, management information needs and interface requirements)

## Factors to Consider When Defining Requirements

All concerned management and user groups must be actively involved in the requirements definition task to prevent problems such as expending resources on a system that will not satisfy the business requirements.

User involvement is necessary to obtain commitment and full benefit from the system. Without management sponsorship, clearly defined requirements and user involvement, the benefits may never be realized.

## Factors to Consider When Acquiring Software

Software acquisition should be based on various factors, such as the:
- Cost differential between development and acquisition
- Availability of generic software
- Time gap between development and acquisition

> **Note:** Ensure that the feasibility study contains documentation that supports the decision to acquire the software.

## 2. Project Design and Development

### Leading Practices for Design and Implementation

This topic contains information on the seven leading practices for designing systems that reflect risk management strategies, which serve as checkpoints to ensure good systems design.

The following table provides a short overview of the seven leading practices and how they help mitigate risk associated with the design and implementation of IS controls.

| Seven Leading Practices for System Design and Implementation | | |
|---|---|---|
| **No.** | **Leading Practice** | **Description and Associated Risk** |
| 1 | Align with the business. | • Identify or create the business opportunity that makes the system worth building.<br>• Ensure that any systems development project directly supports the enterprise in achieving one or more of its goals.<br>• Provide sustained benefit to the enterprise by ensuring that a system continues to support the efficient exploitation of the business opportunity it was built to address. |
| 2 | Use technology to enable change. | • Investigate tasks and activities that may seem impossible at the current time, but, if IT-enabled, would fundamentally and positively change the ways that the enterprise does business. This may include:<br>– Looking for opportunities to create a transformation or value shift in the context of the enterprise business space<br>– Finding ways to do things that provide dramatic cost savings or productivity increases<br>– Thinking of courses that would be least likely to be foreseen or quickly countered or copied by competitors |
| 3 | Leverage existing technology. | • Find ways to incorporate existing systems that have proven to be stable and responsive over time into the design of new systems because the design of systems embodies strategy and the purpose of strategy is to use the means available to the enterprise to best accomplish its goal.<br>• Build new systems on the strengths of older systems, as in the principle of evolutionary process. |
| 4 | Embrace simplicity. | • Use a (simple) mix of technology and business procedures to achieve business objectives, where possible.<br><br>**Rationale:**<br>• This reduces complexity and the risk associated with the work and spreads the cost across multiple objectives.<br>• Using a different technology or process to achieve each different project objective multiplies cost and complexity and reduces the overall probability of project success. |

## Leading Practices for Design and Implementation *(cont.)*

| Seven Leading Practices for System Design and Implementation *(cont.)* | | |
|---|---|---|
| **No.** | **Leading Practice** | **Description and Associated Risk** |
| 5 | Remain flexible. | • Decompose the system design into separate components or objectives and, whenever possible, run the work on different objectives in parallel.<br>• Promote flexibility by preventing the achievement of one objective from becoming dependent on the achievement of another objective.<br><br>**Rationale:** Delays in the work toward one objective will not impact the progress toward other objectives.<br>• Assign personnel to develop the systems that have applicable skills that can achieve a variety of different objectives.<br>• Use the same development technology to achieve several different objectives, making it much easier to shift personnel from one objective to another, as needed, because they use the same skill sets.<br>• Ensure that the project plan foresees, and provides for, an alternative plan in case of failure or delays in achieving objectives as scheduled.<br>• Build the design of the systems to allow for some system features to be dropped from development, if needed, and still be able to deliver substantially. |
| 6 | Build within the enterprise's capability. | • Design and implement controls that are within the enterprise's capability and can be supported over time.<br><br>**Rationale:** Unrealistic goals do not—in the long run—support the enterprise. |
| 7 | Learn from failure. | • Rework and increased effort are an inadequate response to failure. Provide lessons learned to ensure that avoidable challenges during the current project are not repeated in the future.<br><br>**Risk:** Avoidable failures are repeated from project to project. |

## System Design Introduction

While the architecture significantly influences the IS control design, the established requirements must also be addressed. This topic contains information on IS control design.

## System Design Objectives

After the design has been completed, architects should be able to:
- Explain how the software architecture will satisfy the risk response requirements in the system and application.
- Outline the rationale for key design decisions.

**Rationale:** Choices of particular hardware and software configurations may have cost implications of which stakeholders need to be aware and control implications that are of interest to the risk practitioner.

**CRISC Involvement in System Design**

CRISC involvement is focused on whether:
- Risk response requirements are properly communicated and documented.
- An adequate system of controls is incorporated into system specifications and test plans.

Monitoring functions are built into the system, particularly for electronic commerce (e-commerce) applications and other types of paperless environments.

---

**Completion of the System Design**

After the detailed design has been completed and approved, distribute the design to the developers for coding.

The key deliverables coming out of project design include:
- System, subsystem, program and database specifications
- Test plans
- A defined and documented formal software change control process

---

## System Development

**Introduction**

This topic contains information on system development, which uses the detailed design developed previously, to begin coding; thus moving the system one step closer to a final physical software product.

---

**CRISC Involvement in System Development**

While the responsibilities in this system development rest primarily with the programmers and systems analysts who are building the system, the CRISC may supervise the development of specific controls to ensure that they are adequately addressed.

---

## 3. Project Testing

**Introduction**

This section describes system testing, which:
- Is an essential part of the development process that verifies and validates that a program, subsystem or application, and the designed security controls perform the functions for which it has been designed
- Determines whether the units being tested operate without any malfunction or adverse effect on other components of the system
- Uses a variety of development methodologies and organizational requirements to provide for a large range of testing schemes or levels.

> **Note:** Each set of tests is performed with a different set of data and under the responsibility of different people or functions. The risk practitioner can play a role is ensuring that adequate tests are performed to test the functionality of the security controls.

## Testing Types

The following table describes a variety of tests that:
- Relate to the previously mentioned approaches
- Are performed based on the size and complexity of the modified system

<table>
<tr><th colspan="2">Testing Types</th></tr>
<tr><th>Type</th><th>Description</th></tr>
<tr><td>Unit testing</td><td>• Tests an individual program or module<br>• Uses a set of test cases that focus on the control structure of the procedural design<br>• Ensures that the internal operation of the program performs according to specification</td></tr>
<tr><td>Interface or integration testing</td><td>• Uses a hardware or software test that evaluates the connection of two or more components that passes information from one area to another<br>• Takes unit-tested modules and builds an integrated structure dictated by design<br><br>Note: The term “integration testing” also refers to tests that verify and validate the functioning of the application under test with other systems, in which a set of data is transferred from one system to another.</td></tr>
<tr><td>System testing</td><td>• A series of tests designed to ensure that modified programs, objects, database schema, etc., which collectively constitute a new or modified system, function properly<br>• Often performs these test procedures in a nonproduction test/development environment by software developers designated as a test team<br><br>The following table describes specific analyses that may be carried out during system testing.<br>
<table>
<tr><th>Test</th><th>Description</th></tr>
<tr><td>Recovery testing</td><td>Checks the system’s ability to recover after a software or hardware failure</td></tr>
<tr><td>Security testing</td><td>Verifies that the modified/new system includes provisions for appropriate access controls and does not introduce any security holes that may compromise other systems</td></tr>
<tr><td>Stress/volume testing</td><td>Tests an application with large quantities of data to evaluate its performance during peak hours</td></tr>
<tr><td>Volume testing</td><td>Studies the impact on the application by testing with an incremental volume of records to determine the maximum volume of records (data) that the application can process</td></tr>
<tr><td>Stress testing</td><td>Studies the impact on the application by testing with an incremental number of concurrent users/services on the application to determine the maximum number of concurrent users/services the application can process</td></tr>
<tr><td>Performance testing</td><td>Compares the system’s performance to other equivalent systems using well-defined benchmarks</td></tr>
</table></td></tr>
</table>

**Testing Types *(cont.)***

<table>
<tr><th colspan="3">Testing Types (cont.)</th></tr>
<tr><th>Type</th><th colspan="2">Description</th></tr>
<tr><td rowspan="4">Final acceptance testing</td><td colspan="2">• Begins on the modified system:<br>– After the system staff is satisfied with its initial or system tests<br>– During the implementation phase<br>• Incorporates the defined methods of testing into the enterprise's quality assurance (QA) methodology<br>• Proactively encourages QA activities to perform adequate levels of testing on all software development projects<br><br>The following table identifies the two major parts of final acceptance testing.</td></tr>
<tr><th>Type of Test</th><th>Description</th></tr>
<tr><td>QA testing (QAT)</td><td>• Focuses on the technical aspect of the application (documented specifications and the technology employed)<br>• Verifies that the application works as documented by testing the logical design and the technology itself<br>• Ensures that the application meets the documented technical specifications and deliverables<br>• Is performed primarily by the IS department<br><br>Note: The participation of the end user is minimal and on request.<br><br>• Does not focus on functionality testing</td></tr>
<tr><td>User acceptance testing (UAT)</td><td>• Focuses on the functional aspect of the application<br>• Supports the process of ensuring that the system is production-ready<br>• Satisfies all documented requirements<br>• Methods include:<br>– Definition of test strategies and procedures<br>– Design of test cases and scenarios<br>– Execution of the tests<br>– Utilization of the results to verify system readiness</td></tr>
<tr><td colspan="3">Note: Because they have different objectives, do not combine QAT and UAT.</td></tr>
</table>

### Definition of Acceptance Criteria

"Acceptance criteria" refers to the criteria that a deliverable must meet to satisfy the predefined needs of the user.

---

### Use of Production Data in Integrated Test Facilities (ITFs)

Many enterprises rely on ITFs to process test data in production-like systems to confirm the behavior of the new application or modules in real-life conditions, including peak volume and other resource-related constraints.

In this environment, the IS function performs tests with a set of fictitious data in which the client uses extracts of production data to cover the most possible scenarios and some fictional data for scenarios that would not be tested.

> **Note:** When production data are used in a test environment, scramble the data so the confidential nature of data is obscured from the tester and in case the system inadvertently discloses or leaks information due to system defect or misconfiguration.

> **Note:** Such data leakage can occur when the acceptance testing is done by team members who, under usual circumstances, would not have access to such production data.

---

### Certification and Accreditation Process

On completion of acceptance testing, the final step is usually a certification and accreditation process, which:
- Includes evaluating program documentation and testing effectiveness
- Results in a final decision for deploying the business application system

For information security issues, the evaluation process includes reviewing:
- Security plans
- The risk assessments performed and test plans
- The evaluation process results in an assessment of the effectiveness of the security controls and processes to be deployed

**Rationale:** This process generally involves security staff and the business owner of the application and provides some degree of accountability to the business owner regarding the state of the system that needs to be accepted for deployment.

---

### Final Test Report to Management

When the tests are completed, the risk practitioner should issue an opinion to management as to whether the system:
- Meets the business requirements
- Has appropriate controls implemented
- Is ready to be migrated to production

Be sure that this report:
- Specifies the deficiencies in the system that need to be corrected
- Identifies and explains the risk that the enterprise is taking by implementing the new system

---

## 4. Project Implementation

### Implementation Planning

### Introduction

This topic contains information on implementation planning, which is a project in itself and requires a methodology and the adoption of best practices from past experiences.

### Description of Support Structure

Once a project is operational, it requires an efficient support structure for the new system delivered by the project. A support structure requires:
- Setting up roles and naming people to fulfill these roles
- Providing personnel with new skills
- Distributing the workload so that the right people support the right issues
- Developing new processes while respecting the specificities of IT department requirements
- Dedicating an infrastructure for support staff

---

### Major Challenges for Implementation

One of the major challenges is to manage implementation:
- From build to integrate to migrate
- For the phasing-out of the existing system
- For the phasing-in of the new system

Migration must be set up in a step-by-step transition of the affected services.

> **Note:** The implemented processes for a legacy environment:
> - May be different from what may be implemented with the new platform
> - Must be communicated to users and system support staff if there are any changes

---

## End-user Training

### Introduction

This topic contains knowledge on developing a training plan that ensures that end users can become self-sufficient in the operation of the system.

> **Note:** End-user training and the training plan must start early in the development process.

---

## Data Migration

### Introduction

This topic contains information on the data conversions involved with data migration.

---

### Description of Data Conversion

Data conversion is required if the source and target systems utilize different:
- Field formats or sizes
- File or database structures (e.g., relational database, flat files, Virtual Storage Access Method [VSAM])
- Coding schemes
- Hardware and/or operating system (OS) platforms

The object is to convert existing data into the new required format, coding and structure while preserving the meaning and integrity of the data.

---

### Minimizing Data Migration Risk

Carefully plan the data migration, and use the appropriate methodologies and tools to minimize the risk of:
- Disruption of routine operations
- Violation of the security and confidentiality of data
- Conflicts and contention between legacy and migrated operations
- Data inconsistencies and loss of data integrity during the migration process

## Fallback (Rollback) Scenario

### Introduction

This topic contains knowledge regarding how not all new system deployments go as planned. To mitigate the risk of downtime for mission-critical systems, best practices dictate that the tools and applications required to reverse the migration are available prior to attempting the production cutover.

Components have to be delivered that can back out all changes and restore data to the original applications in the case of nonfunctioning new applications.

Some or all of these tools and applications may need to be developed as part of the project.

| Data Conversion Key Considerations | |
|---|---|
| **Consideration** | **Guidelines** |
| Completeness of data conversion | The total number of records from the source database is transferred to the new database (assuming the number of fields is the same). |
| Data integrity | The data are not altered manually, mechanically or electronically by a person, program or substitution or by overwriting in the new system.<br><br>**Note:** Integrity problems also include errors due to transposition and transcription errors and problems transferring particular records, fields, files and libraries. |
| Storage and security of data under conversion | Data are backed up before conversion for future reference or any emergency that may arise out of data conversion program management.<br><br>**Note:** An unauthorized copy or too many copies can lead to misuse, abuse, or theft of data from the system. |
| Data consistency | The field/record called for from the new application should be consistent with that of the original application.<br><br>**Note:** This enables consistency in repeatability of the testing exercise. |
| Business continuity | The new application should be able to continue with newer records as added (or appended) and help in ensuring seamless business continuity. |

## Changeover (Go-live) Techniques

### Introduction

This topic contains information on changeover, which refers to an approach to shift users from using the application from the existing (old) system to the replacing (new) system.

### Description of Changeover (Go Live)

Changeover (go live) is appropriate only after testing the new system with respect to its program and relevant data.

### Types of Changeover Techniques

The following, selected changeover (go-live) techniques are discussed in more detail in this section:
- Parallel changeover
- Phased changeover
- Abrupt changeover

## Parallel Changeover

### Description of Parallel Changeover

This technique involves using both systems during a period of overlap. This includes, in order:
- Running the old system
- Running both the old and new systems in parallel
- Fully changing over to the new system after gaining confidence in the working of the new system

After a period of overlap, the:
- User gains confidence and assurance in relying on the newer system
- Use of the older system is discontinued
- New system becomes totally operational

### Benefits of Parallel Changeover

Parallel changeover:
- Minimizes the risk of using the newer system
- Helps identify problems, issues or any concerns that the user initially comes across in the newer system

### Exhibit 4.3: Parallel Changeover

**Exhibit 4.3** depicts a parallel changeover.

**Exhibit 4.3: Parallel Changeover**

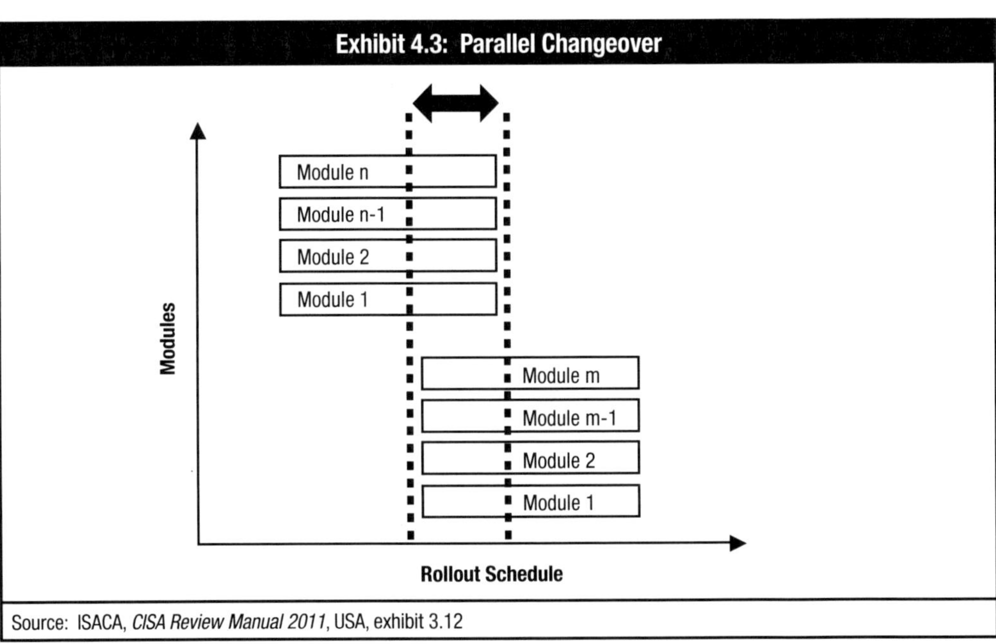


Source: ISACA, *CISA Review Manual 2011*, USA, exhibit 3.12

**Note:** The number (*m* and *n*, respectively) of modules in the new and old systems may be different.

## Phased Changeover

### Phased Changeover Steps

The following table describes how the changeover from the older system to the newer system takes place in a preplanned, phased manner.

| Phased Changeover Steps | |
|---|---|
| **Step** | **Description** |
| 1 | Module 1 of the older system is phased out and replaced by Module 1 of the newer system.<br><br>**Note:** This relationship is not necessarily one to one. Modules 1-3 of the old system could be replaced by module 1 of the new system and so forth. |
| 2 | Module 2 of the older system is phased out and replaced by Module 2 of the newer system. |
| n-1 | Module n-1 of the older system is phased out and replaced by Module n-1 of the newer system. |

### Phased Changeover Risk Factors

Some of the risk factors that may exist in the phased changeover include:

- Resource challenges (both on the IT side—to be able to maintain two unique environments such as hardware, OSs, databases and code—and on the operations side—to be able to maintain user guides, procedures and policies; definitions of system terms, etc.)
- Extension of the project life cycle to cover two systems
- Change management for requirements and customizations to maintain ongoing support of the older system

### Exhibit 4.4: Phased Changeover

**Exhibit 4.4** depicts a phased changeover.

**Exhibit 4.4: Phased Changeover**

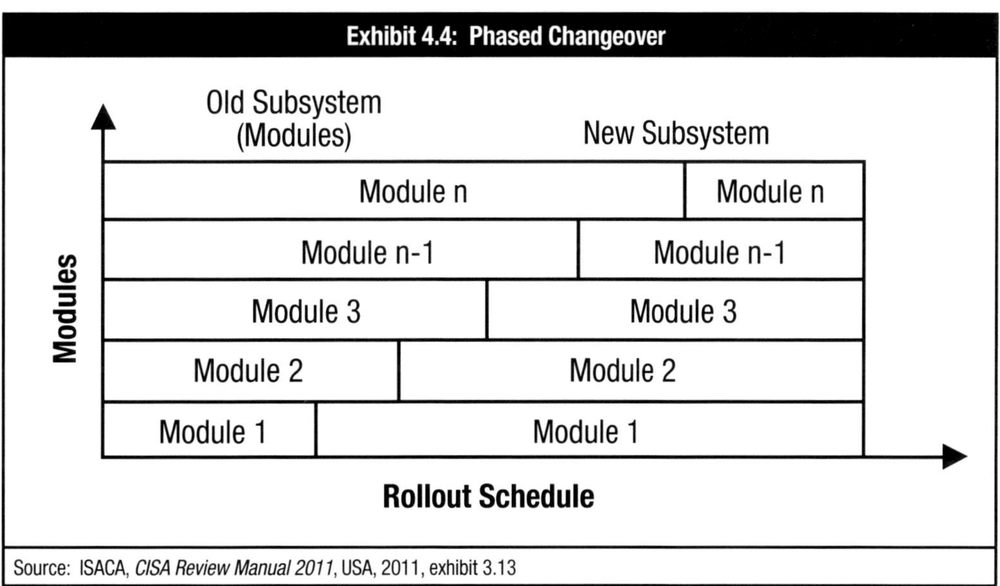


Source: ISACA, *CISA Review Manual 2011*, USA, 2011, exhibit 3.13

## Abrupt Changeover

**Description of Abrupt Changeover**

In this approach, the newer system is changed over from the older system on a cut-off date and time and the older system is discontinued once changeover to the new system takes place.

**Abrupt Changeover Steps**

The following table describes the steps of the abrupt changeover process.

| Abrupt Changeover Steps | |
|---|---|
| **Step** | **Description** |
| 1 | Convert files and programs; perform test runs on the test bed. |
| 2 | Install new hardware, OS, application system and migrated data. |
| 3 | Train employees or users in groups. |
| 4 | Schedule operations and test runs for go live or changeover. |

**Exhibit 4.5: Abrupt Changeover**

**Exhibit 4.5** depicts the abrupt changeover process.

Exhibit 4.5: Abrupt Changeover

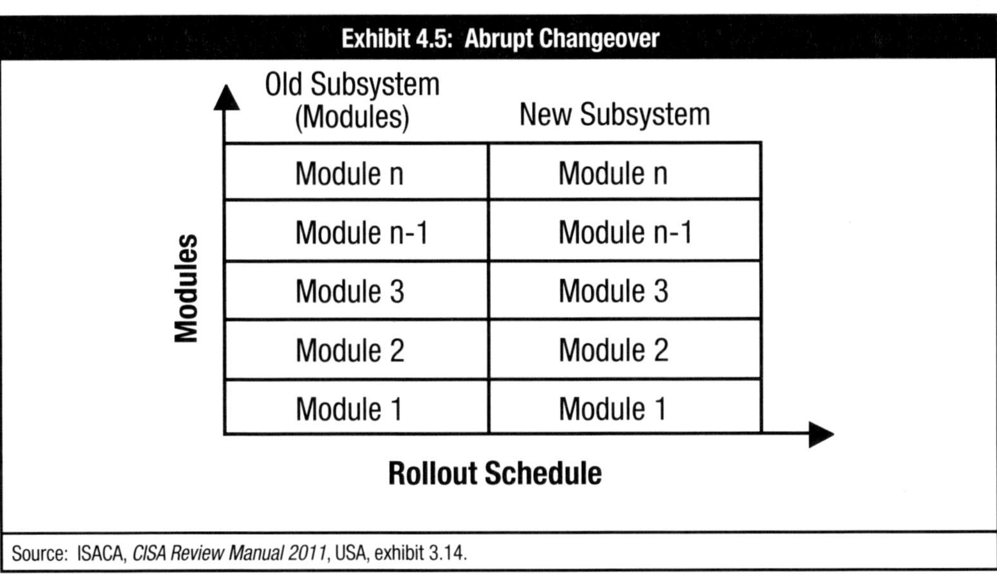


Source: ISACA, *CISA Review Manual 2011*, USA, exhibit 3.14.

**Abrupt Changeover Risk Areas**

Some of the potential abrupt changeover risk areas include:
- Asset safeguarding
- Data integrity
- System effectiveness
- System efficiency
- Change management challenges (depending on the configuration items considered)
- Duplicate or missing records (duplicate or erroneous records may exist if data cleansing is not done correctly)

## Postimplementation Review

### Introduction

This topic contains information on postimplementation review, which is:
- Used following the successful implementation of a new or extensively modified system
- Beneficial to verify that the system has been properly designed and developed and that proper controls have been built into the system

### Postimplementation Review Objectives

A postimplementation review should meet the following objectives:
- Assess the adequacy of the system:
  - Does the system meet user requirements and business objectives?
  - Have controls been adequately defined and implemented?
- Evaluate the projected cost benefits or ROI measurements.
- Develop recommendations that address the system's inadequacies and deficiencies.
- Develop a plan for implementing the recommendations.
- Assess the development project process:
  - Were the chosen methodologies, standards and techniques followed?
  - Were appropriate project management techniques used?

### Example of Key Performance Indicators (KPIs)

A KPI for the implementation process measures the relative success of the changeover compared to desired performance objectives. Success of a changeover is often measured as a percentage of errors, number of trouble reports, duration of system outage, or degree of customer satisfaction.

The use of the KPI indicates to management whether the change control process was managed correctly, with sufficient levels of quality and testing.

**Example:** A KPI may be set at an error rate of 3%. If the changeover results in less than 3% errors, then the objective was satisfied. An error rate of greater than 3 % would indicate an unacceptable level of errors and faults in the process.

### Postimplementation Review Timing

The project development team and appropriate end users perform a post-project review jointly after the project has been completed and the system has been in production for a sufficient time period to assess its effectiveness.

## Project Closeout

### Introduction

This topic contains information on closing a project. Projects should have a finite life: At some point, the project is closed and the new or modified system is handed over to the users and/or system support staff.

## Project Closeout Steps

The following table describes the steps to close a project.

| Project Closeout Steps | |
|---|---|
| **Step** | **Action** |
| 1 | Assign any outstanding issues to individuals responsible for remediation and identify the related budget, if applicable. |
| 2 | Assign custody of contracts, and archive or pass on documentation to those who will need it. |
| 3 | Survey the project team, development team, users and other stakeholders to:<br>• Identify any lessons learned that can be applied to future projects.<br>• Include content-related criteria such as:<br>– Performance fulfillment and project-related incentives<br>– Fulfillment of additional objectives<br>– Adherence to the schedule and costs<br>• Include process-related criteria such as:<br>– Quality of the project teamwork<br>– Relationships to relevant environments |
| 4 | Conduct reviews in a formal process such as a post-project review in which lessons learned and an assessment of project management processes used are documented and referenced, in the future, by other project managers or users working on projects of similar size and scope. |
| 5 | Complete a postimplementation review once the project has been in use (or in "production") for some time—long enough to realize its business benefits and costs—and measure the project's overall success and impact on the business units. |

**Note:** The project sponsor should be satisfied that the system produced is acceptable and ready for delivery.

# *F. Managing Project Risk*

## Introduction

This section contains information on managing the risk associated with an information systems project, such as the failure of the project to deliver expected benefits on time, within budget and to meet the needs and expectations of the stakeholders.

## Relevance

While the SDLC focuses on ensuring benefits realization at a technical level, project management practices—while closely related—focus on managing project-related risk through the management of project scope, time and resources.

The risk practitioner must apply project risk management techniques to ensure that needed IS controls are implemented at a reasonable cost.

- A 2002 Gartner survey found that 20 percent of all expenditures on IT is wasted—a finding that represents, on a global basis, an annual destruction of value totalling about US $600 billion.
- A 2004 IBM survey of Fortune 1000 CIOs found that, on average, CIOs believe that 40 percent of all IT spending brought no return to their organisations.
- A 2006 study conducted by The Standish Group found that only 35 percent of all IT projects succeeded while the remainder (65 percent) were either challenged or failed.

Source: ISACA, *Enterprise Value Governance of IT Investments: The Val IT Framework 2.0*, USA, 2009, page 7

## 1. Key Terms and Principles

### Introduction

This section introduces terms and principles related to project management as well as terms that help relate the process to other key business processes.

### Definition of Project

A project is a structured set of activities concerned with delivering a defined capability (that is necessary, but not sufficient, to achieve a required business outcome) to the enterprise based on an agreed schedule and budget.

### Definition of Program

A program is a structured grouping of interdependent projects that are both necessary and sufficient to achieve a desired business outcome and create value.

**Note:** These interdependent projects could involve, but are not limited to, changes in the nature of the business, business processes, the work performed by people, as well as the competencies required to carry out the work, enabling technology and organizational structure.

### Definition of Portfolio

A portfolio is a grouping of "objects of interest" (investment programs, IT services, IT projects, other IT assets or resources) managed and monitored to optimize business value.

Portfolio management is distinct from project and project management in that the distinct objective is to create maximum value from a grouping of projects and programs.

**Exhibit 4.6: Relationship Among Project, Program and Portfolio Management**

**Exhibit 4.6** describes the relationship among project management, program management and portfolio management.

Exhibit 4.6: Relationship Among Project, Program and Portfolio Management

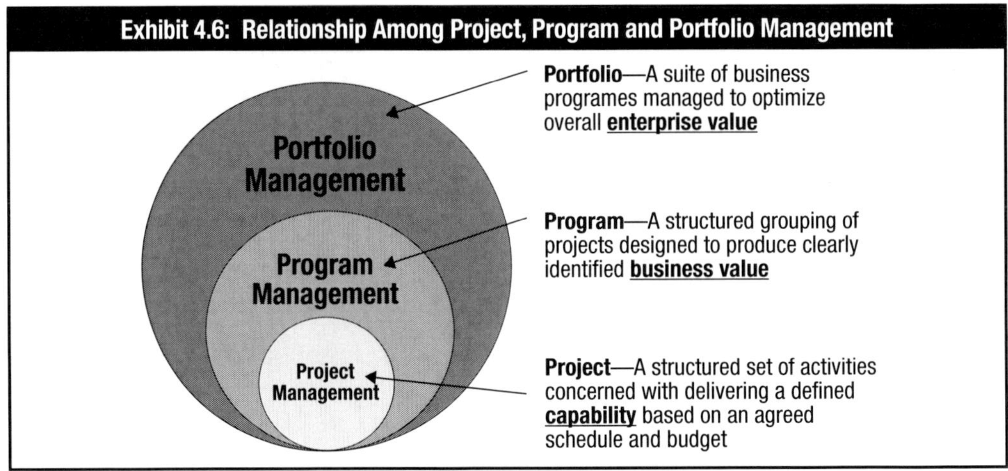


## 2. Overview of Project Risk Management

**Project Management Iron Triangle**

A project is bound by what can be termed the "iron triangle." As shown in **exhibit 4.7**, the iron triangle is defined by three constraints that affect project quality. A change to any one of the three constraints will force a change on the other two. In other words, an increase in scope will force a change in the allocated resources, the project schedule or both.

Exhibit 4.7: Project Management Iron Triangle

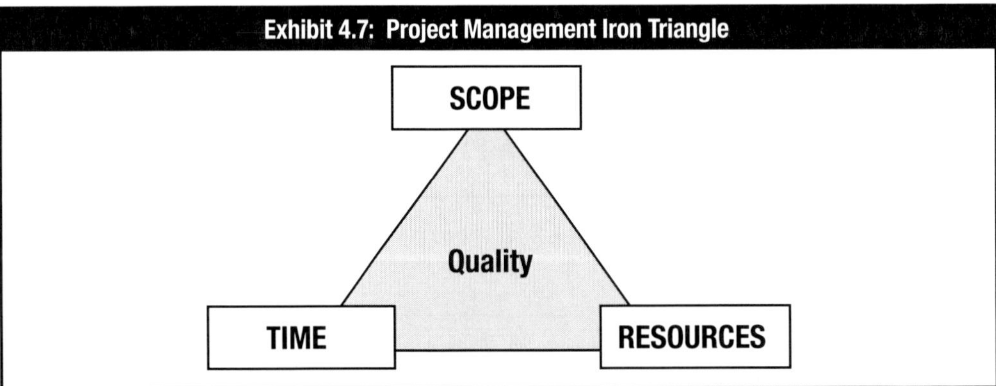


**Key Project Management Phases**

Key phases of the project and program management methodology are:

1. Define a program/portfolio management framework for IT investments.
2. Establish and maintain an IT project management framework.
3. Establish and maintain an IT project monitoring, measurement and management system.
4. Build project charters, schedules, quality plans, budgets, and communication and risk management plans.
5. Ensure the participation and commitment of project stakeholders.
6. Ensure the effective control of projects and project changes.
7. Define and implement project assurance and review methods.

This section revolves around the operational aspects of project management only, specifically items 4, 5, 6 and 7.

### Project Risk

There is always the risk that an IT project will not deliver the expected results and that the investment in the project has been wasted or ineffective. The cause of project failure may be due to many factors such as:
- Changing requirements
- Poorly defined requirements
- New priorities
- Lack of resources or skill
- Lack of oversight for the project
- Problems with new technology
- Financial challenges

## 3. Scope Management

### Introduction

Managing project deliverables requires careful control of project scope, deliverables and schedule. This control is effected through comprehensive documentation of the project, clearly defined requirements and resource management.

### Project Scoping

Project scope is defined by the project owner or sponsor and is often a balance between cost, time and business needs. The scope should be clearly documented early in the project requirements phase of the project life cycle, once the feasibility studies and initial requirements gathering has been completed.

It is extremely important that the scope includes identifying and addressing the security requirements in the initial documentation and budgeting for the information system. A failure to include security costs and deliverables will almost certainly lead to expensive redesign of the project later in the life cycle, or the delivery of a system that is not adequately protected.

The documentation of the project deliverables should be detailed enough to provide an understanding of the complexity and amount of work required to complete the project.

This can often be done through the use of a statement of work (SOW), or a work breakdown structure (WBS). This prevents confusion or disappointment when one party does not meet the expectations of the other party due to misunderstanding of the project deliverables.

Any changes to the scope of the project should be subject to review and approval by a change management board (sometimes called a change control board). This is examined later in this chapter.

### Earned Value

Earned value management (EVM) is a technique for measuring project performance and progress in an objective manner. Earned value has the ability to combine measurements of scope, schedule and cost in a single integrated system. EVM is notable for its ability to provide accurate forecasts of project performance problems. Using the methodology helps improve both scope definition as well as the analysis of overall project performance.

## Project Review and Status Reporting

All projects require regular review and reporting. A project must have milestones that require the careful review of, and reporting on, the project status. These reports will indicate whether a project is on schedule or on budget and highlight any problems that may be developing with project scope, resourcing, deliverables, funding or time.

Such reviews often mandate that predefined deliverables have been produced before the next project phase is entered. Rigorous reviews enable the identification of any current or future project challenges, enable corrective action where necessary and provide management with decision points as to whether to continue, modify or abandon a project.

## Managing Scope Changes

"The only thing certain in life is change."This premise certainly applies to projects and certain changes will have to be made to accommodate new requirements, adjust to changing priorities, or manage unexpected complications with integrating new technology to the existing IT architecture or vice versa.

To manage project change it is important to allow changes where necessary and—at the same time—prevent uncontrolled changes or scope creep. A formal project change management process helps achieve this while minimizing the effect that such changes may have on project schedule, budget, resources or stakeholder expectations.

A formal project change management process ensures that the change is executed according to the approval given, and that the change is documented, thoroughly tested and correctly implemented.

## Change Request Approval Process

While the change management process is the overall process to change a system/software, the change approval process describes how a single change request is submitted, evaluated and approved.

The following table describes the responsibilities related to the change request approval process.

| Change Request Approval Process | | |
|---|---|---|
| **Step** | **Responsibility** | **Description** |
| 1 | Stakeholder | • Initiates a formal project change request that contains clear description of the:<br>– Requested change<br>– Reasons for the change<br>• Submits the change request to the project manager |
| 2 | Project manager | • Judges the impact of each change request on project activities (scope), schedule and budget<br>• Archives copies of all change requests in the project file |
| 3 | Change advisory board | • Evaluates the change request (on behalf of the sponsor)<br>• Decides whether to recommend the change<br>• If accepted, instructs the project manager to update the project plan to reflect the requested change |
| 4 | Project sponsor | • Formally accepts or rejects the updated project plan recommendation by the change advisory board |

## 4. Time Management

### Introduction

This topic contains information on managing time by establishing and scheduling time frames.

### Budgeting vs. Scheduling

Managing time involves two key efforts:
- **Budgeting**—Totaling the human and machine effort involved in each task
- **Scheduling**—Establishing the sequential (or networked) relationship between tasks

Since tasks of a project reflect a self-contained, ordered group of activities, a project can be represented as a network in which tasks are shown as branches connected at nodes immediately preceding and following other tasks.

### Critical Path Method (CPM)

All project schedules have a critical path, which is the set of successive activities that go from the beginning to the end of the project with the shortest possible completion time.

The essential technique for using the critical path method (CPM) is to construct a model of the project that includes the following:
1. A list of all activities required to complete the project (typically categorized within a work breakdown structure [WBS]),
2. The time (duration) that each activity will take to completion
3. The dependencies between the activities

Some activities may be performed concurrently (at the same time as other tasks); others may not be able to start before preceding activities have been completed (consecutively). Identifying such interdependencies between tasks is a key benefit of the CPM.

### Exhibit 4.8: CPM

**Exhibit 4.8** depicts the CPM.

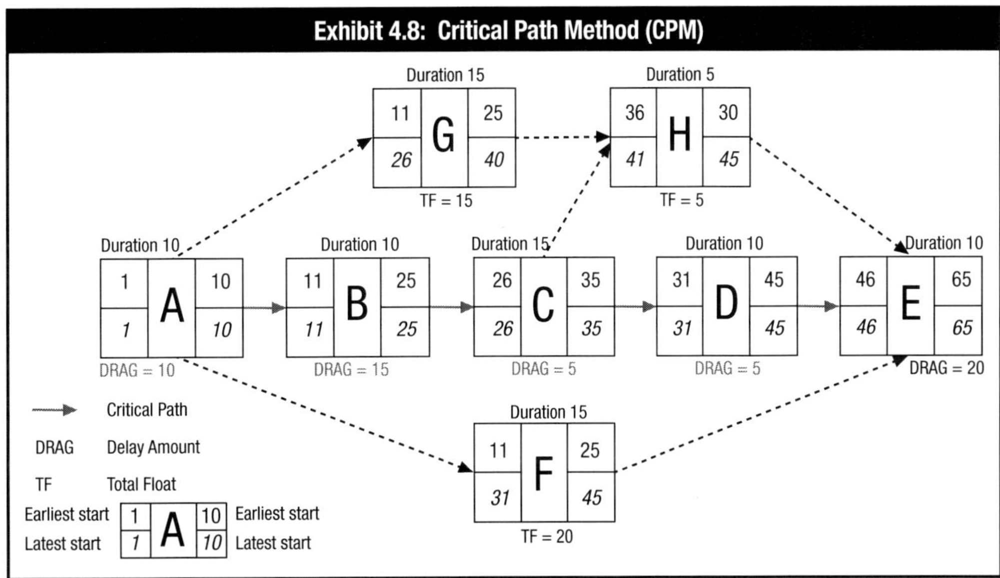

### Definitions

**Slack (float) time**—Activities that are not part of the critical path have some slack time (float time), which is the difference between the:
- Latest possible completion time of each activity that will not delay the completion of the overall project
- Earliest possible completion time based on completion of all predecessor activities

> **Note:** Activities on the critical path have zero float time, and conversely, activities with zero slack time are on the critical path.

**Drag time**—The drag is the calculation of the impact that a delay in one activity will have on the completion date of the project. The drag of an activity is equal to its duration if it does not have another activity in parallel; however, it if has parallel activities, its drag is the lesser of either the activity duration or the total float of the parallel activity.

## 5. Budget Management

### Introduction

This topic contains information on managing the project budget, which focuses on managing resources efficiently to achieve the desired outcome.

### Description of Resource Usage

Resource usage is the process by which the project budget is being spent. If actual spending is in line with planned spending, resource usage must be measured and reported.

Projects incur both fixed and non-fixed costs. A fixed cost such as licenses or office space will often remain relatively unchanged if the time to complete the project is extended or decreased. Salaries may be affected by overtime—more hours required than budgeted for or the cost of recruiting staff with specialized skills.

> **Note:** It is not sufficient to monitor only actual spending.

### Example—Resource Productivity

If a task is planned to take 24 man-hours, then it is implicitly supposed that:
- The resource being deployed is capable of finishing the task within the scheduled time.
- At the same time, results are to be delivered at a satisfactory quality level.

This assumes that every budget and project plan presupposes a certain "productivity" of resources.

### Techniques for Measuring Resource Productivity

Resource usage can be checked with a technique called earned value analysis (EVA), which consists of comparing the following metrics at regular intervals during the project:
- Budget to date
- Actual spending to date
- Estimate to complete and estimate at completion

**Example:** A project task is budgeted to take three eight-hour working days. One work day has passed:
- From a pure time perspective, it will take another 16 hours to complete the task.
- From an earned value perspective, the resource is asked how much of the task he/she was able to complete in the eight hours:
  - If the resource performed a third of the task, he/she is in alignment with projections.
  - If the resource performed 25 percent of the task, he/she is behind projections and will need another day.
  - If the resource performed 50 percent of the task, he/she is ahead of projections and will be able to complete the task in two, instead of three, days.

## 6. Use of Metrics to Support Resource Planning

### Introduction

This topic contains information on gathering metrics to support resource planning, which is required to ensure that day-to-day operations proceed smoothly while identifying and designating the technology and roles and responsibilities required for program development.

**Note:** These metrics help ensure that resource deficiencies are detected and corrected before they impact the performance of overall security program development efforts.

### Collecting Metrics

The following table describes the steps to collect metrics to support resource planning.

| Steps for Resource Planning Metrics | |
|---|---|
| **Step** | **Action** |
| 1 | • Develop metrics for resource utilization to support efforts to maximize program development efforts.<br><br>**Rationale:** Program development activities are subject to the same staffing and organizational dependency issues as any other management process. |
| 2 | • Gather historical data on resource dependencies that may affect the security program.<br>• Use this data in planning periodic activities designed to:<br>– Identify changing security resource requirements in time to devise alternate plans<br>– Meet control objectives that may also change in response to new requirements |
| 3 | • Ensure that all personnel who take lead roles in the performance of critical security functions have a backup that can perform the given function unassisted in the absence of the primary leader.<br><br>**Rationale:** Metrics that count the number of people in critical functions without a trained backup herald a call to action. |
| 4 | • Ensure that lead roles are covered.<br><br>**Note:** If critical elements of the security program are not managed with appropriate accountability, this metric may indicate gaps in the information security program that need to be addressed. |
| 5 | • Where responsibilities for operating technologies are delegated to other departments, collect metrics that provide insight into the resource requirements and planning processes of the responsible enterprise. |

# G. Project Management Tools and Techniques

## 1. General Project Management Techniques

**Introduction**

This topic contains knowledge on general project management practice techniques.

---

**Project Management Knowledge and Practices**

Project management knowledge and practices are best described in terms of their component processes of initiating, planning, executing, controlling and closing a project.

Overall characteristics of successful project planning are that it is a risk-based management process and iterative in nature.

---

**Project Management Techniques and Tools for Controlling Time and Using Resources**

Project management techniques and tools:
- Provide systematic quantitative and qualitative approaches to software size estimating, scheduling, allocating resources and measuring productivity
- Assist the project manager in controlling the time and resources utilized in the development of a system
- Vary from a simple manual effort to a more elaborate computerized process

> **Note:** Base the approach on the project size and complexity.

---

**Exhibit 4.9: Relationships Between Project Management Elements**

Project management should pay attention to the intertwining relationships between project management elements:
- Deliverables
- Duration
- Budget

**Exhibit 4.9** depicts an oversimplified and schematized complex relationship.

**Exhibit 4.9: Relationships Between Project Management Elements**

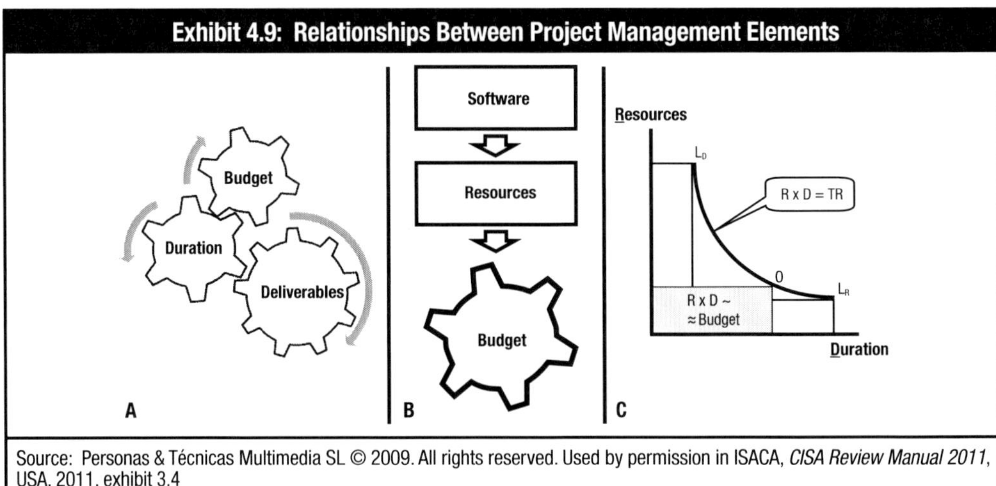


Source: Personas & Técnicas Multimedia SL © 2009. All rights reserved. Used by permission in ISACA, *CISA Review Manual 2011*, USA, 2011, exhibit 3.4

## Project Management Elements

The following table describes the relationships between the project management elements shown in **exhibit 4.8**.

<table>
<tr><th colspan="3">Project Management Elements</th></tr>
<tr><th>Element</th><th colspan="2">Description</th></tr>
<tr><td>Deliverables</td><td colspan="2">• Project duration and budget must be commensurate with the nature and characteristics of the deliverables.<br>• In general, there will be a positive correlation (growing together) between highly demanding deliverables, a long duration and a high budget.<br>• The quality of the deliverables is an important element that is also considered during the management of time and resources:<br>– The parameters for quality of the deliverables may be specified clearly by the project steering committee or project sponsor, or the project manager may have to elicit the parameters from user management.<br>– Regardless of who specifies the parameters for quality, the project manager must have a clear and documented view of the quality expectations for the deliverables of the project steering committee, sponsor and users.</td></tr>
<tr><td>Budget</td><td colspan="2">• Budget is deduced from the resources required to carry out the project by multiplying fees or costs by the amount of each resource.<br>• At the beginning of the project, the required resources are estimated by using techniques of software/project size estimation.<br><br>Note: Size estimation yields a “total resources” calculation.</td></tr>
<tr><td rowspan="5">Resources</td><td colspan="2">• For simplification purposes, there is the assumption that resources are fixed for the duration of the project.<br>• The curve shows resources assigned (R) × duration (D) = total resources (TR, a constant quantity); which is the classic “man × month” dilemma curve.<br>• Any point along the curve meets the condition R × D = TR.<br>• If any point O on the curve is chosen, the area of the rectangle will be TR, proportional to the budget.</td></tr>
<tr><th>If resources are ...</th><th>Then the project will take a ...</th></tr>
<tr><td>Few</td><td>Long time (a point close to LR).</td></tr>
<tr><td>Many</td><td>Shorter time (a point close to LD).</td></tr>
<tr><td colspan="2">LR and LD are two practical limits and result in:<br>• A duration that is too long may not seem reasonable.<br>• The use of too many (human) resources at once would be unmanageable.</td></tr>
</table>

**Note:** There are a few heuristics (rules of thumb) available to choose a convenient combination of assigned resources and project duration.

## 2. Gantt Charts

### Introduction

This topic contains knowledge on constructing Gantt charts, which aid in scheduling the activities (tasks) needed to complete a project.

### Purpose of Gantt Charts

Gantt charts (depicted in **exhibit 4.10**):
- Show when an activity should begin and end along a timeline
- Show which activities:
  - Can be in progress concurrently
  - Must be completed sequentially
- Reflect the resources assigned to each task and by what percent they are allocated
- Compared to a baseline project plan, outline:
  - Which activities have been completed early or late
  - The project progress indicating whether the project is behind, ahead or on schedule
- Track the achievement of milestones or significant accomplishments for the project such as the end of a project phase or completion of a key deliverable

### Exhibit 4.10: Sample Gantt Chart

**Exhibit 4.10** shows a Gantt chart within a work breakdown structure.

**Exhibit 4.10: Sample Gantt Chart**

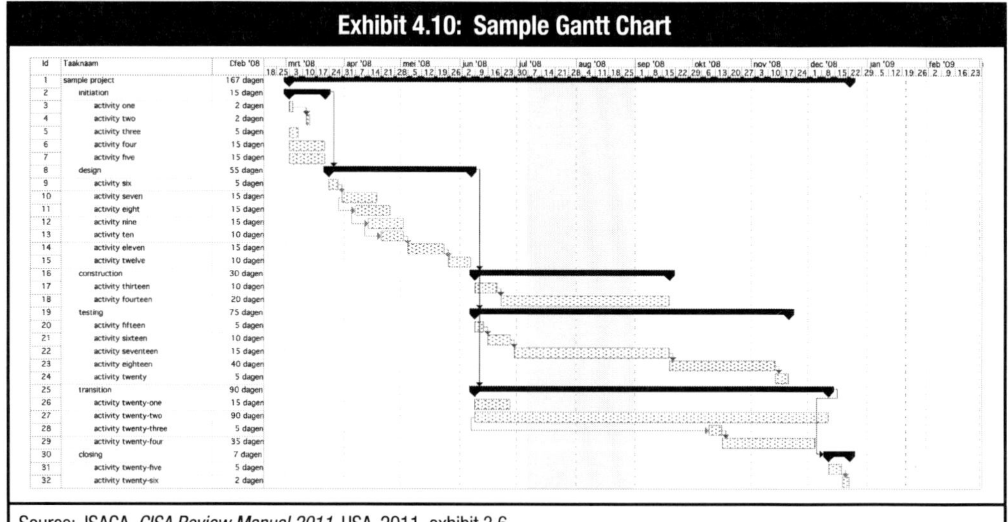


Source: ISACA, *CISA Review Manual 2011*, USA, 2011, exhibit 3.6

## 3. Critical Path Methodology (CPM) and Program Evaluation Review Technique (PERT)

### Introduction

This topic contains information on PERT, which is a CPM-type technique that uses three different estimates for each activity instead of using a single number (as used by CPM).

The advantage of using three different estimates for each activity is that the formula is based on the reasonable assumption that:
- The three time estimates follow a beta statistical distribution.
- Probabilities (with associated confidence levels) can be associated with the total project duration.

### PERT Usage

PERT is often used in system development projects with uncertain durations (e.g., pharmaceutical research or complex software development).

## Identification of All Activities and Related Events/Milestones

When designing a PERT network for system development projects, the first step is to identify all the activities and related events/milestones of the project and their relative sequence.

**Note:** The risk practitioner:
- Must be careful not to overlook any activity
- May prepare many diagrams that provide increasingly more detailed time estimates

**Note:** Some activities such as analysis and design must be preceded by others before program coding can begin. The list of activities determines the detail of the PERT network.

**Example:** An event or result may be the completion of the operational feasibility study or the point at which the user accepts the detailed design.

## Identification of a Critical Path Using PERT

A critical path is the:
- Longest path through the network (only one critical path in a network)
- Route along which the project is shortened (accelerated) or lengthened (delayed)

**Example:** In **exhibit 4.11**, the critical path is A, C, E, F, H and I.

## PERT Time Estimates

The following table describes the three estimates used for completing each task's activity.

| Task Activity Completion Estimates | |
|---|---|
| **Estimate** | **Description** |
| First | This is the most optimistic time, if everything went well. |
| Second | This is the most likely scenario and is based on experience attained from projects similar in size and scope. |
| Third | This is the pessimistic or worst-case scenario. |

## Calculation of the PERT Time Estimate

To calculate the PERT time estimate for each given activity, the following calculation is applied: [Optimistic + Pessimistic + 4(most likely)]/6.

The three PERT estimates are:
- Reduced (applying a mathematical formula) to a single number
- Applied to the classic CPM algorithm

## Exhibit 4.11: PERT Network-based Chart

**Exhibit 4.11** illustrates the use of the PERT network management technique, in which events are points in time or milestones for starting and completing activities (arrows).

**Exhibit 4.11: PERT Network-based Chart**

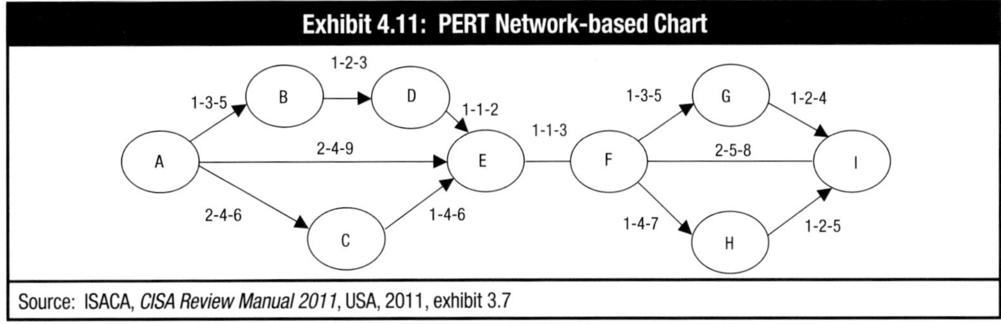


Source: ISACA, *CISA Review Manual 2011*, USA, 2011, exhibit 3.7

## H. Suggested Resources for Further Study

### Suggested Resources for Further Study

In addition to the resources cited throughout this manual, the following resources are suggested for further study:

- ISACA:
  - COBIT 4.1, 2007

    **Note:** The COBIT 4.1 framework is available at no charge from ISACA and can be downloaded at *www.isaca.org/cobit*. The new COBIT 5 framework will be available in 2012.
  - *CISM Review Manual 2012*, 2011
  - *ITAF: A Professional Practices Framework for IT Assurance*, 2008
  - *IT Assurance Guide: Using COBIT*, 2007
  - *Value Management Guidance for Assurance Professionals: Using Val IT 2.0*, 2010
  - *Systems Development and Project Management Audit/Assurance Program*, 2009
  - *The Business Model for Information Security (BMIS)*, 2010
- Barnier, Brian; *The Operational Risk Handbook for Financial Companies: A Guide to the New World of Performance-oriented Operational Risk*, Harriman House Ltd, UK, 2011
- Committee of Sponsoring Organizations of the Treadway Commission (COSO), *Internal Control—Integrated Framework: Guidance on Monitoring Internal Control Systems*, USA, 2009
- IEEE Computer Society, Standard 1074-2006 for Developing a Software Project Lifecycle Process, USA, 2006, *www.ieee.org*
- International Project Management Association (IPMA), *IPMA Competence Baseline (ICB), Version 3.0*, The Netherlands, 2006
- Krutz, Ronald L.; Russell Dean Vines; *The CISM Prep Guide: Mastering the Five Domains of Information Security Management*, Wiley, USA, 2003
- Maizlish, Bryan; Robert Handler; *IT Portfolio Management Step-by-step: Unlocking the Business Value of Technology*, John Wiley & Sons, USA, 2005
- National Institute of Technology and Standards (NIST), *Security Controls in External Environments, NIST Special Publication (SP) 800-53, Revision 3*, USA, 2009
- National Institute of Technology and Standards (NIST), Special Publication 800-53, Revision 3, Online Database, *http://web.nvd.nist.gov/view/800-53/home*
- Office of Government Commerce (OGC), *Projects in Controlled Environments 2 (PRINCE2): Managing Successful Projects With PRINCE2: 2009 Edition*, UK, 2009
- Project Management Institute (PMI), *A Guide to the Project Management Body of Knowledge (PMBOK), 4th Edition*, USA, 2004

# Domain 5–Information Systems Control Monitoring and Maintenance

## A. Chapter Overview

### Introduction

This chapter focuses on monitoring of information systems controls to confirm that controls are well designed and continue to effectively mitigate identified risk over time. The chapter provides information on the implementation of a control monitoring process, reporting of control status to management, automated tools that are available to help enterprises achieve information systems (IS) control monitoring appropriate for their unique needs, and process maturity to help enterprises determine the current state of their control environment and the steps necessary to move toward their desired state.

### Inputs From Other Domains

The IS control monitoring and maintenance process is a distinct subset of *Domain 2—Risk Response*, particularly risk mitigation/reduction.

Information systems control monitoring and maintenance efforts are directly based on and affected by *Domain 4—Information Systems Control Design and Implementation*. Newly implemented IS controls are captured in the controls inventory and key controls are selected for monitoring.

### Relevance

Risk management relies on a controls monitoring process to ensure that IS controls remain effective and efficient over time. Monitoring requires the definition of meaningful performance indicators, systematic and timely reporting of performance, and prompt response to deviations. Monitoring makes sure that the right things are done and are in line with business directions, corporate policies and best practice standards.

### Outputs to Other Domains

Reports on the effectiveness or ineffectiveness of controls are an important input into *Domain 1—Risk Identification, Assessment and Evaluation*. Control effectiveness significantly influences the enterprise's overall risk profile, particularly residual risk.

### Learning Objectives

As a result of completing this chapter, the CRISC candidate should be able to:
- Describe the purpose and levels of a maturity model as it applies to the risk management process.
- Compare different monitoring tools and techniques.
- Describe various testing and assessment tools and techniques.
- Explain how monitoring of IS controls relates to applicable laws and regulations.
- Understand the need for control maintenance
- Establish a process for the ongoing operation and maintenance of controls

## Contents

This chapter contains the following sections:



Process-specific controls for the following processes are addressed in Part II of this manual:

1. Determining the IT Strategy
2. Project and Program Management
3. Change Management
4. Third-party Service Management
5. Continuous Service Assurance
6. Information Security Management
7. Configuration Management
8. Problem Management
9. Data Management
10. Physical Environment Management
11. IT Operations

# *B. Task and Knowledge Statements*

## Introduction

This section describes the task and knowledge statements for Domain 5, which focuses on monitoring and maintaining IS controls to ensure that they function effectively and efficiently.

## Domain 5 Task Statements

The following table describes the task statements for Domain 5.

| No. | Task Statement (TS) |
|---|---|
| TS5.1 | Plan, supervise and conduct testing to confirm continuous efficiency and effectiveness of information systems controls. |
| TS5.2 | Collect information and review documentation to identify information systems control deficiencies. |
| TS5.3 | Review information systems policies, standards and procedures to verify that they address the organization's internal and external requirements. |
| TS5.4 | Assess and recommend tools and techniques to automate information systems control verification processes. |
| TS5.5 | Evaluate the current state of information systems processes using a maturity model to identify the gaps between current and targeted process maturity. |
| TS5.6 | Determine the approach to correct information systems control deficiencies and maturity gaps to ensure that deficiencies are appropriately considered and remediated. |
| TS5.7 | Maintain sufficient, adequate evidence to support conclusions on the existence and operating effectiveness of information systems controls. |
| TS5.8 | Provide information systems control status reporting to relevant stakeholders to enable informed decision making. |

## Domain 5 Knowledge Statements

The following table describes the knowledge statements for Domain 5.

| No. | Knowledge Statement (KS)<br>Knowledge of: |
|---|---|
| KS5.1 | Standards, frameworks and leading practices related to information systems control monitoring and maintenance |
| KS5.2 | Enterprise security architecture |
| KS5.3 | Monitoring tools and techniques |
| KS5.4 | Maturity models |
| KS5.5 | Control objectives, activities and metrics related to IT operations and business processes and initiatives |
| KS5.6 | Control objectives, activities and metrics related to incident and problem management |
| KS5.7 | Security testing and assessment tools and techniques |
| KS5.8 | Control objectives, activities and metrics related to information systems architecture (platforms, networks, applications, databases and operating systems) |
| KS5.9 | Control objectives, activities and metrics related to information security |
| KS5.10 | Control objectives, activities and metrics related to third-party management |
| KS5.11 | Control objectives, activities and metrics related to data management |
| KS5.12 | Control objectives, activities and metrics related to the system development life cycle |
| KS5.13 | Control objectives, activities and metrics related to project and program management |
| KS5.14 | Control objectives, activities and metrics related to software and hardware certification and accreditation practices |
| KS5.15 | Control objectives, activities and metrics related to business continuity and disaster recovery management |
| KS5.16 | Applicable laws and regulations |

**Note:** Knowledge statements 5.6 and 5.8-5.14 are process specific and addressed in Part II of this manual.

# C. The Control Life Cycle

## Introduction

This section provides an overview of the control life cycle as well as the related concepts and principles.

The relative mix and importance of process, application and general controls will be unique to each enterprise.

s

## Exhibit 5.1: The Control Life Cycle

The control life cycle maps out the various phases in the life of a control from the initial selection/design, through development, implementation, maintenance and disposal of the system.

The risk practitioner must ensure that the controls are properly managed throughout each phase of the life cycle.

**Exhibit 5.1** shows the phases of the control life cycle.

**Exhibit 5.1: The Control Life Cycle**

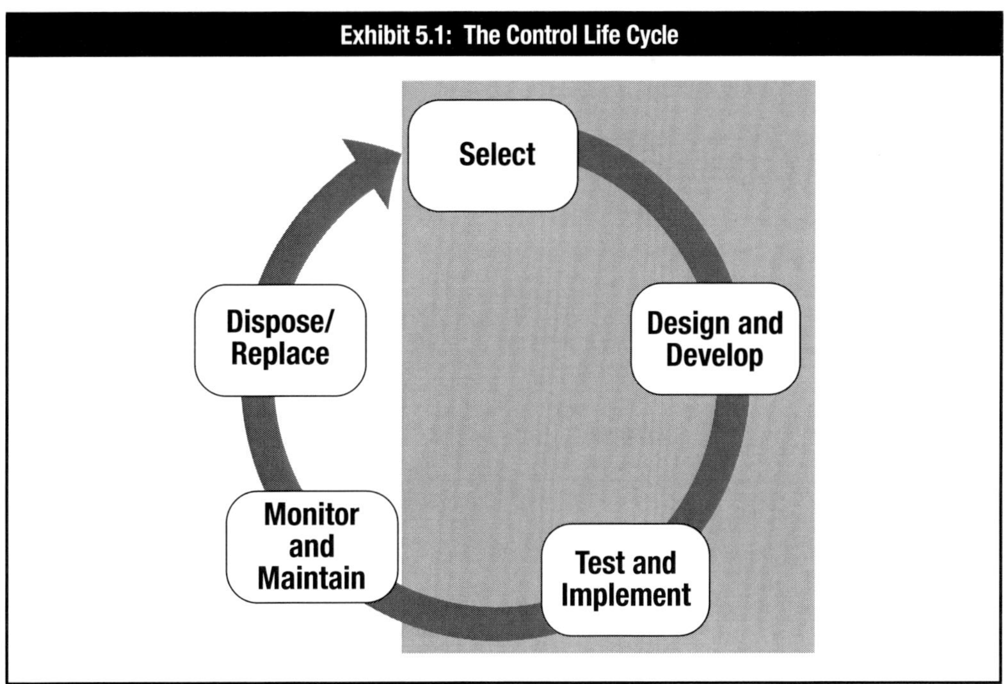


The shaded areas indicate those steps of the control life cycle that are addressed in *Domain 4—Information Systems Control Monitoring and Maintenance.*

## Control Monitoring and Maintenance With the Control Life Cycle

As **exhibit 5.1** shows, the control life cycle is a continuous process that begins with the selection of an appropriate response to identified risk.

Once a control has been selected, it must be designed into the systems and processes of the enterprise, tested, and implemented properly. Those topics are covered in *Domain 4—Information Systems Control Design and Implementation*.

> **Note:** While this chapter focuses on the monitoring and maintenance of IS controls, many principles and concepts can be applied to other control types, such as managerial or operational controls.

# D. Information Systems Control Monitoring and Maintenance

## 1. IS Control Monitoring and Maintenance Process Overview

### Introduction

This section contains an overview of the IS control monitoring and maintenance process as well as the related key concepts and principles.

### Effect of Changes on Controls

Business processes and related information flows change over time to adjust to new technologies, changing market demands, new products and more complex and stringent compliance requirements. The risk practitioner must be attentive to the impact these changes may have on the effectiveness of controls.

These changes can—often quite subtly—affect the efficiency and effectiveness of controls, or even bypass them altogether. Controls that were previously effective become inefficient, redundant or obsolete and have to be removed or replaced.

Control monitoring helps maintain control effectiveness by providing timely feedback on the status of controls to ensure that controls are effectively managed throughout the control life cycle.

### Control Monitoring Objectives

The IS monitoring and maintenance function addresses the following objectives:

- Ensure that key controls are in use, are working correctly, and are effectively mitigating the identified risk.
- Confirm that controls are being properly maintained, updates are being applied, and configurations and exceptions are handled according to policy and procedures.
- Provide assurance for management that the current risk levels are within acceptable levels.
- Management is providing oversight (ownership) of control activities.
- Management review activities and the effectiveness of key controls are reported to stakeholders.
- Processes that require correction are identified, and remedial actions are assigned and completed.

### More Control Is Not Necessarily Better

At some point, the addition of controls begins to detract from the efficiency and profitability of a process without adding an equitable level of risk mitigation.

While controls provide a benefit, there are related costs—both tangible and intangible (opportunity costs). Controls place a burden on the enterprise—a control has a cost to implement (i.e., purchase, install), a cost to maintain (i.e., reset passwords, annual maintenance fees), a cost in productivity (i.e., slower performance), a cost to monitor (i.e., data gathering, analysis and reporting); therefore, the selection of controls should be done carefully with due consideration of the effect of the control on the enterprise.

Controls should only be implemented if there is a clear corresponding risk, which the control helps mitigate. As seen in *Domain 2—Risk Response* and *Domain 4—IS Control Design and Implementation*, risk response selection mechanisms are frequently based on cost-effectiveness. The purpose of *IS Control Monitoring and Maintenance* is to ensure that the cost-benefit analysis calculations remain valid, that residual risk levels are according to forecast expectations and that the enterprise has an appropriate level of control.

An enterprise that attempts to implement too many controls—controls that are perceived as not justifiable, too aggravating, restrictive, confrontational or invasive—will often see their residual risk increase since employees, customers and managers ignore or bypass the controls.

Likewise, more monitoring is not necessarily better monitoring, which is why a focus on key controls allows an enterprise to focus its resources on the controls that matter the most.

## 2. IS Control Monitoring and Maintenance Process Phases

### Introduction

This topic contains the IS control monitoring and maintenance process phases.

### Reliance on Prior Risk Management Activities

IS monitoring and maintenance relies on previous domain tasks to:
- Identify, assess and evaluate risk.
- Respond to risk.
- Design, implement and refine the IS controls.

The IS monitoring function provides evidence to evaluate the operating effectiveness and efficiency of IS controls over time.

### IS Control Monitoring and Maintenance Process Phases

The following table describes how monitoring and maintenance of IS controls ties into the overall risk management process.

| IS Control Monitoring and Maintenance Process Phases | | |
|---|---|---|
| **Phase** | **Description** | **For details, refer to…** |
| 1 | Prioritize risk:<br>• Understand and prioritize risk to organizational objectives.<br>• Identify the significant application components and flow of information through the system.<br>• Understand the functionality of the application by:<br>– Reviewing the application system documentation<br>– Interviewing appropriate personnel | Domain 2, C. The Risk Response Process |
| 2 | Identify controls:<br>• Identify key controls across the internal control systems that address the prioritized risk.<br>• Identify the application control strengths.<br>• Evaluate the impact of the control weaknesses.<br>• Develop a testing strategy by analyzing the accumulated information. | Domain 4 |
| 3 | Identify operational control information:<br>• Identify information that will persuasively indicate whether the internal control system is operating effectively.<br>• Observe and test user performing procedures. | Domain 5, E. Identify and Assess Information |
| 4 | Implement monitoring:<br>• Develop and implement cost-effective procedures to evaluate the data about control operations. | Domain 5, G. Implementing Control Monitoring Processes |
| 5 | Report results. | Domain 5, F. Tools for Monitoring, and Domain 3, C. Essentials of Risk Monitoring<br><br>**Note:** The same core principles apply to risk reporting and control reporting. |
| 6 | Maintain IS controls:<br>• Test and deploy vendor patches as quickly as possible.<br>– Maintain secure configuration of controls according to proscribed baselines. | Part II, Chapter 9, Configuration Management |

# E. Identify and Assess Information

**Introduction**

This section describes the identification and assessment of information to enable the monitoring of IS controls.

## 1. Determining the Operating Effectiveness of the Internal Control System

**Introduction**

This topic contains information on how to identify key controls and the various ways to monitor these controls.

This requires considerable thought because there are various ways in which controls operate that impact how they can be monitored.

**Purpose of IS Control Monitoring**

The ultimate purpose of monitoring is for an evaluator to gather enough information to evaluate whether IS controls are operating effectively to mitigate risk.

**Exhibit 5.2: Direct and Indirect Information**

**Exhibit 5.2** provides a quick reference on the relative confidence that an evaluator can have in the information provided for analysis and the use of direct vs. indirect information for monitoring.

**Exhibit 5.2: Direct and Indirect Information**

| | Direct Information | Indirect Information |
|---|---|---|
| **Ongoing Monitoring** | • Typically most persuasive<br>• Expecially valuable in high-risk areas | • Can enhance monitoring efficiency<br>• Provides support to direct information |
| **Separate Evaluation** | • Primarily used to revalidate conclusions reached through ongoing monitoring | • Typically least persuasive<br>• Can help scope other separate evaluation procedures |

Source: ISACA, *Monitoring Internal Control Systems and IT: A Primer for Business Executives, Managers and Auditors on How to Embrace and Advance Best Practices*, USA, 2010, figure 8

### Use of Direct and/or Indirect Information for Monitoring

The intended role of the monitoring activity is a major determinant in the decision whether to use direct or indirect information or a hybrid approach.

**Example:** Publicly traded enterprises may be required to use monitoring as a key element of their compliance requirements related to financial reporting.

**Rationale:** Because the inherent legal and regulatory risk is higher in those industries, an enterprise typically places more emphasis on the requirement for direct information than it would for operating controls that are not part of an external reporting requirement or subject to quarterly or annual public reporting.

> **Note:** According to paragraph 101 of volume II of the Committee of Sponsoring Organizations of the Treadway Commission (COSO) *Internal Control—Integrated Framework: Guidance on Monitoring Internal Control Systems* (USA, 2009):
>
> *Some control monitoring tools perform what is often referred to as "continuous controls monitoring." These tools complement normal transaction processing by checking every transaction, or selected transactions, for the presence of certain anomalies (e.g., identifying transactions that exceed certain thresholds, analyzing data against predefined criteria to detect potential controls issues such as duplicate payments, electronically identifying segregation of duties issues). Many of these tools serve more as highly effective control activities (detecting individual errors and targeting them for correction before they become material) than they do as internal control monitoring activities. Regardless, if they operate with enough precision to detect an error before it becomes material, they can enhance the efficiency and effectiveness of the whole internal control system and may be key controls whose operation should be monitored.*

## 2. Control Monitoring Tasks

### Considerations for Monitoring IS Controls

Throughout the IS control monitoring process, the risk practitioner must consider the following questions:

- How significant or meaningful is the risk that is being addressed?
- How directly does the IS control address a defined risk when considered against other potential controls?
- What is the acceptable fault tolerance of the control, or is the control working or not working? What is the nature of the control? Is it manual or automated; is it detective or preventive?
- Does the effectiveness of a control potentially depend on other controls (e.g., aspects of security or application development)?

## Defining Key Controls

Key controls are the controls that provide the best gauge of the maturity and effectiveness of the control environment and the risk management effort. A key control is one which, if breached, would have the largest impact on the business, often by affecting more than one system, department or product line.

Working with the business owners, the risk practitioner will determine the key controls that should be monitored most closely and use those controls to report on how effectively the controls are working to mitigate risk. Many controls, logs and countermeasures may be built into the systems and networks of the enterprise at various points, and each of these controls may be important in the event of a crisis; however, some controls are much more important than others and those are the controls that should be used in reporting on control effectiveness.

**Note:** The ultimate definition of key controls, as well as the relative mix and importance of process, application, and general controls, will be unique to each enterprise. There are several sources of information on controls that can be used as tools in this process including frameworks such as those provided by ISACA, ITIL or the International Organization for Standardization (ISO).

## Defining Key Control Elements

A critical first step in this process is clearly defining the elements that make up a given key control. Although many key controls may have only a single element, others may have multiple elements.

The purpose of this exercise is to ensure that all elements of a key control are ultimately being considered when determining monitoring options.

The purpose of determining control elements is that it can often be found that a control may appear to be set correctly, but in practice is not working effectively. The purpose of risk monitoring is not just to ensure that a control is merely in place or set correctly, but also to ensure that the control is working effectively in its real-world environment.

**Example:** A gate across a private driveway, for example, may be set to open only to people with the correct card, but on examination it is found that people can drive around the gate and still gain access if they do not have the correct card. A control review that only checked that the gate works correctly with the correct card would not identify the fact that the risk of unauthorized access is not being mitigated by the control.

## Examples of Control Elements

The following table describes examples of single- and multiple-element IS controls.

<table>
<tr><th colspan="2">Example IS Control Elements</th></tr>
<tr><th>Control</th><th>Description</th></tr>
<tr><td>Network IDs are established only after department head approval.</td><td>This control consists of a single element: the signature of the department head.</td></tr>
<tr><td>System settings require users to change passwords every 60 days.</td><td>This control likely has multiple elements: It is based on the system parameter being set at 60 days and the systems successfully enforcing the control parameters over time.<br><br>Consideration: The password control system should be checked to see if the setting is set correctly to 60 days and, in addition, a proper control review would confirm that passwords HAVE been changed in the past 60 days and that the users are not able to bypass the controls.<br><br>Note: When defining a monitoring plan, it is important to consider both aspects of the control.</td></tr>
<tr><td>All programming changes are subjected to comprehensive testing based on the significance of the change.</td><td>This control has multiple, complex elements:<br>• The competence of the people defining the test plans<br>• The determination of the “significance” of a change<br>• The determination of “comprehensive” for the tests being performed<br>• The completion of the tests prior to implementation<br><br>Consideration: This control seems to be quite challenging to monitor because the selection criterion (significant change) and the related control attribute (comprehensive testing) are not measurable.</td></tr>
<tr><td>Access controls grant system privileges on a need-to-know basis.</td><td>Access controls always have at least two elements, as described in the following table.
<table>
<tr><th>Element</th><th>Description</th></tr>
<tr><td>Programming logic</td><td>The system configuration, settings and processes within the application or infrastructure resource that:<br>• Identify and authenticates users/other resources<br>• Allow or disallow activity based on certain constraints, such as user ID or a media access control (MAC) address</td></tr>
<tr><td>Provisioning</td><td>The process to grant, maintain, change or revoke access privileges, based on changes in the environment or the user/resource</td></tr>
</table></td></tr>
</table>

## Tasks Associated With IS Control Monitoring

The following table lists the tasks associated with IS control monitoring and maintenance.

| Tasks Associated With IS Control Monitoring | |
|---|---|
| **Task** | **Task Name/Description** |
| 1 | Determine whether a specific control is already being monitored. |
| 2 | Determine the types of information available for monitoring. |
| 3 | Determine the feasibility of using automated tools. |

## Task 1—Determine Whether a Specific Control Is Already Being Monitored

Determine whether the control is already being monitored through a routine business process or other control activity, which is not uncommon.

## Direct or Indirect Control Information

Some common IS processes, if implemented properly, may provide management with direct or indirect information about the operation of certain controls.

The following table describes examples of IS processes and the type of information (direct or indirect) each provides about the operation of certain controls.

| IS Processes and Information Types | | | |
|---|---|---|---|
| **IS Process** | **Indirect** | **Direct** | **Description** |
| Access recertification | X | | Verify whether management is reviewing and approving all user IDs on an annual basis. |
| Security log monitoring | X | | Indirect information based on review of logs gathered from systems or network components. Can be used to identify security controls such as failed access attempts |
| Portfolio management and/or steering committees | X | | An indirect means by which management assures itself that IT development efforts and IT resources are being devoted to those activities that have been approved by management based on review of steering committee minutes or approval documentation |
| Independent review of program development | X | X | Independent feedback on controls relating to testing and user sign-off |
| Change review board | X | X | Evidence that management is monitoring the controls relating to change request sign-off and testing |
| Post-implementation reviews of program changes | X | | Indirect information about the effectiveness of internal controls over the development process, including interviews with users and management |
| Problem management | X | | Indirect information about the effectiveness of several different IS processes that may ultimately be determined to be the source of incidents |
| Performance management | X | | Indirect information about the effectiveness of internal control processes based on user feedback, or direct information based on system generated statistics |
| Recovery testing | | X | Direct evidence that the redundancy or backup controls work effectively |

## Task 2—Determine the Types of Information Available for Monitoring

The following table provides guidance to determine whether to use direct vs. indirect information for monitoring.

> **IMPORTANT:** If a key control has a history of being ineffective or is relatively new, ensure that direct information about the control is the focus until the control reaches a sufficient level of maturity and reliability.

| Using Direct vs. Indirect Information for Monitoring | | |
|---|---|---|
| **Information Type** | **Intended Use** | **Consideration** |
| Direct information | • To provides assurance that a control is operating<br>• To focus on the elements of a control<br>• To ensure that all sources are being considered<br><br>**Note:** A monitoring approach that utilizes direct information would typically require the use of reperformance, examination or observation. | In this process, it is important to recognize that the:<br>• Term "information" could easily be replaced with the audit term "evidence"<br>• Evidence of many controls, especially those requiring judgment and expertise, is not frequently captured in information systems.<br><br>**Results:** The direct information that supports many controls will need to be derived through observation by the evaluator. |
| Indirect information | • To point out control failures, thereby allowing an evaluator to gain an inference about whether a control is operating (when there is no indication that controls have failed)<br><br>**Note:** There needs to be a strong correlation between the key control (or the elements of the key control) and the information used for monitoring its effectiveness. | • The examples following this table are by no means exhaustive. Indirect information sources vary by enterprise.<br>• Performance indicators found in the COBIT framework (*www.isaca.org/cobit*) can provide an excellent source for determining potential indirect monitoring measures. |

## Example 1—Use of Direct Information

An enterprise resource planning (ERP) system typically requires that sales can be made only for products that have been previously defined in the ERP system. The price that a customer pays and the credit limit of the customer are defined in the ERP system through business processes that exist within the enterprise.

In such a case, management typically needs some form of direct information that the controls over:
• Adding customers, inventory items, prices and credit limits work as intended
• Periodically verifying physical inventory levels are in place

> **Note:** Management may be able to utilize indirect information that comes from its operations to satisfy itself that other controls are working.

## Example 2—Use of Indirect Information

The following table describes examples of indirect information.

| Examples of Indirect Information |
|---|
| Assuming that the enterprise periodically compares finished goods inventory levels to the perpetual inventories in its ERP system, the lack of any significant differences between perpetual levels and actual levels provides indirect information that its billing controls are operating.<br><br>**Note:** This does not provide any information about the propriety of cutoff, for which management needs to have direct information. |
| Reports that provide information about any unusual deviations and individual product margins (whereby, the price of an item sold is compared to its standard cost) provide indirect information that controls over billing and pricing are operating. |
| The monitoring of the cause of credit memos can also provide information about billing controls, especially in situations in which it is possible to identify whether credit memos were issued for shipping or pricing disputes.<br><br>**Note:** To make a proper inference about these controls, management would have to satisfy itself that there is no adverse deterioration in aging of its receivables for its customers. |
| Reports that show orders that were rejected for credit limitations provide indirect information that credit checking aspects of the system are working as intended.<br><br>**Note:** To the extent that an "override" can be used to allow a credit limit exception, these reports can also be used by management to satisfy itself that the process of approving these overrides is operating. |

## Task 3—Determine the Feasibility of Using Automated Tools

Determine the feasibility of using automated monitoring tools, which may add value to the monitoring process in situations in which the information that supports a conclusion that a control is in place and operating resides directly or indirectly in electronically stored data.

A control to verify that only authorized changes can be made to system applications, configurations or files may use a tool that compares historical file sizes to current file sizes to detect any changes and alert the administrator to a potentially unauthorized change.

Monitoring tools can focus on many, but not all, dimensions of internal control.

## Software-based Monitoring Tools

The following list describes variations in software tools:

- Some tools focus on a very narrow set of control issues; others cover a broader range.
- Software vendors use varying names to describe what their tools provide.
- Software tools that do the same basic thing can be called by different names, and conversely, tools with the same basic names can do entirely different things.
- Some tools are designed to focus on monitoring or auditing internal controls; others are integrated into IS operating processes, but provide essential information that is relevant to monitoring.
- There are several software-based tools that can assist the enterprise in gathering, analyzing and reporting on control data. Many of these tools will draw data from multiple systems and aggregate the data into common files or databases. The tool will then correlate activity on different systems and seek to discover related events (or activity) that span more than one system.

## Considerations for Selecting Monitoring Software

Other things to consider when selecting monitoring software tools include:
- Many tools are likely to be considered multipurpose in that they are designed to provide a form of operational process or control.
- Depending how they are used, these tools can also provide direct or indirect information to management that controls are in place and operating.
- The presence of a tool should not be an assumption that the tool is being used.
- To be considered effective for controls monitoring, there needs to be a focus on communicating results/issues from these tools to those individuals responsible for acting on findings.

## Types of Monitoring Tools

There are various types of software tools that can be used to perform different types of control monitoring. These tools are organized into the following groups based on the focus of the tool as it relates to monitoring internal control:
- Transaction data
- Conditions
- Changes
- Processing integrity
- Error management

**Reference:** For details on each of these tools, see the applicable examples in Domain 5, E. Tools for Monitoring.

## Limitations of Monitoring Tools

Although automated monitoring tools can be highly effective in a number of situations, they are not without limitations and can generally CANNOT:
- Are not without their limitations
- Generally cannot:
  - Determine the propriety of the accounting treatment afforded individual transactions because this must be determined based on the underlying substance of the transaction itself
  - Address whether an individual transaction was accurately entered into the system; rather, they can deal only with whether the transactions met internal standards for acceptable transactions (for example, it was valid)
  - Determine whether all relevant initial transactions were entered into systems in the proper period because this is typically dependent on human activity

## F. Tools for Monitoring

**Introduction**

This section describes tools for monitoring IS controls.

### 1. Tools for Monitoring Transaction Data

**Tools for Monitoring Transaction Data**

Tools used in this fashion tend to deal with:
- IS controls designed to focus on the integrity of transaction processing
- Processed data, which confirms the IS controls related to the completeness of processed data

The following table lists different uses of tools focusing on transaction data.

> **IMPORTANT:** Tools cannot evaluate whether all transactions were entered into a system, but they can evaluate the completeness of transactions being moved from one system to another.

**Tools for Monitoring Transaction Data**

| Tool Use | Description |
|---|---|
| Comparing transaction data against defined rule sets | The intention is to identify instances in which the controls over a process or system are not working as intended.<br><br>This category of tools for monitoring information can be used to:<br>• Highlight exceptions and/or anomalies<br>• Analyze unusual trends in activities, values and volumes<br>• Compare balances or details between two systems or between distinct parts of a process |
| *Ad hoc* reporting | Tools in this category can take the form of *ad hoc* programs that are run periodically against the defined:<br>• Population (either a sample of the data population or the entire population)<br>• Programs that are implemented into a processing environment to continuously monitor a specific set of transactions |
| Data correlation across multiple sources | Given their ability to be used in a variety of scenarios and to correlate data and information from multiple sources, these tools can be used in multiple situations to:<br>• Highlight significant manual or unusual automated journal entries so that financial management can ensure that such entries were proper and approved.<br>• Search for unusual or duplicate payments so that management can ensure that controls over disbursements are working effectively.<br>• Analyze unusual activity as part of management's fraud prevention activities.<br>• Determine the frequency of supervisory overrides.<br>• Validate that all transactions fall within a specific control range. |

## 2. Tools for Monitoring Conditions

### Tools for Monitoring Conditions

Software tools in this category:
- Examine specific settings or parameters that control how an application or infrastructure resource is configured
- Compare the configuration information to either baseline information, a prior analysis or both to determine whether they are consistent with the enterprise's expectations
- May operate periodically (frequently described as scanning-based)
- Can be embedded in the process as either software or hardware (frequently described as agent-based)

**Rationale:** This increases the speed and effectiveness of the monitoring process while simultaneously allowing it to be performed on a more frequent or even continuous basis.

The following table describes where tools for monitoring conditions can best be applied.

| Application of Tools for Monitoring Conditions | |
|---|---|
| **Situation** | **Example** |
| A large number of conditions across different systems | Numerous parameter settings that affect internal control in an integrated enterprise resource planning (ERP) system or multiple different "instances" or versions of the same ERP package that support different business units |
| Few parameters across many systems | Analysis of security parameters across relevant servers in a global network |
| Complex conditions or a high volume of records | A large number of users of a single application or multiple applications to be evaluated for appropriate segregation of responsibilities as defined by application access rights |

These types of tools deal with integrity controls. There are also dimensions of these controls that support access and authorization-related controls. This is particularly true for tools that analyze security-related parameters because these parameters are relevant to ensuring that only those authorized to utilize resources can gain access.

## 3. Tools for Monitoring Changes

### Tools for Monitoring Changes

The following table describes an example of change-monitoring tools. They can be considered an extension of the tools that focus on "conditions."

| Examples of Tools for Monitoring Changes | |
|---|---|
| **Topic** | **Description** |
| Tool focus | The basic difference is that these tools:<br>• Usually operate on a continuous basis (i.e., they are agent-based)<br>• Are specifically designed to identify and report changes to critical resources, data or information, thereby making it possible to verify that changes are appropriate and authorized<br><br>Change within IT resources is pervasive, continuous and unavoidable, so the purpose of these controls is to make sure that the changes are authorized and implemented correctly. |
| Preventive vs. detective change control detection | When controlling change is considered a key control, enterprises have some form of change control that includes both a:<br>• Preventive control that will only permit authorized personnel to make changes<br>• Detective control, whereby all changes are recorded, reviewed and potentially approved by someone independent of those making changes |
| Database structure changes | Changes that have been made to database structures can be logged by the database environment itself and then subsequently reviewed by management. |

### Considerations for Monitoring Changes

The following table provides examples of considerations for monitoring changes.

| Examples of Considerations for Monitoring Changes | |
|---|---|
| **Topic** | **Description** |
| Control mechanism for changes to programming or ERP configuration | Certain ERP environments provide their own control mechanism for controlling programming changes or changes to the ERP configuration. To the extent that this exists, the reports of these changes can be used both as a detective control and a means by which management monitors controls. |
| Recording changes | Although this approach to controlling change works effectively, it is dependent on the ability of the resource being changed to provide an effective means to record changes and, thereby, ensure that any changes can be reviewed and approved. The risk practitioner will need to ensure that the recording logs cannot be altered or deleted by users or administrators. |
| Essential considerations | The following must be considered:<br>• Not all IT resources provide the capability for recording changes.<br>• In large IT environments, the number of individual resource components that would need to be analyzed on a detective basis can be overwhelming.<br>• Using native logging capabilities of some resources may affect system performance unacceptably.<br>• In certain high-risk areas, management may not wish to use the native features of certain resources because of the potential impact that such features could have on systems performance and the simplicity of turning off these features. |

**Considerations for Monitoring Changes *(cont.)***

| Examples of Considerations for Monitoring Changes *(cont.)* | |
|---|---|
| **Topic** | **Description** |
| Category application with appropriate conditions | When these conditions are present, tools in this category can:<br>• Become mechanisms for both control and monitoring<br>• Identify changes that have been made to infrastructure resources, databases, application programs, and security rights and permissions<br>• Provide visibility for all changes so that, ultimately, they can be validated independently, which is, in essence, monitoring that the underlying change control process is working<br>• Provide alerts when certain types of mission-critical changes are being made so that there is transparency throughout the enterprise and necessary actions can be taken on a timely basis<br>• Provide a verifiable audit history for direct information of control functionality over time<br><br>As with tools that focus on conditions, these largely focus on integrity and authorization-related controls. |

## 4. Tools for Monitoring Process Integrity

**Tools for Monitoring Process Integrity**

The following table describes an example of automated tools to evaluate processing integrity that:
- Are designed to verify and monitor the completeness and accuracy of the various situations
- May occur in the overall IT process

| Tools for Monitoring Process Integrity | |
|---|---|
| **Topic** | **Description** |
| Tool focus | Automated tools focus on balancing and controlling data as they progress through processes and systems. |
| Activity performance | Tools in this category can:<br>• Reconcile financial totals and/or transaction/record counts from one file or database to another file or database within the same or between different application or operating systems (OSs)<br>• Ensure data file, record and field accuracy as data move across systems and processes<br>• Monitor information from source systems and/or data warehouses to the general ledger (GL) |
| Design to maintain an audit trail | Because these systems operate independently of transaction processing, they can be designed to maintain an audit trail of key information for monitoring or trending studies. |
| Control mechanism | Within an enterprise, these tools are typically considered a control mechanism.<br><br>Depending on how management uses them, these tools could be considered mechanisms for monitoring controls as they relate to information processing. |

## 5. Tools for Error Management and Reporting

### Tools for Error Management and Reporting

Error management tools are designed to detect transactions that do not meet defined criteria so that they can be corrected and reprocessed. Considerations are presented in the following table.

**Examples:**
- An automotive parts supplier may receive a technically valid electronic data interface message describing an authorized shipping schedule; however, the message may have invalid order identification.
- A telecommunications provider may receive message information from its telephone switching systems on customers whose information has not made it through the process of being added to the billing system.

The fact that invalid transactions are rejected is frequently considered an application control. These systems frequently capture the transactions in an area where they can later be reprocessed after correcting the cause of the error.

Management's monitoring of the volume and resolution of activity in these systems or accounts provides both direct and indirect information that the control is working effectively.

> **IMPORTANT:** These capabilities typically are part of an existing information system rather than an add-on vendor solution.

| Considerations for Error Management Tools | |
|---|---|
| **Topic** | **Description** |
| Best approach | Determining the best approach for an enterprise should be driven by the importance of the control and the related risk that control is designed to mitigate. There are trade-offs with respect to these solutions that need to be considered carefully by management in the process of determining whether tools of this nature will work in their environment. |
| Parameter specifics | Many controls that are built into systems are controlled through the configuration of a specific set of parameters within both IT infrastructure resources and application systems. |

## 6. Continuous Monitoring

### Introduction

Continuous monitoring:
- Is becoming important in today's electronic business (e-business)
- Provides a method to collect evidence on system reliability while normal processing takes place
- Allows for monitoring the operation on a continuous basis
- Gathers selective monitor evidence through the computer

> **Note:** If the selective information collected by the computer technique is not deemed serious or material enough to warrant immediate action, store the information in separate monitor files for verification by the monitor at a later time.

## Relevance

Continuous monitoring techniques are important tools:

- When they are used in time-sharing environments that process a large number of transactions, but leave a scarce paper trail
- Because they improve the security of a system by permitting IS monitors to evaluate operating controls on a continuous basis without disrupting the enterprise's usual operations
- Because they report a system misuse in a timely fashion

**Results:**

- This reduces the time lag between the misuse of the system and the detection of that misuse.
- IS monitoring allows management to gain greater confidence in a system's reliability due to the realization that failures, improper manipulation and lack of controls will be detected on a timely basis.

## Benefits

The continuous monitor approach:

- Cuts down on paperwork
- Reduces monitoring cost and time
- Leads to the conduct of an essentially paperless monitoring process
- Improves accountability of local management for the enforcement and monitoring of controls

## Selection of Continuous Monitoring Tools and Techniques

The selection and implementation of continuous monitoring tools depends, to a large extent, on the complexity of an enterprise's computer systems and applications and the IS monitor's ability to understand and evaluate the system with and without the use of continuous monitoring techniques.

The risk practitioner must recognize that continuous monitoring techniques are not a cure for all control problems and that the use of these techniques provides only limited assurance that the information processing systems examined are operating as they were intended to function. Continuous monitoring does not replace the need for periodic reevaluation of controls or the requirement for audits.

## Automated Monitoring Techniques

The following table describes five types of automated evaluation techniques applicable to continuous monitoring.

<table>
<tr><th colspan="2">Types of Automated Evaluation for Continuous Monitoring</th></tr>
<tr><th>Technique</th><th>Description</th></tr>
<tr><td>Systems Control Audit Review File and Embedded Audit Modules (SCARF/EAM)</td><td>The technique involves embedding specially written monitor software in the enterprise's host application system so that the application systems are monitored on a selective basis.</td></tr>
<tr><td>Snapshots</td><td>This technique:<br>• Takes what may be termed as "pictures" of the processing path that a transaction follows, from the input to the output stage<br>• Tags transactions by applying identifiers to input data and recording selected information about what occurs for the subsequent review</td></tr>
<tr><td>Monitor hooks</td><td>This technique:<br>• Embeds hooks in application systems to function as red flags if a suspicious condition occurs<br>• Induces IS monitors to act before an error or irregularity gets out of hand</td></tr>
<tr><td>Integrated test facility (ITF)</td><td>In this technique:<br>• Dummy entities are set up and included in a monitor's production files.<br>• The IS monitor can make the system either process live transactions or test transactions during regular processing runs and have these transactions update the records of the dummy entity.<br>• The operator enters the test transactions simultaneously with the live transactions that are entered for processing.<br>• The monitor then compares the output with the data that have been independently calculated to verify the correctness of the computer-processed data.</td></tr>
<tr><td>Continuous and intermittent simulation (CIS)</td><td>In this technique:<br>• During a process run of a transaction, the computer system simulates the instruction execution of the application.<br>• This simulator follows the table below as each transaction is entered to decide whether the transaction meets predetermined criteria.
<table>
<tr><th>If the transaction ...</th><th>The simulator ...</th></tr>
<tr><td>Meets the predetermined criteria,</td><td>Monitors the transaction.</td></tr>
<tr><td>Does not meet the predetermined criteria,</td><td>Waits until it encounters the next transaction that meets the criteria.</td></tr>
</table></td></tr>
</table>

## Exhibit 5.3: Concurrent Monitoring Tools—Advantages and Disadvantages

**Exhibit 5.3** describes the relative advantages and disadvantages of the various concurrent monitoring tools.

**Exhibit 5.3: Concurrent Monitoring Tools—Advantages and Disadvantages**

| | SCARF/EAM | ITF | Snapshots | CIS | Audit Hooks |
|---|---|---|---|---|---|
| Complexity | Very high | High | Medium | Medium | Low |
| Useful when: | Regular processing cannot be interrupted. | It is not beneficial to use test data. | An audit trail is required. | Transactions meeting certain criteria need to be examined. | Only select transactions or processes need to be examined. |

Source: ISACA, *CISA® Review Manual 2011*, USA, 2011, exhibit 3.33

# G. Implementing Control Monitoring Processes

**Introduction**

The section describes the implementation of monitoring processes.

The monitoring process requires an enterprise to design and implement ongoing monitoring procedures and/or separate evaluations that are needed to gather and analyze persuasive information to support conclusions about the effectiveness of internal control.

## 1. Determine Monitoring Method and Frequency

**Factors to Determine Monitoring Method and Frequency**

The management of an enterprise uses a unique judgment after factoring in:
- Its objectives
- Its risk
- Its controls
- The persuasiveness of information that is available about its controls

**Controls Already Monitored**

Certain controls may already be monitored as part of an existing process.

In these situations, activities may be limited to ensuring that the frequency of monitoring is sufficient for management's purpose.

**Questions to Determine Control Monitoring Method**

The following table outlines the questions that need to be asked and answered to determine monitoring method.

| Questions to Determine Control Monitoring Method | |
|---|---|
| **Question** | **Description** |
| What is the nature of the control? | • The answer depends on the specific elements of the control, as discussed previously.<br>• Generally, automated controls require less frequent monitoring of their automated aspects.<br>• Once the control is verified to be working properly, automated aspects are unlikely to change—assuming that effective change management controls are in place. |
| How will monitoring be used? | • A properly designed and executed monitoring program helps support both internal needs to:<br>– Ensure that objectives and external assertions are being achieved<br>– Provide persuasive information that an internal control operated effectively at a point in time or during a particular period<br>• This duality of purpose ultimately should play a role in decisions made as to the frequency and method of evaluation and to who should be the evaluator. |

## Questions to Determine Control Monitoring Method *(cont.)*

<table>
<tr><th colspan="3">Questions to Determine Control Monitoring Method (cont.)</th></tr>
<tr><th>Question</th><th colspan="2">Description</th></tr>
<tr><td>Who is most appropriate to evaluate a given control?</td><td colspan="2">• The answer provides needed input to the decision about how monitoring should best be performed.<br>• There are several criteria that must be considered that relate to the competence and objectivity of the “evaluator” of an internal control.<br><br>Example: If management decides that the chief information officer (CIO) should evaluate a control, it is most likely that a decision will also be made that the CIO’s monitoring would be part of an ongoing management process, not a separate evaluation.</td></tr>
<tr><td rowspan="4">What is the mix of direct and indirect information?</td><th>When monitoring using ...</th><th>Then ...</th></tr>
<tr><td>Direct information,</td><td>Perform monitoring on whatever schedule is deemed sufficient by management.</td></tr>
<tr><td>Indirect information,</td><td>Apply additional separate evaluations as they may be required more frequently.</td></tr>
<tr><td colspan="2">The specifics for each monitoring procedure need to be developed for both ongoing and separate evaluations.<br><br>This activity focuses largely on the development of activities and reporting responsibilities and the training and change management activities for those impacted.</td></tr>
<tr><td>Can the evaluation be an ongoing process?</td><td colspan="2">• To the extent that controls can be monitored through an ongoing process, it is more likely that control issues will be highlighted on a timely basis.<br>• Not all monitoring activities, however, lend themselves to being easily implemented as part of an ongoing process.<br>• This decision ultimately needs to be made by management for each unique situation.</td></tr>
</table>

## 2. Select and Implement Automated Monitoring Tools

### Acquisition of Monitoring Tools

The selection and implementation of monitoring tools should follow the same basic technology acquisition and implementation processes that an enterprise uses for any of its business systems.

### Selection Criteria for Monitoring Tools

The following table provides several criteria to consider when determining whether and how to use a given tool.

**IMPORTANT:** Each of the criteria listed below must be considered in terms of the benefit of automating the monitoring process compared to alternatives.

| Criteria for Tool Use | |
|---|---|
| **Criterion** | **Description** |
| Sustainability | • It is important that monitoring software be able to change at the same speed as technology applications and infrastructure to be effective over time.<br>• If monitoring tools do not change at the same pace as the underlying applications, there is a risk that new or changed controls in the application will not be monitored. |
| Scalability | • Monitoring tools have to be able to keep up with the growth of an enterprise and meet anticipated growth in process, complexity or transaction volumes. |
| Customizability | • Many software products come with rules that have been defined by the vendor and may not meet management's business requirements.<br>• For software to be effective, it must be customizable to the specific needs of an enterprise.<br>• Ideally, the ability to customize should be built into the software so that the end users can adapt the software—the cost of custom programmer changes to parameters or rule sets can be prohibitive and impact scalability. |
| Ownership | • Someone in the enterprise must "own" the tool and must be able to effectively identify organizational changes and opportunities that may require modifications to the tool.<br>• Effort takes time and a certain level of expertise for the enterprise to maximize the benefits from its investment in the tool. |
| Impact on performance | • Because monitoring tools must operate in a manner that coexists with existing systems and data, it is important to understand the impact of a monitoring tool on base system's performance and capacity.<br>• Embracing a monitoring tool that adversely impacts business processing capability is not sustainable in most enterprises. |
| Usability of existing tool | • If an operating tool currently in use can be leveraged to perform monitoring activities, there are fewer considerations to be made than when an enterprise is acquiring a new tool. Tools that may integrate or be readily interoperable with other tools may be a preferred choice. |
| Tool complexity | • Tool complexity and challenging user interfaces may affect its usage. How much training would be required by the administrators or users of the tool? |
| Transferability | • This refers to the ease of shifting the use of the tool from one user (group) to another. |
| Cost-benefit (return on investment [ROI]) | • Enterprises must understand the total cost of ownership (TCO) when purchasing monitoring tools.<br>• Software licensing costs are only a very minor part of the total cost of software ownership. |

## 3. Clarify Reporting Requirements and Expectations

### Clarify Reporting Requirements and Expectations

Because the controls being monitored often cross organizational objectives and business functions, one of the most important aspects of this activity is clearly defining how exceptions should be reported and to whom.

Report exceptions or deficiencies identified in the monitoring process to the individual:
- Who is in a position to take corrective actions
- With overall responsibility for:
  - Controls in a given area
  - A given set of objectives

**Note:** This requires clear ownership for risk and control monitoring, The risk practitioner must ensure that there is a declared owner for risk and for ensuring that the controls are being operated and monitored effectively.

### Considerations for Correcting Deficiencies

It is worth noting that the process of correcting deficiencies may be considered a management activity rather than an element of internal control. Classification aside, corrective actions should be undertaken when control deficiencies are severe. It is crucial that:
- The right people receive the information necessary to enable corrective action.
- Management retains sufficient oversight to gain assurance that the corrective action has been taken.

### Considerations for Prioritization of Corrective Actions

The following table describes several factors that may influence an enterprise's prioritization of identified exceptions and deficiencies as described in the Committee of Sponsoring Organizations of the Treadway Commission (COSO) 2009 *Internal Control—Integrated Framework: Guidance on Monitoring Internal Control Systems* (USA).

| Monitoring Factors and Considerations | |
|---|---|
| **Factor** | **Consideration** |
| Errors | The likelihood that the deficiency will result in an error or other adverse event:<br>• The fact that a deficiency has been identified means that there is at least some likelihood that an error could occur.<br>• The greater that likelihood, the greater the severity of the control deficiency will be. |
| Other controls | • The effectiveness of other, compensating controls:<br>• The effective operation of other controls may prevent or detect an error that results from an identified deficiency before that error can materially affect the enterprise.<br>• The presence of such controls, when monitored, can provide support for reducing the severity of a deficiency. |
| Deficiencies and organizational objectives | The potential effect of a deficiency (vulnerability) on organizational objectives:<br>• As the effect of an identified deficiency increases, its severity increases. |

## Considerations for Prioritization of Corrective Actions *(cont.)*

| Monitoring Factors and Considerations *(cont.)* | |
|---|---|
| **Factor** | **Consideration** |
| Effect of deficiencies on other objectives | The potential effect of the deficiency on other objectives:<br>• Beyond consideration of the factors listed previously, enterprises may consider the effect of a deficiency on their overall operating effectiveness or efficiency.<br><br>**Example:** An identified deficiency may prove to be immaterial in relation to the financial reporting objective, but it may cause inefficiencies that warrant correction in relation to operational objectives. |
| Multiple deficiencies | The aggregating effect of multiple deficiencies:<br>• When multiple deficiencies affect the same or similar risk factors, their mutual existence (the aggregated effect of the deficiencies) increases the likelihood that the internal control system may fail, thus increasing the severity of the identified deficiencies. |

## Prioritizing Deficiencies

Determining who prioritizes the deficiencies is a matter of judgment. Enterprises likely will consider the:
- Size and complexity of the enterprise
- Nature and importance of the underlying risk
- Experience and authority of the people involved in the monitoring process

The prioritization of identified deficiencies should be performed by appropriately competent and objective personnel that hold positions of responsibility within the enterprise.

## Determining Where Controls Are Needed

Those responsible for implementing controls should focus on root causes of problems when:
- Analyzing a control failure resulting from monitoring
- Identifying where controls are needed

## Introduction to the Cause-and-effect Diagram

A cause-and-effect diagram:
- Is generally used to explore all the potential or real causes (or inputs) that result in a single effect (or output)
- Can help identify root causes and areas where there may be problems
- Assists in comparing the relative importance of the different causes to ensure effective prioritization and coordination of remediation efforts

> **Note:** Arranging causes according to their level of importance or detail results in a depiction of relationships and hierarchy of events.

## Building a Cause-and-effect Diagram

The following table outlines the steps to successfully build a cause-and-effect diagram.

| Steps to Build a Cause-and-effect Diagram | |
|---|---|
| **Step** | **Action** |
| 1 | Agree on the effect or problem statement before beginning. |
| 2 | Identify major categories of failures.<br><br>**Note:** This may be equipment, policies, procedures and people for an IT process, but it can include other categories as well, depending on the circumstances of the environment. |
| 3 | Link the potential or observed control failures to the categories. |
| 4 | Discuss the control failure points with the project team. |
| 5 | Revise the monitoring process and repeat testing as necessary. |

## Cause-and-effect Diagram Process Considerations

When following the cause-and-effect diagram process, there are additional considerations, including the need to:

- Be concise.
- Think carefully about what could be causes and add them to the tree for each node.
- Pursue each line of causality back to its root cause.
- Consider grafting relatively empty branches onto others or splitting up branches that become crowded.
- Identify which root causes are most likely to merit further investigation.

## Exhibit 5.4: Cause-and-effect Diagram Based on Software Change Failure

**Exhibit 5.4** depicts an example of a cause-and-effect diagram based on the failure of a software change.

**Exhibit 5.4: Cause-and-effect Diagram Based on Software Change Failure**

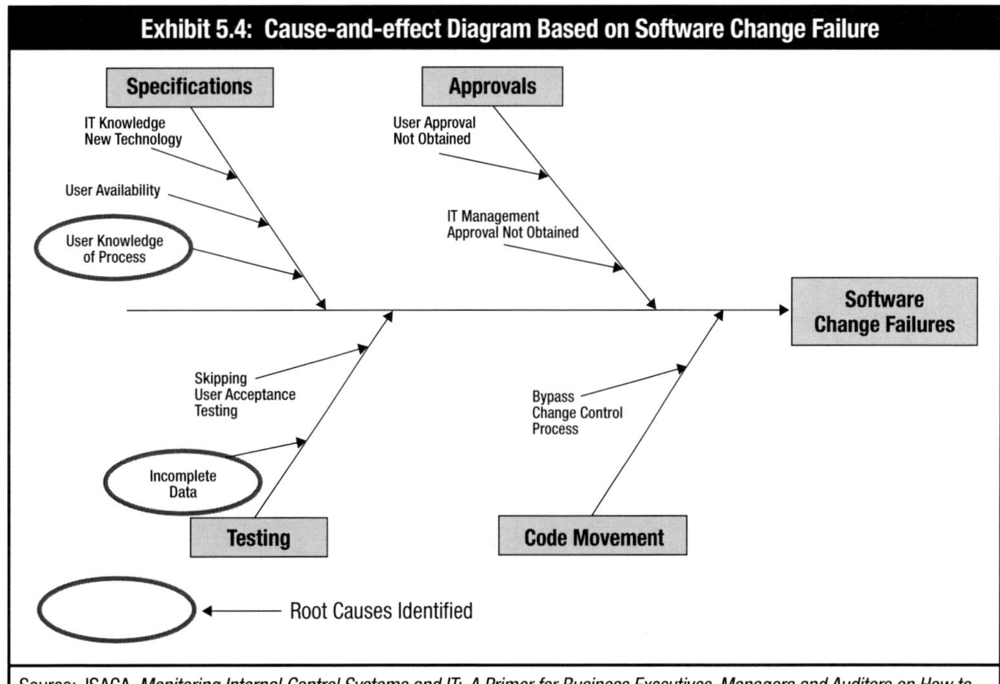


Source: ISACA, *Monitoring Internal Control Systems and IT: A Primer for Business Executives, Managers and Auditors on How to Embrace and Advance Best Practices*, USA, 2010, figure 10

# H. Implementing Control Maintenance Processes

**Introduction**

This section describes the implementation of control maintenance processes.

The maintenance process requires an enterprise to update, maintain, adjust and reconfigure controls according to changing risk levels and emerging threats or vulnerabilities. The maintenance of controls is done either on a regular basis to keep the controls operating correctly—including applying patches, new rules etc., and on an exception basis to respond to incidents or vulnerabilities discovered through control monitoring.

## 1. Determine Control Maintenance Requirements

**Control Maintenance Objectives**

Effective risk management requires that the IS controls be maintained over time to ensure that they continue to provide effective risk reduction and protect the enterprise from changing or emerging risk.

**Control Maintenance Activities**

Control maintenance includes activities such as:
- Patch management
- Configuration management
- Exception management
- Documentation

**Maintenance of Vendor Supplied Controls**

Some controls deployed across the enterprise will be provided by hardware and software vendors. Examples of these controls may include:
- Intrusion Detection and Intrusion Prevention Systems (IDS/IPS)
- Firewalls
- Routers
- Access Control Systems
- Anti-malware (Antivirus) Systems

Maintenance of these controls starts with an inventory of what controls the enterprise has, and where the controls are located.

The enterprise is responsible for ensuring that relevant vendor released patches are deployed as quickly as possible across the enterprise. However all changes must be tested (as much as possible) prior to deployment since changes may not work properly or have sometimes disabled other associated systems or processes.

Vendor patches may be released in order to repair a vulnerability in the system, update malware or attack signatures, or provide enhanced functionality.

A critical challenge with deploying patches across multiple devices is to ensure that all systems are updated, that the rollout of the patch does not congest networks or unduly impact production activity.

## Maintenance of IS Control Configurations

Many information system controls have customized settings that regulate the operation of the control according to the requirements of the enterprise's security policy.

IS control configurations are not static, but will often require tuning or modification to meet changing risk scenarios or emerging threats. The effectiveness of IS controls is maintained through adjusting the control configuration settings; however, it is critically important to ensure that any changes to control configuration are subject to a change control or configuration management procedure that will prevent unauthorized or untested changes and will help detect any errors in the new configuration prior to implementation.

These control configuration settings are often found in access control lists (ACLs), configuration tables, or in settings that can only be altered by a person with administrator-level access.

The risk practitioner needs to ensure that the changes to these settings (the configuration of the device), are carefully tested, controlled and documented.

Without proper controls over changes to control configurations, changes are made without proper approval or documentation and the enterprise cannot determine whether the control is operating correctly or not. Poor management of the rules may impact the performance of the control.

## Exception Handling

It is not uncommon to encounter a situation where a user or process requires an exception to a policy that is being enforced by an IS control. In such an instance, the exception must be reviewed, approved and documented prior to granting the exception. The exception may be required for business reasons or to enable testing of a new application. The exception should be evaluated for its potential impact on risk levels prior to approval and implementation.

Exceptions to the rules that govern the operation of control should only be made when necessary, and as infrequently as possible. Too many exceptions may be an indicator that the policy is not appropriate or that the control is not set up correctly or does not align with business priorities.

When an exception is required, the request should be submitted formally, in writing, and then require approval by a manager with the appropriate level of authority.

Exceptions to policy (and the needed changes to the configuration of the IS control) must be documented and reviewed on a periodic basis to ensure that the exception is still required, and removed once it is no longer needed.

## Maintenance of Control Documentation

Current and concise IS control documentation is an important input into the risk management process. The documentation should reflect the current configuration of individual controls as well as the control architecture. This is why it is common to hear the documentation described as "living"—they are always being kept alive and accurate.

## 2. Implement Control Maintenance Process

### Control Maintenance Process

Change management is the process that controls changes to IS controls and production systems. The objective of change management is to ensure that unauthorized changes are not permitted, authorized changes are reviewed for impact on risk prior to implementation, all changes are logged and validated to ensure that the change was implemented properly in the authorized manner, and that the effect of the change is monitored for effectiveness.

The flow chart below outlines a process to manage changes to control configuration.

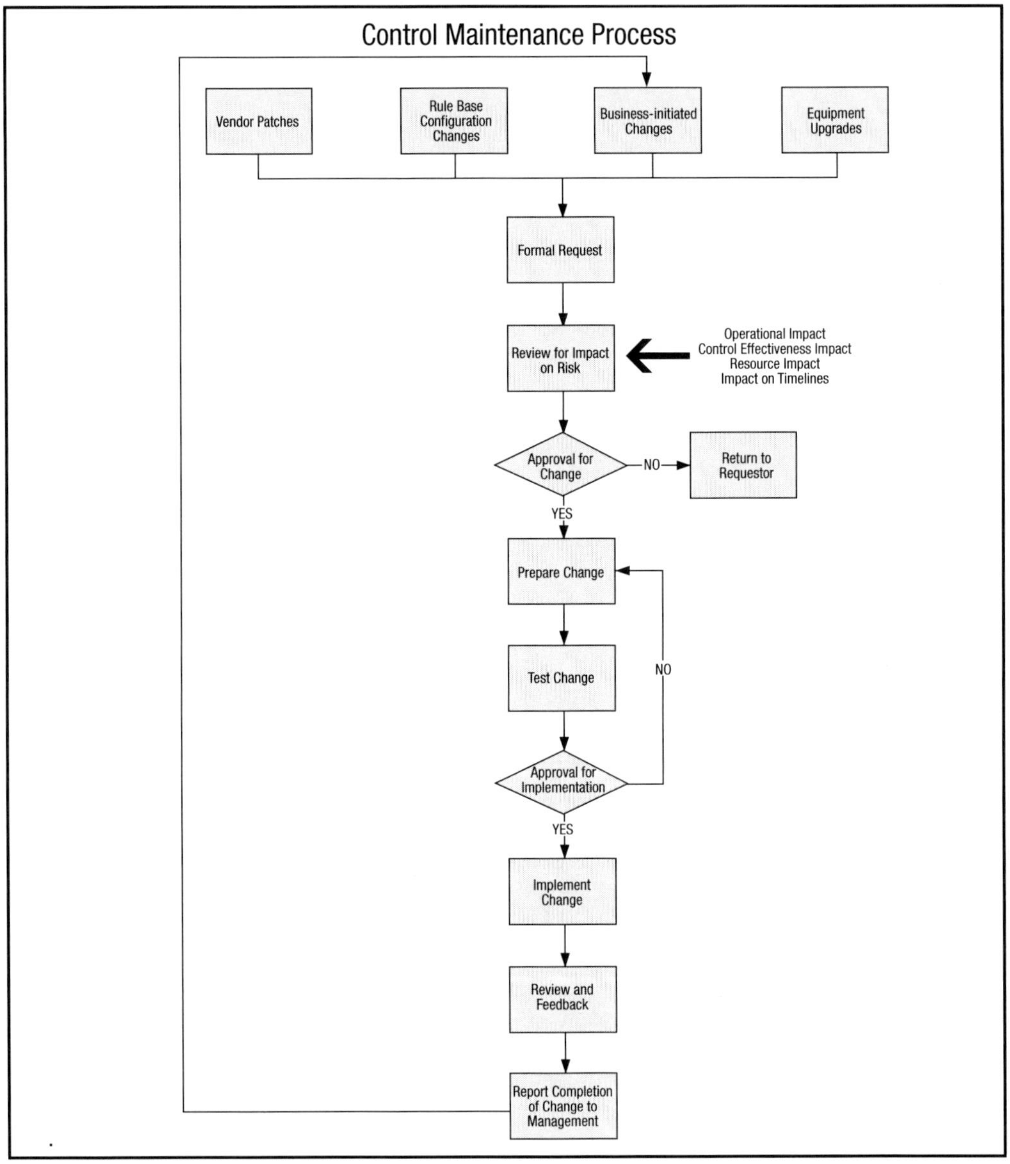

Control Maintenance Process

Vendor Patches
Rule Base Configuration Changes
Business-initiated Changes
Equipment Upgrades
Formal Request
Review for Impact on Risk
Operational Impact
Control Effectiveness Impact
Resource Impact
Impact on Timelines
Approval for Change
NO
Return to Requestor
YES
Prepare Change
Test Change
NO
Approval for Implementation
YES
Implement Change
Review and Feedback
Report Completion of Change to Management

The Configuration Management Database (CMDB) and all other relevant documentation should be updated to reflect the changes made.

- Establishing a formal control maintenance process should avoid the problem of undocumented or improper changes to controls. The formal process mandates testing, review and documentation of all changes both prior to, and following the implementation of the proposed change.

## Monitoring and Maintenance of IS Controls

During the previous phases of the risk management process, IS controls were designed and implemented. It is important that regular reviews of those controls be conducted to ensure that the IS controls are meeting their objectives and effectively mitigating the risks—as intended when the controls were selected and installed.

Reviews may be conducted on a regular scheduled basis or on an "as-needed" basis—perhaps as mandated by an incident, an emerging threat or as a result of a KRI threshold being reached.

The results of the IS control reviews should be documented and reported to management. The results may be used for further risk assessment, audit, investigation or compliance reporting as required.

## Periodic Control Testing

IS control maintenance is a subset of overall risk monitoring and should be conducted in conjunction with other control monitoring. An effective risk response on one that integrates all forms of control—administrative, technical (typical IS controls) and physical control measures into an interoperable and complete response solution.

IS controls may be network based, application based, hardware based, or user-based. The tests may be conducted by internal review and testing teams (including security, audit, and IT staff) or external parties (such as penetration testing specialists, compliance auditors, or external auditors).

The purpose of the control testing is to ensure that the controls are:
- Implemented as designed
- Operating correctly
- Achieving the desired result

As issues are discovered in the testing process, the controls must be maintained and adjusted to address the issues and resolve outstanding vulnerabilities.

## I. Suggested Resources for Further Study

### Suggested Resources for Further Study

In addition to the resources cited throughout this manual, the following resources are suggested for further study:

- ISACA:
  - COBIT 4.1, 2007
    **Note:** The COBIT 4.1 framework is available at no charge from ISACA and can be downloaded at *www.isaca.org/cobit*. The new COBIT 5 framework will be available in 2012.
  - *COBIT and Application Controls: A Management Guide*, 2009
  - *Monitoring Internal Control Systems and IT: A Primer for Business Executives, Managers and Auditors on How to Embrace and Advance Best Practices*, 2010
- National Institute of Technology and Standards (NIST), *Guide for Assessing the Security Controls in Federal Information Systems and Organizations, Building Effective Security Assessment Plans*, Special Publication 800-53A, Revision 1, USA, 2010

# Part II—Risk Management and Information Systems Control in Practice

## Overview

### Part II Introduction

Part II of the *CRISC Review Manual 2012* provides information on how the risk management theory and concepts that were introduced in Part I apply to specific processes.

It particularly addresses those knowledge statements from Domains 1, 2 and 3 that relate to risk, controls, control objectives, activities and metrics related to selected business and IT processes.

### Part II Contents

Part II contains the following chapters:



### Individual Chapter Content

Each chapter first identifies the knowledge statements addressed within the chapter. It then provides an overview of the specific process as well as:
- An explanation of its importance to achieving business objectives
- Process-related key concepts and principles
- A high-level process overview
- Examples of common risk
- Selected key risk indicators (KRIs)
- Examples of common IS controls supporting the process
- A description of the practitioner's perspective
- Suggested reading materials and references

## Individual Chapter Structure

Each chapter contains the following sections:

A. Chapter Overview
B. Related Knowledge Statements
C. Key Terms and Principles
D. Process Overview
E. Risk Management Considerations
F. Information Systems Control Design, Monitoring and Maintenance
G. The Practitioner's Perspective
H. Suggested Resources for Further Study

# 1. Determining the IT Strategy

## A. Chapter Overview

### Introduction

This chapter provides an overview of the process for aligning the IT strategic technological direction with risk management and:
- Explains its importance in achieving business objectives
- Outlines a high-level process overview
- Introduces related key concepts
- Presents examples of common risk
- Lists selected key risk indicators (KRIs)
- Provides examples of common IS controls supporting the process
- Describes the practitioner's perspective
- Offers suggested reading materials and references

### Learning Objectives

The CRISC candidate should have a general understanding of the process for determining how the IT strategy aligns with the business and how to integrate IT strategy with risk management.

### Contents

This section contains the following topics:

## B. Related Knowledge Statements

**Contents**

The following table lists the applicable knowledge statements from the CRISC job practice.

| No. | Knowledge Statement (KS)<br>Knowledge of: |
|---|---|
| KS1.8 | Threats and vulnerabilities related to business processes and initiatives |
| KS1.22 | Threats and vulnerabilities associated with emerging technologies |
| KS2.5 | Organizational risk management policies |
| KS4.4 | Control practices related to business processes and initiatives |
| KS5.5 | Control objectives, activities and metrics related to IT operations and business processes and initiatives |

## C. Key Terms and Principles

| | |
|---|---|
| **Introduction** | This section introduces terms and principles related to the determination of the IT strategy as well as terms that help relate the process to other key business processes. |
| **Definition of Strategic Planning** | The process of deciding on the enterprise's objectives, on changes in these objectives, and the policies to govern their acquisition and use |
| **Definition of IT Strategic Plan** | A long-term plan—i.e., three- to five-year horizon—in which business and IT management cooperatively describe how IT resources will contribute to the enterprise's strategic objectives (goals) |
| **Definition of IT Tactical Plan** | A medium-term plan—i.e., six- to 18-month horizon—that translates the IT strategic plan direction into required initiatives, resource requirements and ways in which resources and benefits will be monitored and managed |

## D. Process Overview

### Introduction

Enterprises spend significant amounts of money every year on IT resources, including applications, infrastructure, information and people.

However, studies have shown that many IT investments fail to meet organizational expectations and that the positive, expected results of these investments are not being realized.

The risk practitioner has a role in ensuring that the IT strategy is closely aligned with the overall business strategy and that the direction in which the enterprise is headed is reflected in the IT investment strategy and subsequent project selection.

### Relevance

The information services function sets the IT strategy according to business requirements. Determining the IT strategy leads to the creation of a technological infrastructure plan and an architecture design that sets and manages clear and realistic expectations of what technology can offer in terms of products, services and delivery mechanisms. The strategic plan is regularly updated and encompasses core aspects such as systems architecture, technological direction, acquisition plans, IT standards, migration strategies and contingency planning. This enables timely responses to changes in the competitive environment, economies of scale for IS staffing and investments, and improved interoperability of platforms and applications.

Traditional IT strategy has often focussed on business or IT functions and ignored the security and assurance requirements of the enterprise. In addition, where security controls were considered in the design of the system, they were frequently built on a system-by-system basis instead of through an enterprisewide approach. This leads to incomplete security models, inconsistency between the security of various systems, inefficiencies in the creation and operation of a security program and difficulties in accurate reporting according to compliance standards.

### Process Objectives

Key objectives of the methodology for determining the IT strategy are to:
- Create and maintain a technology infrastructure plan that is closely aligned with business objectives and reflects the enterprise's risk appetite and tolerance levels.
- Create and maintain technology standards.
- Publish technology standards.
- Monitor technology evolution.
- Define (future/strategic) use of new technology.
- Comply with regulations and standards of due care and risk management.

### Process Phases

Key phases of the methodology of determining the IT strategy are:
1. Understand the business strategy.
2. Align existing technologies with the strategy; identify gaps; assess existing technologies to determine capacity, throughput, etc.
3. Assess alternatives to address gaps (strategic, operational, technical).
4. Develop and publish technology standards.
5. Develop and publish a technology infrastructure plan.
6. Establish and maintain a process for monitoring compliance with the plan and standards.
7. Establish and maintain a process for assessing new technologies.

## Risk Associated With IT Strategy Development

The risk associated with the development of IT strategy primarily stems from the difficulty of learning how to understand the true requirements of the business and aligning the IT strategy effectively with those requirements. It is common to find that the business cannot properly describe its needs or requirements, and the IT analysts are not always adept at helping the enterprise determine or draw out its requirements. This leads to misalignment and ineffective solutions that serve to frustrate both parties and undermine the benefits that should have been realized by developing a strong strategic plan.

Another challenge in defining strategy is the scope of the problem. Many IT projects are developed on a system-by-system or project-by-project basis, or are focused on one department or division of the enterprise and do not consider the needs of other departments.

There is a need for more enterprisewide security solutions—a top-down approach that understands the strategic view of the enterprise and then can build components that are standardized and consistent across all systems, departments and IT projects.

Enterprises change at a rapid pace, and the strategic plans for the enterprise must be flexible enough to adapt to those changes. A failure to be flexible enough, even at the strategic level, can result in projects that are only effective as solutions for yesterday's business needs and do not reflect the priorities and requirements of the enterprise today.

## The Development of a Strategic Plan

The development of a strategic plan starts with a strategic overview of long-term (3-5 years) goals and mission of the enterprise. This requires input from several sources, especially from top management.

The risk is that many so-called strategic plans are too narrow and concerned with current problems and, while considered strategic, are really only tactical or mid-range plans based on solving the "known" problems of today and yesterday. A strategic plan needs to consider the operating environment (context) of the enterprise several years into the future. The operational context will consider political influences, proposed regulations, staffing issues, new technology, competition, financial constraints and economic forecasts. A strategic plan cannot be considered strategic unless it starts out by describing the world in which the enterprise will operate some time into the future.

Other sources of information that the risk practitioner will search for in the review of the strategic plan are mission statements, press releases and sources that indicate the culture of the enterprise. The organizational culture may influence how quickly the enterprise may want to deploy new technology, whether the enterprise is willing to push the edge of systems development or wait to follow the lead of other enterprises, and whether the enterprise is committed to growth or stability.

The risk practitioner should also ensure that the development of the strategic plan considers:
- Employee input
- Human resources and employee retention and development
- Customer expectations
- Competitive overview
- Legal requirements
- Market conditions and analysis
- Supply chain
- Vendor developments and new technology

# E. Risk Management Considerations

## Introduction

This section discusses risk management practices related to determining the IT strategy.
The following points are addressed:
- Risk factors
- Generic IT risks and their potential root causes
- Key risk indicators (KRIs)

> **Note:** The lack of strategic planning for technological direction is, in itself, a risk. An absence of strategic planning poses the risk of:
> - Improper oversight and governance of IT investments
> - Purchase of noncompliant or noninteroperable technologies
> - Purchase of unnecesary licenses, equipment, training and software
> - Inadequate forecasting of bandwidth, memory and processing needs

Therefore, the lack of a strategic plan often will result in a lack of technology investment, disconnect between business direction and technology capabilities, acquisition of unnecessary technology or inappropriate system capacity.

## Risk Factors

Examples of factors affecting the IT strategy are:
- **Enterprise size and complexity**—Size, complexity, geographic range, etc., all affect the ability to implement and maintain an IT strategic plan:
  - Infrastructure: insourced, facilities managed, outsourced
  - Organizational structure: employees, consultants, hybrid
  - Operational environment: regional, international, competitive, regulated, government, military, not-for-profit
  - Control environment: mature, immature, resilient, *ad hoc*
  - Applications source: in-house, purchased off the shelf, customized
  - IT environment: centralized, decentralized, cloud, virtual, outsourced
- **Risk management capabilities**—A defined risk culture and risk management program with effective identification, management and evaluation of risk
- **Portfolio management capability**—The diligence with which business projects are approved, managed and implemented

## Generic Risk Scenarios

Generic risk scenarios related to the IT strategy are:
- Technological acquisitions that are inconsistent with strategic plans
- Uncontrolled acquisition, use and proliferation of IS assets
- Inappropriate IT infrastructure to meet organizational requirements
- Increased costs due to uncoordinated and unstructured acquisition plans
- Deviations from the approved technological direction
- Licensing violations
- Regulatory requirements that are not met
- Difficulty in determining compliance due to inadequate information
- Failure to maximize the use of emerging technological opportunities to improve business and IT capability
- Technical incompatibilities or maintenance issues
- Incompatibilities between technology platforms and applications
- Increased support, replacement and maintenance costs
- Reliance on unsupported technology

## KRIs

Examples of KRIs are the:
- Percent of applications operating on nonstandard, unpatched or unapproved operating platforms
- Number of requests for exceptions to policies or standards
- Number of hardware/software purchase approvals without IT technology board review
- Percent of systems that are not in compliance with technology standards
- Number of technology platforms in use across the enterprise
- Alignment of IT strategy with IT projects
- Number of unapproved changes to systems, configurations, networks or applications
- Effectiveness of change control to prevent scope creep or uncontrolled changes to IT projects
- Feedback of satisfaction levels of users to deployed projects
- Standardization of systems with the overall IT architecture
- Regularity of reviews/updates of system documentation and the technology infrastructure plan

# F. Information Systems Control Design, Monitoring and Maintenance

## Introduction

This section provides an overview of common controls relating to determining the IT strategy. The following points regarding IS control are addressed:
- Objectives and activities
- Metrics
- Monitoring practices

## IS Control Objectives and Activities

Key control activities for the following aspects of planning the technological direction are described in this section:
- Planning the technological direction
- The technology infrastructure plan
- Monitoring future trends and regulations
- Technology standards
- The IT architecture board
- Enterprise security architecture planning

## Determining the IT Strategy

Determining the IT strategy includes the following: • Perform a strengths, weaknesses, opportunities and threats (SWOT) analysis of all current critical and significant IT assets on a regular basis.
- Follow up on market evolutions and relevant emerging technologies.
- Identify the latest developments in IT that could have an impact on the success of the business.
- Establish the appropriate technological risk appetite (e.g., pioneer, leader, early adopter, follower).
- Identify what is needed in terms of technological direction for business systems architecture, migration strategies and contingency aspects of infrastructure components.
- Identify security needs and ensure that security requirements are addressed throughout the IT systems development life cycle and enterprise IT security plan.

## Technology Infrastructure Plan

The technology infrastructure plan includes:
- A clear linkage to the IT strategic and tactical plans
- Ongoing assessments of the current vs. planned information systems, resulting in a migration strategy or road map to achieve the future state
- Transitional and other costs, complexity, technical risks, future flexibility, value and product/vendor sustainability
- The need to identify changes in the competitive environment, economies of scale for IS staffing and investments, and improved interoperability of platforms and applications
- Ensuring that controls are in place to provide the information necessary for compliance with regulations

## Monitoring Future Trends and Regulations

Monitoring future trends and regulations includes the following:

- Assign adequately skilled staff members to routinely monitor technological developments, competitor activities, infrastructure issues, legal requirements and regulatory environment changes, and provide relevant information to senior management (including consulting third-party experts to obtain their opinions and to confirm findings and proposals of internal staff).
- Ascertain that the IT department maintains membership in vendor user groups, subscribes to technical journals and maintains a research budget.
- Evaluate new technologies in the context of their potential contribution to the realization of broader business goals and targets using established criteria, e.g., return on investment (ROI) or the ability to achieve market leadership.
- Ensure that the enterprise's legal counsel monitors legal and regulatory conditions in all relevant geographic locations and informs the IT steering committee of any changes that may impact the technology infrastructure plan.
- Monitor ongoing and evolving security conditions, threats, exposures and attack methods, and ensure that the security of information and information systems is adequate to meet industry best practices.

## Technology Standards Objectives

Technology standards have the following common objectives:

- Corporate technology standards are approved by the IT architecture board and communicated throughout the enterprise by using a technology forum.
- Management establishes and maintains an approved list of vendors and system components that conform to the technological infrastructure plan and technology standards.
- A process is established to prevent the acquisition of nonconforming systems or applications.
- Technology guidelines are put in place to effectively support the enterprise's technological solutions.
- Monitoring and benchmarking processes are implemented to enforce compliance with the standards—e.g., by measuring noncompliance with technology standards.
- Technology standards are updated as part of a periodic review of the technological infrastructure plan and stakeholders are involved in the development and approval of migration strategies and change plans, taking into consideration impacts on personnel and operations.
- The IT department's recruiting and training practices are aligned with the technology standards.
- Technology solutions are interoperable, consistent and provide defense-in-depth (layered defense) across the enterprise.

## IT Architecture Board Responsibilities

Those fulfilling the role of an IT architecture board should:

- Provide architecture guidelines and advice on their application.
- Agree on, and formally document, the board's role and authority, including the overall IT architecture design and alignment with the information architecture.
- Put a process in place to monitor and benchmark instances of noncompliance to technology standards.
- Meet regularly and take minutes that include actions, assignments of responsible parties, time lines and tasks.

## IS Control Metrics for Monitoring

The following table provides an overview and examples of IS control metrics for monitoring the IT strategy.

| Overview and Examples of IS Control Metrics for Monitoring the IT Strategy | | | | | |
|---|---|---|---|---|---|
| **Attribute** | **Target** | **Current Period** | **Prior Period** | **Prior Period − 1** | **Prior Period − 2** |
| Percent of applications operating on nonstandard, unpatched or unapproved operating platforms | | | | | |
| Number of requests for exceptions to policies or standards | | | | | |
| Number of hardware/software purchase approvals without IT technology board review | | | | | |
| Percent of systems that are not in compliance with technology standards | | | | | |
| Number of technology platforms in use across the enterprise | | | | | |
| Alignment of IT strategy with IT projects | | | | | |
| Number of unapproved changes to systems, configurations, networks or applications | | | | | |
| Effectiveness of change control to prevent scope creep or uncontrolled changes to IT projects | | | | | |
| Feedback of satisfaction levels of users to deployed projects | | | | | |

## G. The Practitioner's Perspective

**Introduction**

Setting the IT strategy of the enterprise is an integral part of IT's strategic planning responsibility. It is in this function that the decision whether (and how) to utilize technology to advance the enterprise's objectives is defined and integrated into an overall strategic plan. The following section provides an overview of how the five CRISC domains (listed below) relate to planning the technological direction in practice:

- Domain 1—Risk Identification, Assessment and Evaluation
- Domain 2—Risk Response
- Domain 3—Risk Monitoring
- Domain 4—Information Systems Control Design and Implementation
- Domain 5—Information Systems Control Monitoring and Maintenance

**Domain 1**

Risk Identification, Assessment and Evaluation:

- The risk identification, assessment and evaluation process is key to the determination of technological direction. In alignment with the enterprise strategy, the IT technology board or an equivalent IT management group determines the future IS direction. It is critical that the IS direction be based on the effective management of risk to information systems, data, business processes and assets of the enterprise. The enterprise must be committed to implementing a cost-effective, risk-based information systems control and investment strategy.
- The risk identification process related to technology considers the consequences of specific technology acquisitions, such as:
  - Interoperability with existing technologies
  - Obsolescence
  - Future costs of nonstandard solutions
  - Adherence to known and anticipated regulatory requirements
  - Failure to embrace new technologies
  - Security implications
  - Other tangible and intangible factors
  - Availability, integrity and confidentiality of information and information systems
- The process should include interaction with core business management and IT functions (systems development, operations, technical support, information security, etc.) to ensure alignment with business requirements.
- The risk identification, assessment and evaluation process provides the risk practitioner with a listing of critical systems, current risk, risk levels and an indication of acceptable risk levels.

**Domain 2**

Risk Response:

- Risk that has been identified in the previous domain must now be managed through the selection of a suitable response plan. The core risk response options are:
  - Risk acceptance (where the cost or impact of a risk is less than the cost to mitigate or reduce the risk effectively)
  - Risk avoidance (to choose to discontinue or not enter an area of exceptional uncontrolled risk)
  - Risk transference (to purchase insurance or spread the risk between various parties)
  - Risk mitigation (to reduce the risk level through the implementation of administrative, physical or technical controls)
- It is essential to identify risk in conjunction with the development of the IT strategy because once the strategy has been accepted and projects have started, it is expensive to initiate changes since investments in new technology have been made.
- Risk response is handled at a senior IT management level based on the available options and the level of risk deemed acceptable for the enterprise.

| | |
|---|---|
| **Domain 3** | Risk Monitoring:<br>• Key risk indicators (KRIs) are symptoms of a breakdown in the IT strategy, either due to misalignment with the business strategies, lack of a cohesive technical plan, or poor internal controls. KRIs can be used to monitor IT activities, but the most telling risk monitoring process will be how well the integration of IT and business strategy progresses and that the identified risk to information and information systems is managed. |
| **Domain 4** | Information Systems Control Design and Implementation:<br>• The IS control objectives and activities described previously provide an excellent resource of best practice processes to ensure an effective technological direction management function.<br>• The practitioner should periodically review the control design against stated policy and control requirements. Where the design effectiveness of the control is of questionable or negative value, enhancement or replacement of the control should be considered. |
| **Domain 5** | Information Systems Control Monitoring and Maintenance:<br>• The KRIs defined previously are best suited for a senior management audience. KRI metrics should be given to C-level management as an overview of the effectiveness of risk management efforts and IT planning integration with the business. Failure to achieve adequate KRIs requires additional analysis because the issues may be with IT management, business management or some combination.<br>• Annual evaluation of the maturity of controls provides a barometer of controls in their current state, comparison to previous periods, and the target maturity level. Target maturity levels should be agreed on by stakeholders and evaluated by senior management as part of the IT annual review.<br>• Maturity assessment can be performed as a self-assessment, with oversight by an objective third party, peer review or independent assessment (internal audit or external provider). |

## H. Suggested Resources for Further Study

**Suggested Resources for Further Study**

In addition to the resources cited throughout this manual, the following resource is suggested for further study:

- ISACA:
  - COBIT 4.1, 2007
    **Note:** The COBIT 4.1 framework is available at no charge from ISACA and can be downloaded at *www.isaca.org/cobit*. The new COBIT 5 framework will be available in 2012.
  - *The Risk IT Framework*, 2009
  - *The Risk IT Practitioner Guide*, 2009
  - *Enterprise Value: Governance of IT Investments, The Val IT Framework 2.0*, 2008
  - *Implementing and Continually Improving IT Governance*, 2010
- Lientz, Bennet P.; Lee Larssen; *Risk Management for IT Projects: How to Deal With Over 150 Issues and Risks*, Butterworth-Heinemann, USA, 2006
- Sherwood, John; Andrew Clark; David Lynas; *Enterprise Security Architecture: A Business-Driven Approach*; CMP Books, USA, 2005

**Page intentionally left blank**

# 2. Project and Program Management

## A. Chapter Overview

### Introduction

This chapter provides an overview of the project and program management process and:
- Explains its importance to achieving business objectives
- Introduces related key concepts
- Presents examples of common risk
- Lists selected key risk indicators (KRIs)
- Provides examples of common IS controls supporting the process
- Describes the practitioner's perspective
- Offers suggested reading materials and references

### Learning Objectives

Project and program management is a major risk element. The CRISC candidate should have a comprehensive understanding of project management processes and how project management interrelates with risk management.

### Contents

This chapter contains the following topics:

# B. Related Knowledge Statements

## Contents

The following table lists the applicable knowledge statements from the CRISC job practice.

| No. | Knowledge Statement (KS)<br>Knowledge of: |
|---|---|
| KS1.14 | Threats and vulnerabilities related to project and program management |
| KS4.10 | Controls related to project and program management |
| KS5.13 | Control objectives, activities and metrics related to project and program management |

## C. Key Terms and Principles

**Introduction**

This section introduces terms and principles related to project and program management as well as terms that help relate the process to other key business processes.

**Definition of Project**

A structured set of activities concerned with delivering a defined capability (that is necessary, but not sufficient, to achieve a required business outcome) to the enterprise based on an agreed-on schedule and budget

**Definition of Program**

A structured grouping of interdependent projects that are both necessary and sufficient to achieve a desired business outcome and create value

> **Note:** These interdependent projects could involve, but are not limited to, changes in the nature of the business, business processes, the work performed by people as well as the competencies required to carry out the work, enabling technology and organizational structure.

**Definition of Portfolio**

A grouping of "objects of interest" (investment programs, IT services, IT projects, other IT assets or resources) managed and monitored to optimize business value

Portfolio management is distinct from project and project management in that the distinct objective is to create maximum value from a grouping of projects and programs.

**Exhibit 2.1: Relationship Among Project, Program and Portfolio Management**

Exhibit 2.1: Relationship Among Project, Program and Portfolio Management

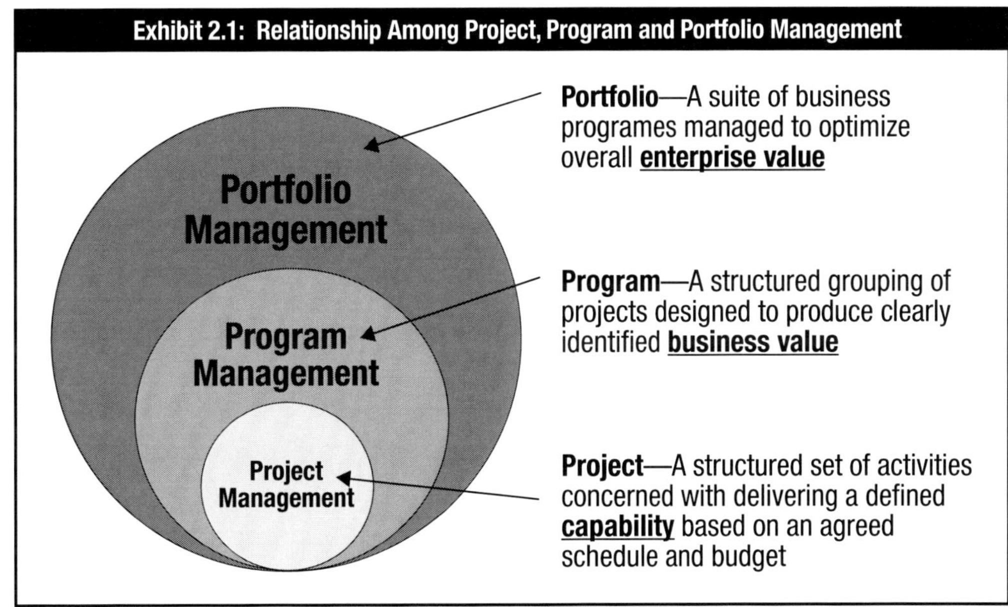

## D. Process Overview

**Introduction**

This section introduces the project management process and its importance to the achievement of business objectives.

**Process Objectives**

A project portfolio management framework for the management of all IT projects is critical to ensure correct prioritization and coordination. The framework includes a master plan, assignment of resources, definition of deliverables, approval by users, a phased approach to delivery, quality assurance (QA), a formal test plan, and testing and postimplementation review after installation to ensure project risk management and value delivery to the business.

This approach reduces the risk of unexpected costs, project failures and project cancellations. It improves communication to the business and end users, increases end-user involvement, ensures the value and quality of project deliverables, and maximizes the contribution of IT-enabled investment programs.

**Relevance**

Ineffective project and program management is a great risk and one of the most frequent causes of business process failure.

**Process Phases**

Key phases of the project management methodology are:

1. Define a program/portfolio management framework for IT investments.
2. Establish and maintain an IT project management framework.
3. Establish and maintain an IT project monitoring, measurement and management system.
4. Build project charters, schedules, quality plans, budgets, and communication and risk management plans.
5. Ensure the participation and commitment of project stakeholders.
6. Ensure the effective control of projects and project changes.
7. Define and implement project assurance and review methods.

# E. Risk Management Considerations

## Introduction

This section discusses risk management practices related to project and program management. The following points are addressed:
- Risk factors
- Generic threats and vulnerabilities
- Key risk indicators (KRIs)

## Risk Factors

Examples of factors affecting the project and program management process are:
- **Enterprise size and complexity**—Size, complexity, geographic range, etc., all affect the ease with which project and program management can be implemented and maintained.
- **Corporate culture**—Industry, traditions and management style have a high impact on risk, risk appetite, risk tolerance and risk awareness.
- **Operational model:**
  - Control environment: mature, immature, resilient
  - Applications-source: in-house, purchased off the shelf, customized
  - Application portfolio: managed, *ad hoc*
- **Risk management capabilities**—The maturity of enterprise risk management (ERM) and the decisions to approve and manage projects
- **Project management capability**—The experience within the business to manage projects
- **Project-specific factors**—Scope, time line, budget, criticality, complexity, purpose, and the level of impact on existing business and IT processes
- **Portfolio management capability**—The diligence with which business cases are assessed and the project portfolio is managed

## Generic Threats and Vulnerabilities

Generic threats and vulnerabilities related to the project and program process are:
- Program management framework:
  - Inappropriate project prioritization
  - Disorganized and ineffective approach to project programs
  - Misalignment of project and program objectives
- Project management framework:
  - Different project management approaches within the enterprise
  - Lack of compliance with the enterprise's reporting structure
  - Inconsistent tools for project management
- Project management approach:
  - Confusion and uncertainty caused by different project management approaches within the enterprise
  - Lack of compliance with the enterprise's reporting structure
  - Failure to respond to project issues with optimal and approved decisions
- Stakeholder commitment:
  - Unclear responsibilities and accountabilities for ensuring cost control and project success
  - Insufficient stakeholder participation in defining requirements and reviewing deliverables
  - Reduced understanding and delivery of business benefits
- Project scope statement:
  - Misunderstanding of project objectives and requirements
  - Failure of projects to meet business and user requirements may lead to misunderstanding of the impact of this project with other related projects
- Project phase initiation:
  - Lack of alignment of projects to the enterprise's vision
  - Wrong prioritization of projects
  - Undetected deviations from the overall project plan
  - Poor resource allocation (e.g., inadequate skill match)

## Generic Threats and Vulnerabilities *(cont.)*

- Integrated project plan:
  - Undetected errors in project planning and budgeting
  - Lack of alignment of projects to the enterprise's objectives and to other interdependent projects
  - Undetected deviations from the project plan
- Project resources:
  - Gaps in skills and resources jeopardizing critical project tasks
  - Inefficient use of resources
  - Contract disputes with outsourced resources
- Project risk management:
  - Undetected and unforeseen project risk
  - Lack of mitigating actions for identified risk
- Project quality plan:
  - Project deliverables failing to meet business and user requirements
  - Gaps in expected and delivered quality within projects
  - Inefficient and fragmented approach to quality assurance (QA)
  - Implemented system or changes adversely impact existing systems and infrastructure
- Project change control:
  - Lack of control over project scope, cost and schedule
  - Lost business focus
  - Inability to manage resources
- Project planning of assurance methods:
  - Unreliable assurance activities
  - Ineffective and/or inefficient assurance activities
  - Accreditation and implementation delays
- Project performance measurement, reporting and monitoring:
  - Ineffective reporting on project progress and unidentified issues
  - Lack of control over project progress
  - Loss of focus on customer expectations and business needs
- Project closure:
  - Undetected project management weaknesses
  - Missed opportunities from lessons learned

## KRIs

Examples of KRIs are the:

- Number of projects not satisfying the business case
- Percentage of projects not within the project timeline
- Number of projects missing project milestones
- Percentage of projects not following project management standards and practices
- Percentage of projects not on budget
- Percentage of projects not meeting internal control requirements
- Percentage of projects not meeting business process requirements
- Number of third-party issues affecting delivery of the project
- Number of projects in which business subject matter experts do not participate in the project
- Number of projects in which the project management office (PMO) does not provide project oversight

## F. Information Systems Control Design, Monitoring and Maintenance

### Introduction

This section provides an overview of common controls related to the project and program process. The following points regarding IS control are addressed:
- Objectives and activities
- Metrics
- Monitoring practices

### IS Control Objectives and Activities

Key control activities of the project and program management process are described in this section for the following aspects:
- The program management framework
- The project management framework
- The project management approach
- Stakeholder commitment
- The project scope statement
- Project phase initiation
- The integrated project plan
- Project resources
- Project risk management
- The project quality plan
- Project change control
- Project planning of assurance methods
- Project performance measurement, reporting and monitoring
- Project closure

### Key Control Activities Related to the Program Management Framework

Key control activities for the program management framework may include the following:
- Define and document the program, including all of the projects required to achieve the program's expected business outcomes; specify the required resources, including funding, project managers, project teams, IT resources and business resources where applicable; and gain formal approval of the document from key business and IT stakeholders
- Assign accountability (clearly and unambiguously) for each project, including achieving the benefits, controlling the costs, managing the risks and co-coordinating the project activities
- Determine the interdependencies of multiple projects in the program and develop a schedule for their completion that will enable the overall program schedule to be met
- Determine the program stakeholders (inside and outside the enterprise) and establish and maintain appropriate levels of coordination, communication and liaison with these parties
- Verify periodically with the business that the current program as designed will meet business requirements and make adjustments as necessary; review progress of individual projects and adjust the availability of resources as necessary to meet scheduled milestones

### Key Control Activities Related to the Project Management Framework

Key control activities for the project management framework may include ensuring that:
- The project management framework is consistent with, and is an integral component of, the enterprise's program management framework
- The project management framework includes:
  - Guidance on the role and use of the program or project office
  - A change control process for recording, evaluating, communicating and authorizing changes to the project scope, project requirements or system design
  - Requirements for integrating the project within the overall program
- The project management method covers, at a minimum, the initiating, planning, executing, controlling and closing project stages; and checkpoints and approvals

**Key Control Activities Related to the Project Management Approach**

Key control activities for the project management approach may include the following:
- Establish a project management governance structure prior to project initiation that is appropriate to the project's size, complexity and risks, including legal, regulatory and reputational risks
- Assign each IT project one or more sponsors with sufficient authority to manage execution of the project within the overall program
- Define the responsibility and accountability of the program sponsor, the project manager, and, as necessary, the steering committee and project management office
- Track the execution of a project and put in place mechanisms such as regular reporting and stage reviews that are the responsibility of the project manager to complete in a timely manner

**Key Control Activities Related to Stakeholder Commitment**

Key control activities for stakeholder commitment may include the following:
- Obtain the commitment and participation of key stakeholders, including management of the affected user department and key end users in the initiation, definition and authorization of a project
- During project initiation, outline ongoing key stakeholder commitment and roles and responsibilities for the duration of the project life cycle. Ongoing involvement includes, but is not limited to, project approval, project phase approval, project checkpoint reporting, project board representation, project planning, product testing, user training, user procedures documentation and project communication material development.

**Key Control Activities Related to the Project Scope Statement**

Key control activities for the project scope statement key may include the following:
- Provide a clear, written statement defining the nature, scope and business benefit of every project to the stakeholders to create a common understanding of project scope
- Ensure that key stakeholders and program and project sponsors within the enterprise and IT agree on and accept the requirements for the project, including definition of project success (acceptance) criteria and key performance indicators (KPIs)
- Ensure that the project definition describes the requirements for a plan that identifies internal and external project communications
- With the approval of stakeholders, maintain the project definition throughout the project, reflecting changing requirements

**Key Control Activities Related to Project Phase Initiation**

Key control activities for project phase initiation may include the following:
- Gain approval and sign-off from designated managers and customers of the affected business and IT functions for the deliverables produced in each project phase
- Base the approval process on clearly defined acceptance criteria agreed on by key stakeholders prior to work commencing on the project phase deliverable
- Assess whether the project is on schedule, within budget and aligned with the agreed-on scope; assess identified variances; and identify the impact on the project plan and realization of expected benefits
- Evaluate the project at agreed-on major review points and make formal "stop/go" decisions based on predetermined critical success criteria

## Key Control Activities Related to the Integrated Project Plan

Key control activities for the integrated project plan may include the following:

- Develop a project plan that provides information to enable management to control project progress. The plan should include details of project deliverables, required resources and responsibilities, clear work breakdown structures and work packages, estimates of resources required, milestones, key dependencies, and identification of a critical path. Interdependencies of resources (e.g., key personnel) and deliverables with other projects should be identified.
- Maintain the project plan and any dependent plans to ensure that they are up to date and reflect actual progress and material changes
- Ensure that there is effective communication of project plans and progress reports among all projects and with the overall program and that any changes made to individual plans are reflected in the other plans

## Key Control Activities Related to Project Resources

Key control activities for project resource activities may include the following:

- Identify resource needs for the project and clearly map out appropriate roles and responsibilities, with escalation and decision-making authorities agreed on and understood
- Identify required skills and time requirements for all individuals involved in the project phases in relation to defined roles and staff the roles based on available skill information (e.g., IT skills matrix)
- Utilize experienced project management and team leader resources with skills appropriate to the size, complexity and risk of the project
- Consider and clearly define the roles and responsibilities of other involved parties, including finance, legal, procurement, human resources, internal audit and compliance
- Clearly define and agree on the responsibility for procurement and management of third-party products and services, and manage the relationships

## Key Control Activities Related to Project Risk Management

Key control activities for project risk management may include the following:

- Establish a formal project risk management framework that includes identifying, analyzing, responding to, mitigating, monitoring and controlling risks
- Assign the responsibility for executing the enterprise's project risk management framework within a project to appropriately skilled personnel. Consider allocating this role to an independent team, especially if an objective viewpoint is required or a project is considered critical.
- Perform the project risk assessment of identifying and quantifying risks continuously throughout the project, and manage and communicate risks appropriately within the project governance structure
- Reassess project risks periodically, including at entry into each major project phase and as part of major change request assessments
- Identify risk and issue owners for responses to avoid, accept or mitigate risks
- Maintain and review a project risk register of all potential project risks; maintain a log of all project issues and their resolution; analyze the log periodically for trends and recurring problems to ensure that root causes are corrected

## Key Control Activities Related to the Project Quality Plan

Key control activities for the project quality plan may include the following:

- Identify ownership and responsibilities, quality review processes, and success criteria and performance metrics, all to provide quality assurance for the project deliverables
- Define requirements for independent validation and verification of the quality of deliverables in the plan

## Key Control Activities Related to Project Change Control

Key control activities for project change control may include the following:

- Establish a standard change request form and a request process requiring documentation of the requested change and the expected benefits of the change. The program management team should designate the individuals (business stakeholders, IT personnel) authorized to make project change requests.
- Review change requests; estimate the potential effects on the project, including resource requirements and impact on the schedule; and document the estimated project impact in the change request
- Review the completed change request and document the approval or denial of the request by key stakeholders, including the business project sponsor and IT project manager
- Consider and approve at the program level all approved project change requests based on an assessment of the effect the change will have on the other projects. If the requested change should not be implemented, share the reasons with the requesting project management team so they can evaluate alternative approaches.
- Update the project and program plans for all approved changes and communicate approved changes to all business and IT stakeholders in a timely manner

## Key Control Activities Related to Project Planning of Assurance Methods

Project planning of assurance methods key control activities may include the following:

- Define the assurance tasks required to ensure compliance with internal controls and security requirements that impact the systems or processes in the scope of the project, include key compliance stakeholders in the definition and approval of assurance tasks
- Determine and document how the assurance tasks will be performed, include appropriate subject matter specialists (e.g., audit, security or compliance) in the process

## Key Control Activities Related to Project Performance Measurement, Reporting and Monitoring

Key control activities for project performance measurement, reporting and monitoring include the following:

- Establish and use a set of project criteria as part of the program management framework, including, but not limited to, scope, schedule, quality, cost and level of risk
- Measure project performance against key project performance criteria analyze deviations from established key project performance criteria for cause and assess positive and negative effects on the program and its component projects. Report to identified key stakeholders the progress for the program and component projects, deviations from established key project performance criteria, and positive and negative effects on the program and its component projects.
- Monitor changes to the program and review existing key project performance criteria to determine whether they still represent valid measures of progress; document and submit any necessary changes to the program's key stakeholders for their approval before adoption; and communicate revised criteria to project managers for use in future performance reports
- Recommend, implement and monitor remedial action, when required, in line with the program and project governance framework

## Key Control Activities Related to Project Closure

Key control activities for project closure may include the following:
- Define and apply key steps for project closure, including postimplementation reviews that assess whether a project attained desired results and benefits
- Plan and execute a postimplementation review to determine whether a project delivered expected benefits and to improve the project management and system development process methodology
- Identify, assign, communicate and track any uncompleted activities required to achieve planned program project results and benefits
- Collect, from project participants and reviewers, the lessons learned and key activities that led to delivered benefits; analyze the data; and make recommendations for improving the project management method for future projects

> **IMPORTANT:** Metrics are only as reliable as the data they are based on. The risk practitioner must ensure that a data validation process is in place.

## Example of IS Control Metrics for Monitoring

The following table provides an overview and examples of IS control metrics for monitoring the project management process:
- Establish a tracking and reporting system to document and communicate the status of projects:
- Capture the number of projects approved and rejected, and communicate the status of approved, active and completed projects.

| Monthly Project Management Dashboard Report | | | | | |
|---|---|---|---|---|---|
| **Attribute** | **Target** | **Current Period** | **Prior Period** | **Prior Period − 1** | **Prior Period − 2** |
| Number of concurrent projects | | | | | |
| Number of projects completed, abandoned, postponed | | | | | |
| Number of projects in the pipeline | | | | | |
| Number of projects by business priority | | | | | |
| Number of projects waiting on resources | | | | | |
| Number of projects awaiting business case modification/redefinition | | | | | |
| Ratio of projects completed to those planned/actual | | | | | |
| Ratio of specific projects to those budgeted/actual for time, scope and expenses | | | | | |
| Number of project problem reports opened and closed | | | | | |
| Number of project problem reports remaining open from the previous period | | | | | |
| Number of project problem reports escalated to the executive sponsor or higher management | | | | | |
| Number of escalated problem reports remaining open from the previous period | | | | | |

## Example of IS Control Metrics for Monitoring

The following table provides an overview and examples of IS control metrics for monitoring the project management process:

- Establish and maintain a project management office (PMO) with responsibilities for tracking and reporting project status, and investigating project issues.

| Quarterly Project Status Report—High-visibility Projects | | | | | | | | | |
|---|---|---|---|---|---|---|---|---|---|
| Project | Budget (US $ Million [M]/ Billion [B]) | Spent (US $ Million) | Percent Complete | Business Case/ Governance | Project Management | Budget Management | Internal Control | Business Process | Third-party Management |
| A | 200 M | 100 M | 60 | OK | OK | OK | Alert | OK | OK |
| B | 145 M | 155 M | 70 | OK | Problems | Problems | OK | OK | Problems |
| C | 1.2 B | 5 M | 5 | OK | Alert | Alert | OK | OK | Alert |
| D | 120 M | 100 M | 75 | Problems | Problems | Problems | OK | OK | OK |
| E | 75 M | 50 M | 50 | Problems | Problems | Problems | Problems | Problems | N/A |

## Example of Status Report Comparison

Project A:

| Monitoring Consideration | Status |
|---|---|
| Business case/governance | Defined and managed |
| Project management | On target, milestones met |
| Budget management | Within budget for completion percentage |
| Internal control | Internal audit performed a review of the project and identified internal control issues requiring attention and potential redesign. |
| Business process | The business is satisfied with the deliverables. |
| Third-party management | Consultants are providing the agreed-on assistance. |

Project B:

| Monitoring Consideration | Status |
|---|---|
| Business case/governance | Defined and managed |
| Project management | The project is behind schedule; see *third-party management* and *budget management.* |
| Budget management | There is a budgetary overrun due to the failure to manage third-party consultants. |
| Internal control | Internal control was reviewed by internal audit, with no significant findings. |
| Business process | The business is satisfied with the deliverables. |
| Third-party management | Consultants have had significant personnel turnover, thus extending the project. The project team did not escalate the issue quickly enough to minimize the effect on the budget. The issue was escalated. |

**Example of Status Report Comparison *(cont.)***

Project C:

| Monitoring Consideration | Status |
|---|---|
| Business case/governance | Defined and managed |
| Project management | Is a new project with a large capital outlay. At the present time, there are no significant issues, but intensive oversight is required due to the size of the budget. |
| Budget management | Requires intensive oversight due to the size of the budget |
| Internal control | Internal control processes are on target. |
| Business process | Business processes are on target. |
| Third-party management | Significant consultant involvement requires intensive oversight. |

Project D:

| Monitoring Consideration | Status |
|---|---|
| Business case/governance | The business case has changed and the project value has been diminished. The business is considering changing key elements of the project. |
| Project management | Project management failed to adjust the project schedule to changes in the business case. The entire project may require major modification. |
| Budget management | Budget changes were undetermined due to the business case re-definition. |
| Internal control | Internal control processes are on target, but may require changes if there is a new business case. |
| Business process | Business processes are on target. |
| Third-party management | Consultants are providing the agreed-on assistance. |

### Example of Status Report Comparison *(cont.)*

Project E:

| Monitoring Consideration | Status |
|---|---|
| Business case/governance | Risk assessment had not been performed prior to the start of the project nor has there been a risk assessment by the project team. The executive sponsor is not actively involved. |
| Project management | The project team is not following the project development methodology. Business subject matter experts are not involved in the day-to-day processes. |
| Budget management | The actual expenses exceeding the budget are based on the percent of completion. The budget requires oversight for a potential overrun. |
| Internal control | The information security function is not involved in the process, but has identified a failure to comply with the information security architecture. |
| Business process | The business is not satisfied with the deliverables. |
| Third-party management | There are no consultants. |

## G. The Practitioner's Perspective

### Introduction

It is important to understand the relationship between risk and project management. Since poorly managed projects are one of the most significant risk factors to the enterprise, a properly conducted project will reduce the likelihood and impact of project-related risk. This section provides an overview of how the five CRISC domains relate to project and program management in practice:
- Domain 1—Risk Identification, Assessment and Evaluation
- Domain 2—Risk Response
- Domain 3—Risk Monitoring
- Domain 4—Information Systems Control Design and Implementation
- Domain 5—Information Systems Control Monitoring and Maintenance

### Domain 1

The risk identification, assessment and evaluation process identifies those risk scenarios that would preclude the achievement of a project's objectives or where a project portfolio does not optimize the overall investment.

The initial identification of risk related to project management should be integrated into the project management process and overseen by an internal senior-level governance body such as an IT/business steering committee or executive committee. The project risk assessment and evaluation can be performed as part of the project initiation phase and should be repeated for material changes affecting the risk profile during each phase or milestone review.

The practitioner should focus attention on any change that precludes successful completion of the project. These include, but are not limited to, changes in the business case, architectural changes (operating system, equipment, database design, network design), staffing, third-party participation, internal control requirements, and organizational issues.

The project risk process should require the executive sponsor's approval and a report to the steering committee responsible for development.

### Domain 2

The risk response will be based on the risk identification, assessment and evaluation. Project management has an inherent risk due to the dynamic nature of the controls. Most project risk is a result of the project management process being intentionally or inadvertently circumvented in the course of operations. Since these issues may be a result of scheduling and/or cost overruns, skipped tasks or hurried activities, a pre-planned response process for high-frequency and/or high-impact events should be established, documented and appropriate practice drills implemented.

Risk responses require a formal approach to issues, opportunities and events to ensure that solutions are cost-effective and in alignment with business objectives. The following should be considered:
- When preparing the risk response, identify the risk in business terms: loss of productivity, disclosure of confidential information, lost opportunity costs, etc.
- Understand the business risk appetite.
- Take an integrated approach—business, IT and project management need to provide a cohesive response.
- Risk responses requiring an investment should be supported by a well-thought-out business case that justifies the expenditure, outlines alternatives and describes the justification for the alternative selected.

**Domain 3**

Risk monitoring provides the communication between the affected stakeholders. Most problems occur suddenly and are a result of human error or hardware failure. When solving problems, many times the underlying symptoms are ignored and each problem is approached as an individual incident. The function of risk monitoring is to minimize this occurrence through effective key risk indicators (KRIs). The KRIs cited in this document are generic. The identification of meaningful KRIs is essential to transparency and effective communication with stakeholders and requires attention.

The KRI should describe the risk issue, provide a trend line to identify performance patterns and measure remediation processes.

An issue monitoring process should be implemented to provide IT and the stakeholders with an understanding of open issues, an estimated issue closure date, a means to prioritize and track issue resolution, and status reports to interested parties.

**Domain 4**

Risk frameworks require effectively designed controls to address inherent risk. After the application of controls, the risk framework provides a residual risk level that is within levels and costs that are acceptable to the stakeholders.

Control design must be a part of the system development life cycle (SDLC) process. Subject matter experts from the business (management and operations), IT development, IT operations, internal control and information security should have a role.

Requirements related to project management should be defined during the project requirements definition and should take into consideration the SDLC and architectural review processes. The cost associated with processes required to achieve the requirements should be defined and accepted by the business. This may include tollgate reviews.

Automated tools should be considered for reporting project status.

KRIs designed to monitor the key project-related processes should be included in the control design.

**Domain 5**

Effective monitoring of project management should be built into an effective control monitoring system, e.g., a project management office (PMO).

Timely reporting using KRIs defined previously should be distributed to stakeholders within affected IT and business units.

Quarterly review of high-profile projects should be prepared. Dashboard reporting can be effective in highlighting issues and facilitating the identification of trends. These reports should be shared with senior IT and business management. Some enterprises also share these reports with the board of director's audit committee.

Annual evaluation of the maturity of the project management controls provides a barometer of controls in their current state, comparison to previous periods, and the target maturity level. Target maturity levels should be agreed on among stakeholders.

Maturity assessment can be performed as a self-assessment, with oversight by an objective third party, peer review or independent assessment (internal audit or external provider).

## H. Suggested Resources for Further Study

### Suggested Resources for Further Study

In addition to the resources cited throughout this manual, the following resources are suggested for further study:

- ISACA:
  - COBIT 4.1, 2007
    **Note:** The COBIT 4.1 framework is available at no charge from ISACA and can be downloaded at *www.isaca.org/cobit*. The new COBIT 5 framework will be available in 2012.
  - *The Risk IT Framework*, 2009
  - *The Risk IT Practitioner Guide*, 2009
  - *Enterprise Value: Governance of IT Investments, The Val IT Framework 2.0*, 2008
  - *Systems Development and Project Management Audit/Assurance Program*, 2009
  - *Implementing and Continually Improving IT Governance*, 2010
- Office of Government Commerce (OGC):
  - *Projects in Controlled Environments 2 (PRINCE2): Directing Successful Projects With PRINCE2*, UK, 2009
  - *Projects in Controlled Environments 2 (PRINCE2): Managing Successful Projects With PRINCE2*, UK, 2009
- Project Management Institute (PMI), *A Guide to the Project Management Body of Knowledge (PMBOK), 4th Edition*, USA, 2008

**Page intentionally left blank**

# 3. Change Management

## A. Chapter Overview

### Introduction

This chapter provides an overview of the change management process and:
- Explains its importance to achieving business objectives
- Outlines a high-level process overview
- Introduces related key concepts
- Presents examples of common risk related to change management
- Lists selected key risk indicators (KRIs)
- Provides examples of common IS controls supporting the change management process
- Offers suggested reading materials and references

### Learning Objectives

Because the change management process has a high level of inherent risk, the CRISC candidate should have a comprehensive understanding of the change management process itself and how it mitigates the risk related to implementing changes.

### Contents

This chapter contains the following topics:

## B. Related Knowledge Statements

**Contents**

The following table lists the applicable knowledge statements from the CRISC job practice.

| No. | Knowledge Statement (KS)<br>Knowledge of: |
|---|---|
| KS1.13 | Threats and vulnerabilities related to the system development life cycle |
| KS1.14 | Threats and vulnerabilities related to project and program management |
| KS1.16 | Threats and vulnerabilities related to management of IT operations |
| KS4.12 | Controls related to management of IT operations |
| KS5.5 | Control objectives, activities and metrics related to IT operations and business processes and initiatives |

## C. Key Terms and Principles

### Definition of Change Management

Effective change management—within the IT service domain—helps ensure that changes to the IT infrastructure are applied using standardized methods and procedures to minimize the number and impact of any related incidents upon service. Change management comprises that aspect of the overall IT governance framework, which is enterprise sensitive, dealing with IT configuration, release and problem management issues.

> **Note:** Changes in the IT infrastructure may arise as part of project, program or service improvement initiatives, based on IT services-related problems or in response to changing legislative requirements.

### Exhibit 3.1: Relationship Among Change Management, Release Management, Problem Management and Configuration Management

**Exhibit 3.1** describes the relationship among Change Management, Configuration Management, Release Management and Problem Management.

**Exhibit 3.1: Relationship Among Change Management, Release Management, Problem Management and Configuration Management**

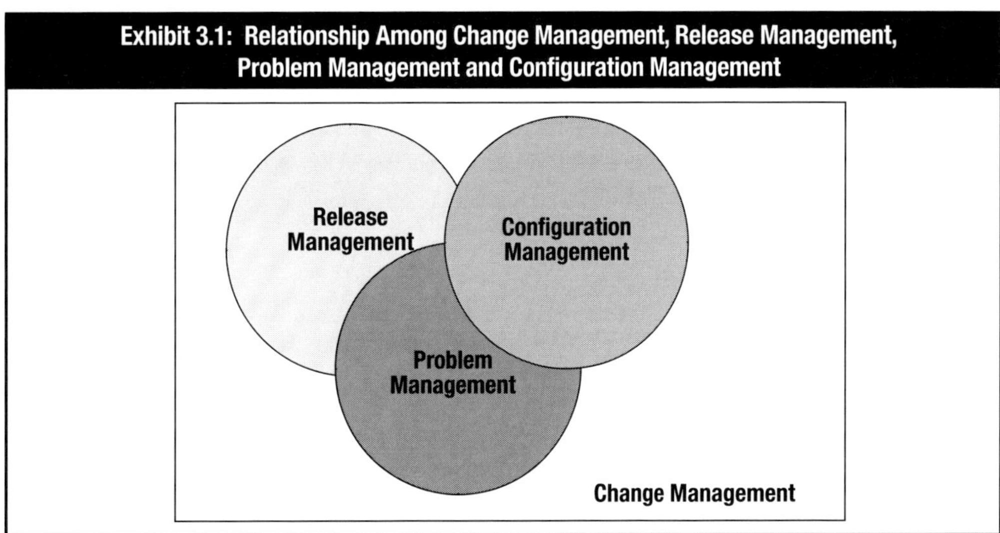


Change Management should specifically be identified as the governance framework of processes and controls within which configuration management (baseline definition and maintenance control), release management (software and hardware release controls) and problem management (issue resolution) take place.

### Definition of Configuration Management

The control of changes to a set of configuration items over a system life cycle

Configuration management focuses on the processes relating to the establishment and maintenance of hardware and software baselines, hardware versions and models, software versions (commercial off-the shelf software [COTS]) and release baselines (changes to internally developed systems) and related service agreements

### Definition of Release Management

Release management focuses on the software release life cycle processes, ranging from the initial development of a piece of software to its eventual release, and creating updated versions of the released version to help improve the software or to fix existing software bug(s). This terminology can also be applied to the hardware acquisition, deployment, upgrade and retirement processes.

### Definition of Problem Management

Systematic assessment and resolution through a defined set of IT change and control activities, of technical and procedural issues arising from the use of IT systems in a production environment.

## *D. Process Overview*

**Introduction**

This section introduces the change management process and its importance to the achievement of business objectives.

**Relevance**

Changes to systems, networks, projects, controls and business processes pose a serious risk to the enterprise. As a result of the change, errors may be introduced into the system or process and system and security control functionality may be impacted. The change management process helps mitigate the risk that changes to information systems might negatively impact the stability or integrity of the production environment.

**Process Phases**

Key phases of change management are to:
1. Develop and implement a process to consistently record, assess and prioritize change requests.
2. Assess the impact of each change on the enterprise and IT.
3. Prioritize requested changes based on business needs and resource availability.
4. Ensure that all changes follow the approved change management process.
5. Authorize changes for implementation once testing is completed successfully.
6. Ensure that standard operating procedures are developed for the change rollout process, including backup, rollback and data migration procedures.
7. Provide a history of change content, authorization and effective date by change request and program or module.
8. Manage and disseminate relevant information regarding changes to stakeholders.

## *E. Risk Management Considerations*

### Introduction

This section discusses risk management practices related to change management. The following points are addressed:
- Risk factors
- Generic threats and vulnerabilities
- Key Risk Indicators (KRIs)

### Risk Factors

Examples of factors affecting risk related to change management are:
- **System complexity**—Size, complexity, geographic range, etc., all affect the ease with which change management can be implemented and maintained.
- **Organizational structure**:
  - Infrastructure: insourced, facilities managed, outsourced (specific facility), outsourced (cloud)
  - Organizational structure: insourced, outsourced (consultants), hybrid
  - Control environment: mature, immature, resilient
  - Applications-source: in-house, purchased off the shelf, customized
  - Application portfolio: managed, *ad hoc*
  - Hardware/software purchasing: centralized, decentralized
  - Geographic location: single location, multiple locations, international, global
- **System criticality**—The process used to assess the effect that each change has on the core business processes of the enterprise
- **Change management capability**—Maturity of the change management process and the ability to monitor, document and authorize change activities
- **Rate of change**—The volatility of the systems and the rate of change to business processes, systems and procedures

### Generic Threats and Vulnerabilities

Generic threats and vulnerabilities related to the change management process are:
- Unauthorized changes resulting in compromised security and unauthorized changes or access to business processes and data
- The inability to research or document previous changes to understand root cause or processing dates affected by changes
- The inability to monitor application changes affecting business processes and data
- Emergency changes are not documented; the potential for unauthorized changes to be implemented as part of emergency changes
- Configuration documentation that does not reflect the current system configuration
- Failure to meet compliance requirements (access control, system availability, etc.)
- Changes resulting in bypassing of security controls
- Increased dependence on key individuals
- Reduced system availability
- Unintended processing side effects
- Adverse effects on capacity and performance of the infrastructure
- Change does not meet change objectives

### KRIs

Examples of KRIs are the:
- Number of disruptions or data errors caused by an inaccurate assessment of the impact of changes
- Number of incidents of application code rework caused by incorrect change specifications
- Ratio of emergency fixes to total changes
- Number of change requests in the backlog
- Time between release of vendor patch and rollout of change
- Failure to document changes via a change management system
- Changes not formally tracked, reported or authorized
- Changes promoted to production without management approval

# F. Information Systems Control Design, Monitoring and Maintenance

## Introduction

This section provides an overview of common controls related to the change management process. The following points regarding IS control are addressed:
- Objectives and activities
- Metrics
- Monitoring practices

## IS Control Objectives and Activities

Key control activities for the following aspects of the change management process are described in this section:
- Change standards and procedures
- Impact assessment, prioritization and authorization
- Emergency changes
- Change status tracking and reporting
- Change closure and documentation

## Key Control Activities Related to Change Standards and Procedures

Set up formal change management procedures to handle, in a standardized manner, all requests (including maintenance and patches) for changes to applications, procedures, processes, networks and network configurations, system and service parameters, and the underlying platforms:
- Develop, document and promulgate a change management framework that specifies the policies and processes regarding:
  - Roles and responsibilities
  - Classification and prioritization of all changes based on business and security risk
  - Assessment of impact
  - Authorization and approval of all changes by the business process owners and IT
  - Tracking and status of changes
  - The impact on data integrity (e.g., all changes to data files being made under system and application control rather than by direct user intervention)
- Establish and maintain version control over all changes.
- Implement roles and responsibilities that involve business process owners and technical IT functions, with an appropriate segregation of duties.
- Establish record management practices and audit trails to record key steps in the change management process. Ensure timely closure of changes. Elevate and report to management changes that are not closed in a timely fashion.
- Consider the impact of contracted services providers (e.g., of infrastructure, application development and shared services) on the change management process, integration of organizational change management processes with change management processes of service providers, and the impact of the organizational change management process on contractual terms and service level agreements (SLAs).

**Key Control Activities Related to Impact Assessment, Prioritization and Authorization**

Assess all requests for change in a structured way to determine the impact on the operational system and its functionality. Ensure that changes are categorized, prioritized and authorized. Develop a process to allow business process owners and/or IT staff to request changes to infrastructure, systems or applications. Develop controls to ensure that all such changes are processed only through the formal change request management process:

- Categorize all requested changes according to systems or areas affected (e.g., infrastructure, operating systems, networks, application systems, purchased/packaged application software).
- Prioritize all requested changes so that the change management process identifies both the business as well as technical and security needs for the change. Consider legal, regulatory and contractual reasons for the requested change.
- Assess the impact of all requests in a structured fashion. The assessment process should address impact analysis on infrastructure, systems and applications. Consider security, legal, contractual and compliance implications of the requested change and also interdependencies among changes. Involve business process owners in the assessment process, as appropriate.
- Estimate the resources and time required to develop, test and implement the change and the impact on current workload.
- Determine the acceptable window of downtime for rollout of the change
- Each change should be formally approved by business process owners and IT technical stakeholders, as appropriate.

**Key Control Activities Related to Emergency Changes**

Establish a process for defining, raising, testing, documenting, assessing and authorizing emergency changes that do not follow the established change process:

- A documented process exists within the overall change management process to declare, assess, authorize and record an emergency change.
- Emergency changes are processed in accordance with the emergency change element of the formal change management process.
- Ensure that emergency access arrangements for changes are appropriately authorized, documented and revoked after the change has been applied.
- Conduct a postimplementation review of all emergency changes, involving all concerned parties. The review should consider implications for aspects such as further application system maintenance and impact on development.

**Key Control Activities Related to Change Status Tracking and Reporting**

Establish a tracking and reporting system to document rejected changes, communicate the status of approved and in-process changes, and complete changes. Make certain that approved changes are implemented as planned:

- Establish a process to allow requestors and stakeholders to track the status of requests throughout the various stages of the change management process.
- Categorize change requests in the tracking process (e.g., rejected, approved but not yet initiated, approved and in process, and closed).
- Implement change status reports with performance metrics to enable management review and monitoring of both the detailed status of changes and the overall state (e.g., aged analysis of change requests). Ensure that status reports form an audit trail so that changes can subsequently be tracked from inception to eventual disposition.
- Monitor open changes so that all approved changes are closed in a timely fashion, depending on priority.

## Key Control Activities Related to Change Closure and Documentation

Whenever changes are implemented, update the associated system and user documentation and procedures accordingly:

- Ensure that documentation—including operational procedures, configuration information, application documentation, help screens and training materials—follows the same change management procedure and is considered to be an integral part of the change.
- Consider an appropriate retention period for change documentation and pre- and post-change system and user documentation.
- Update business processes for changes in hardware or software to ensure that new or improved functionality is used.
- Subject documentation to the same level of testing as the actual change.

## IS Control Metrics for Monitoring

The following table provides an overview and examples of IS control metrics for monitoring the change management process.

| Monthly Change Management Monitoring | | | | | |
|---|---|---|---|---|---|
| Attribute | Target | Current Period | Prior Period | Prior Period −1 | Prior Period −2 |
| Total incident/problem reports for period | | | | | |
| Number of disruptions or data errors caused by an incomplete impact assessment of changes | | | | | |
| Number of incidents of application code rework caused by inadequate change specifications | | | | | |
| Percent of emergency fixes to total changes | | | | | |
| Aging of open change requests | | | | | |
| Percent of change requests not properly approved or documented | | | | | |
| Failure to document changes via a change management system | | | | | |
| Changes not formally tracked, reported or authorized | | | | | |
| Changes promoted to production without management approval | | | | | |
| Percent of workstations discovered to have unauthorized or unlicensed software during the period | | | | | |

## G. The Practitioner's Perspective

### Introduction

Change management is one of the highest risk areas within IT operations. Business processes formerly performed manually by employees are revised and executed using programs implemented by the change management process. The introduction of unauthorized or incorrectly designed business processes can undermine the reliability and integrity of the business processes. The change management process is the primary control point for ensuring the integrity of business processes and the rollout of the change. A failure to implement the change correctly could result in data or processing integrity errors, interruption in service or financial penalties. The following section provides an overview of how the five CRISC domains (listed below) relate to change management in practice:
- Domain 1—Risk Identification, Assessment and Evaluation
- Domain 2—Risk Response
- Domain 3—Risk Monitoring
- Domain 4—Information Systems Control Design and Implementation
- Domain 5—Information Systems Control Monitoring and Maintenance

### Domain 1

The risk identification, assessment and evaluation process is key to the determination of which changes are implemented. The risk process:
- Determines the effect the change will have on core business activities, security and the supporting infrastructure
- Evaluates alternatives to the proposed change and its effect on the core business processes
- Manages interrelationships with other business processes and potential consequences and changes to the interfacing systems
- Prioritizes proposed changes
- Provides a basis for scheduling resources and implementations

The risk practitioner should focus attention on how changes are initially identified and then entered into a change management tracking system, and on the quality of the initial risk assessment performed prior to the change project being assigned to a development team.

### Domain 2

The risk response will be based on the risk identification, assessment and evaluation. Risk response requires a formal approach to issues, opportunities and events. This will ensure that solutions are in alignment with the business objectives, do not negatively impact other processes, and are cost-effective. The following should be considered:
- When preparing the risk response, identify the risks in business terms: loss of productivity, disclosure of confidential information, lost opportunity costs, violation of regulatory compliance, etc.
- Understand the business risk appetite, acceptable service interruptions, confidentiality of data, compliance requirements, etc.
- Keep the business stakeholders apprised of identified risks and how IT will respond to these risks.
- Ensure that steps are taken to identify and investigate any unauthorized changes

### Domain 3

Risk monitoring has the potential to provide management with an early warning of change management issues. The key risk indicators (KRIs) described previously should identify negative trends in the change process and in the change project, which require management oversight and follow-up. An issue monitoring process should be implemented to provide IT and the stakeholders with an understanding of open issues, estimated issue closure dates, a means to prioritize and track issue resolution, and status reports to interested parties.

## Domain 4

The IS control objectives and activities described previously provide an excellent resource of leading practice processes to ensure an effective change management function.

The risk practitioner should periodically review the control design against stated policy and control requirements. Where the design effectiveness of the control is of questionable or negative value, enhancement or replacement of the control should be considered.

## Domain 5

IS control monitoring and maintenance requires the following:

- Timely reporting using the KRIs defined previously should be distributed to stakeholders within IT and the business units affected.
- Annual evaluation of the maturity of controls provides a barometer of controls in their current state, comparison to previous periods, and the target maturity level. Target maturity levels should be agreed on by stakeholders and evaluated by senior IT and business operations management as part of the IT annual review.
- Maturity assessment can be performed as a self-assessment, with oversight by an objective third party, peer review, or independent assessment (internal audit or external provider).
- Change-related issue monitoring should be reported and evaluated routinely.

## *H. Suggested Resources for Further Study*

### Suggested Resources for Further Study

In addition to the resources cited throughout this manual, the following resources are suggested for further study:

- ISACA:
  - COBIT 4.1, 2007
    **Note:** The COBIT 4.1 framework is available at no charge from ISACA and can be downloaded at *www.isaca.org/cobit*. The new COBIT 5 framework will be available in 2012.
  - *The Risk IT Framework*, 2009
  - *The Risk IT Practitioner Guide*, 2009
  - *Enterprise Value: Governance of IT Investments, The Val IT Framework 2.0*, 2008
  - *Change Management Audit/Assurance Program*, 2009
  - *Implementing and Continually Improving IT Governance*, 2010
- Kouns, Jake; Daniel Minoli; *Information Technology Risk Management in Enterprise Environments: A Review of Industry Practices and a Practical Guide to Risk Management Teams*, Wiley-Interscience, USA, 2010
- Office of Government Commerce (OGC), *ITIL: IT Service Management,* Version 3, UK, 2007, *http://itil.osiatis.es/ITIL_course/it_service_management/change_management/overview_change_management/overview_change_management.php*
- Westerman, George; Richard Hunter; *IT Risk: Turning Business Threats Into Competitive Advantage*, Harvard Business School Press, USA, 2007

**Page intentionally left blank**

# 4. Third-party Service Management

## A. Chapter Overview

### Introduction

This chapter provides an overview of the third-party service management process and:
- Explains its importance to achieving business objectives
- Outlines a high-level process overview
- Introduces related key concepts
- Presents examples of common risk
- Lists selected key risk indicators (KRIs)
- Provides examples of common IS controls supporting the process
- Describes the practitioner's perspective
- Offers suggested reading materials and references

### Learning Objectives

The CRISC candidate should have a general understanding of the third-party service management process and how third-party service management affects risk management.

### Contents

This chapter contains the following topics:

## B. Related Knowledge Statements

### Contents

The following table lists the applicable knowledge statements from the CRISC job practice.

| Knowledge Statement (KS) | |
|---|---|
| **No.** | **Knowledge of:** |
| KS1.11 | Threats and vulnerabilities related to third-party management |
| KS4.7 | Controls related to third-party management |
| KS5.10 | Control objectives, activities and metrics related to third-party management |

## C. Key Terms and Principles

**Introduction**

This section introduces terms and principles related to third-party service management as well as terms that help relate the process to other key business processes.

**Definition of Service Level Agreement (SLA)**

An agreement, preferably documented, between a service provider and the customer(s)/user(s) that defines minimum performance targets for a service and how they will be measured

**Definition of Operational Level Agreement (OLA)**

An internal agreement covering the delivery of services that support the IT organization in its delivery of services

**Software as a Service (SaaS)**

A service model where the delivery of software-based services is done from a centralized location to the users who can access the software from a browser or thin client without having to install the software on their local machines. Often used for financial systems and customer relationship management (CRM) systems. The client buys a license from a SaaS vendor and then can access the software over the Internet and store data on the vendor's systems.

**Infrastructure as a Service (IaaS)**

IaaS is a hardware provisioning model where an enterprise outsources the supply, maintenance and administration of the equipment used to support operations, including storage, hardware, servers and networking components. The vendor or service provider retains ownership of the equipment and is responsible for housing, running and maintaining it. The client typically pays for the service on a per-use or per license basis.

**Platform as a Service (PaaS)**

The client purchases the right to access equipment from a third party over the Internet. This can include storage services, hardware, network capacity and operating systems. The client usually is provided access to virtual servers.

## D. Process Overview

### Introduction

This section introduces the third-party service management process and its importance to the achievement of business objectives.

### Relevance

As enterprises focus on their core business and look to reduce cost, outsourcing of various processes may be an appropriate alternative.

The need to ensure that services provided by third parties (suppliers, vendors and partners) meet business requirements requires an effective third-party management process. This process requires the:
- Clear definition of roles, responsibilities and expectations in third-party agreements
- Specification of process inputs and outputs
- Definition of boundaries and interfaces
- Ability to monitor and provide assurance of third-party compliance with contractual and regulatory commitments

Effective management of third-party services minimizes the business risk associated with nonperforming or noncompliant business partners.

### Process Phases

Key phases of the methodology of third-party service management are:
1. Identify and categorize third-party service relationships.
2. Define and document supplier management processes.
3. Establish supplier evaluation and selection policies and procedures.
4. Identify, assess and mitigate supplier risk.
5. Monitor supplier service delivery.
6. Evaluate long-term goals of the service relationship for all stakeholders.

| **Note:** The phases are not necessarily sequential. |
|---|

## E. Risk Management Considerations

### Introduction

This section discusses risk management practices related to the third-party service management process. The following points are addressed:
- Risk factors
- Generic threats and vulnerabilities
- Key risk indicators (KRIs)

### Common Types of Third Party Services

There are many types of third-party services available, each with it s own risk and benefits. The risk practitioner must carefully examine the conditions of the third-party agreement as a part of the risk management process.

Typical third-party services include:
- Software as a Service (SaaS)
- Platform as a Service (PaaS)
- Infrastructure as a Service (IaaS)
- Storage and processing-based systems—cloud computing (or computing via the Internet) has been depicted as a "cloud" in diagrams and is technology that allows many companies to share a network of servers and applications from a vendor. The advantage of cloud computing is the flexibility to quickly adapt to changing performance needs or requirements without the end user needing to purchase more equipment. The additional capacity is provided via the services available by the vendor over the cloud through access to the servers and hardware that the vendor provides.
- Disaster recovery services—also see chapter five in this Part II of the manual, Continuous Service Assurance—many enterprises have chosen a vendor-provided hot site that can be used for disaster recovery. Such hot sites are fully equipped data centers that are available in the event that a client has a catastrophe that disables its computing services. A client can purchase the right to access a hot site on a subscription basis.

### Risk Factors

Examples of factors affecting risk related to third-party service management are:
- **Criticality and uniqueness of the third-party relationship**—The reliance the business puts on a single service provider (e.g., sole, single, preferred or generic supplier)
- **Service provider attributes**—Geographic location, proximity, similarity of business and service provider organizational model and culture, financial health, language, etc.
- **Service provider portfolio management capability**—The diligence, with which service providers are identified, classified and tracked
- **Type of third-party service**—Systems development, systems infrastructure, IT operations, facilities management, applications, services
- **Level of integration**—Integration of the service provider into the enterprise
- **Existence and monitoring of service level agreements (SLAs)**—The level and thoroughness with which service levels are established, monitored and resolved
- **Regulatory requirements**—Limitations on third-party usage and transmission or storage of data across international borders

## Generic Threats and Vulnerabilities

Generic threats and vulnerabilities related to the third-party service management process are:
- Financial losses and reputational damage resulting from service interruption
- The inability to challenge costs and service quality
- The inability to optimize the choice of suppliers
- Inadequate service quality
- Inefficient and ineffective usage of supplier management resources
- Noncompliance with regulatory and legal obligations
- Unresolved problems and issues
- Security and other incidents
- A supplier that is not responsive or committed to the relationship
- An undetected service degradation
- Unidentified significant and critical suppliers
- Unclear roles and responsibilities leading to miscommunication, poor service and increased cost
- Instances where the provider goes bankrupt or may be bought by a competing firm

## KRIs

The risk practitioner can use KRIs to monitor and report on the status of risk associated with third-party service providers. Some examples of KRIs are:
- An excessive number of user complaints due to contracted services
- A purchase expenditure that is not subject to competitive procurement
- SLAs not meeting or exceeding agreed-on metrics
- An excessive number of formal disputes with suppliers
- Excessive or repeated supplier invoice disputes
- Major suppliers not subject to clearly defined requirements and service levels
- A low level of business satisfaction with the effectiveness of communication from the supplier
- A low level of supplier satisfaction with the effectiveness of communication from the business
- An excessive number of significant incidents of supplier noncompliance per time period
- Suppliers without a dedicated relationship manager

## F. Information Systems Control Design, Monitoring and Maintenance

### Introduction

This section provides an overview of common controls related to the third-party service management process. The following points regarding IS controls are addressed:
- Objectives and activities
- Metrics
- Monitoring practices

### IS Control Objectives and Activities

Key control activities for the third-party service management process are:
- Identification of all supplier relationships
- Supplier relationship management
- Supplier risk management
- Supplier performance management

### Key Control Activities Related to Identification of All Supplier Relationships

Identification of key control activities related to supplier relationships may include:
- Define and regularly review criteria to identify and categorize all supplier relationships according to the supplier type, significance and criticality of service. The list should include a category describing vendors as preferred, non-preferred or not recommended.
- Establish and maintain a detailed register of suppliers, including name, scope, purpose of the service, expected deliverables, service objectives and key contact details.

### Key Control Activities Related to Supplier Relationship Management

Key control activities related to supplier management may include:
- Define and formalize roles and responsibilities through service level agreements (SLAs) for each service supplier.
- Assign relationship owners for all suppliers and make them accountable for the quality of service(s) provided.
- Document the supplier relationship managers and communicate the information within the enterprise.
- Establish and document a formal communication process between the enterprise and the service providers.
- Ensure that contracts with key service suppliers provide for a review of supplier internal controls by management or independent third parties.
- Regularly review the reports between the enterprise and the service suppliers.
- Register incidents caused by suppliers and report them using the enterprise's internal incident management process.
- Periodically review and assess supplier performance against established and agreed-on service levels while clearly communicating suggested changes to the service supply contracts or recommendations regarding new suppliers.

### Key Control Activities Related to Supplier Risk Management

Key control activities relating to supplier risk management may include:
- Identify and monitor suppliers in accordance with the enterprise's established risk management process.
- Identify and document in the contract supplier risks (and remedies) associated with the suppliers' inability to fulfill the contractual agreements.
- Consider remedies when defining the contract—including software escrow agreements, exit strategies, alternative suppliers or standby agreements—in the event of supplier failure.
- Review all contracts for legal and regulatory requirements. Ensure that contracts include which legal jurisdiction would apply in case of any dispute.

## Key Control Activities Related to Supplier Performance Management

Key control activities related to supplier performance management may include:
- Define and document criteria to monitor service suppliers' performance.
- Ensure that suppliers regularly report on agreed-on performance criteria.
- Invite users to provide feedback for assessment of supplier performance and quality of service.
- Evaluate the costs and market conditions for the service levels by benchmarking against alternative suppliers, and identify the potential for improvement.
- Define arbitration procedures to consult an arbitration committee in the event that any action must be undertaken.

## G. The Practitioner's Perspective

### Introduction

The use of third-party service providers presents a unique risk to the enterprise. Outsourcing business processes leaves the enterprise vulnerable to risks beyond its organizational borders. The third party may provide core business services that are not easily replaced if the relationship becomes contentious or service levels are unsatisfactory. For this reason, understanding and managing risk is essential. The following section provides an overview of how the five CRISC domains (listed below) relate to third-party service management in practice:

- Domain 1—Risk Identification, Assessment and Evaluation
- Domain 2—Risk Response
- Domain 3—Risk Monitoring
- Domain 4—Information Systems Control Design and Implementation
- Domain 5—Information Systems Control Monitoring and Maintenance

### Domain 1

The decision to outsource (utilize third-party providers) and the subsequent transfer of processing responsibility must follow a systematic process, similar to a systems development life cycle (SDLC).

Risk identification needs to be comprehensive, involving IT, business units, legal, audit, internal control and information security subject matter experts. All relevant issues—for example, scope of services; ability to deliver according to acceptable service levels; cultural, language and regulatory requirements; and conditions for the transfer of personally identifiable or sensitive information—need to be included in the process.

The risk assessment process should be performed several times during the outsourcing process, for example, during:

- The outsourcing decision
- Vendor selection
- Contract negotiation

The risk assessment process should be performed at the time of contract renewal, when material regulatory changes occur and business processes change, and on a scheduled, periodic basis.

### Domain 2

The risk response will be based on the result of the risk identification, assessment and evaluation phase.

If the third-party relationship is in the development stage, risk response will focus on working with the negotiation team or the project team responsible for servicer selection to ensure that the risk is addressed in the contract and that there are provisions in place to audit and review the performance of the third-party service provider.

If the third-party relationship has been established, the risk response should include identifying the relationship manager assigned to the service provider. Using the issue monitoring system, any issues should be defined and appropriate risk mitigation processes scheduled. The remediation process should be monitored until closure.

**Domain 3**

The function of risk monitoring is to continue to evaluate the risk factors that could indicate a change in risk levels and thereby ensure that the actual risk levels that are caused by the use of third-party agreements are within the acceptable risk levels of the enterprise and that the risk mitigation controls are working effectively and, as designed, to mitigate risk. The process of continuous or periodic monitoring will minimize the development of unacceptable levels of risk through the use of effective key risk indicators (KRIs). The identification of meaningful KRIs is essential to transparency and effective communication with stakeholders and requires attention to the needs of the business and the risk culture of the enterprise.

KRIs should describe the risk, provide a trend line to identify performance patterns, and measure remediation processes.

The issue monitoring process described in risk response should be the primary vehicle to ensure that risks are being adequately mitigated.

**Domain 4**

Requirements related to third-party management should be defined during the requirements definition phase of the SDLC. As the risk associated with the use of a third-party service provider is identified, the appropriate control requirements should be documented. As the controls are designed, the costs associated with the controls (or the level of residual risk to be accepted) required to achieve the risk management requirements should be defined and accepted by the enterprise. This may include designing audit and detection systems or services, additional monitoring and problem and change management processes, etc.

Service level agreement (SLA) and relationship management controls need definition and implementation. These are the primary oversight processes required for third-party management.

Automated tools should be designed to report when problems related to third parties are cited and should also report on the corresponding remediation process performed.

KRIs designed to monitor the key third-party processes should be included in the control design.

**Domain 5**

The KRIs described previously are effective monitoring tools. Timely reporting using KRIs defined previously should be distributed to stakeholders within IT and the business units affected.

The relationship manager needs to be integrated with business and IT management because many early warning signs are recognizable to the relationship manager before the trend lines of the monitoring system generate an alert.

Annual evaluation of the maturity of controls provides a barometer of controls in their current state, comparison to previous periods, and the target maturity level. Target maturity levels should be agreed on by stakeholders and evaluated by senior management as part of the IT annual review.

Maturity assessment can be performed as a self-assessment, with oversight by an objective third-party, peer review, or independent assessment (internal audit or external provider).

## *H. Suggested Resources for Further Study*

### Suggested Resources for Further Study

In addition to the resources cited throughout this manual, the following resources are suggested for further study:

- ISACA:
  - COBIT 4.1, 2007
    **Note:** The COBIT 4.1 framework is available at no charge from ISACA and can be downloaded at *www.isaca.org/cobit*. The new COBIT 5 framework will be available in 2012.
  - *The Risk IT Framework*, 2009
  - *The Risk IT Practitioner Guide*, 2009
  - *Enterprise Value: Governance of IT Investments, The Val IT Framework 2.0*, 2008
  - *Implementing and Continually Improving IT Governance*, 2010
- International Organization for Standardization (ISO):
  - ISO 27001, *Information security management—Specification with guidance for use*, Switzerland, 2005. This is the replacement for BS7799-2. It is intended to provide the foundation for third-party audit and is harmonized with other management standards, such as ISO/IEC 9001 and 14001.
  - ISO/IEC 27002:2005, *Code of practice for information security management, Appendix 1*, Switzerland, 2005
- National Institute of Technology and Standards (NIST), *Security Controls in External Environments*, Special Publication (SP) 800-53, Revision 3, Section 2.4, Appendix 2, USA, 2009

Page intentionally left blank

# 5. Continuous Service Assurance

## A. Chapter Overview

### Introduction

This chapter provides an overview of the continuous service assurance process and:
- Explains its importance to achieving business objectives
- Outlines a high-level process overview
- Introduces related key concepts
- Presents examples of common risk
- Lists selected key risk indicators (KRIs)
- Provides examples of common IS controls supporting the process
- Describes the practitioner's perspective
- Offers suggested reading materials and references

### Learning Objectives

Risk is an ever-changing element of business and the environment in which the enterprise operates. The risk practitioner should have a general understanding of the continuous service assurance process and how continuous service management interrelates with risk management.

### Contents

This chapter contains the following topics:

## B. Related Knowledge Statements

**Contents**

The following table lists the applicable knowledge statements from the CRISC job practice.

| No. | Knowledge Statement (KS)<br>Knowledge of: |
|---|---|
| KS1.15 | Threats and vulnerabilities related to business continuity and disaster recovery management |
| KS4.11 | Controls related to business continuity and disaster recovery management |
| KS5.15 | Control objectives, activities and metrics related to business continuity and disaster recovery management |

## C. Key Terms and Principles

| | |
|---|---|
| **Introduction** | This section introduces terms and principles related to continuous service assurance as well as terms that help relate the process to other key business processes. |
| **Availability** | A state where information is accessible when required by the business process, now and in the future |
| **Definition of Recovery Point Objective (RPO)** | The RPO is determined based on the acceptable data loss in case of a disruption of operations. It indicates the earliest point in time to which it is acceptable to recover the data. The RPO effectively quantifies the permissible amount of data loss in case of interruption. (See also Recovery Time Objective [RTO].) |
| **Definition of Recovery Strategy** | An approach by an enterprise that will ensure its recovery and continuity in the face of a disaster or other major outage |
| **Definition of Recovery Testing** | A test to check the system's ability to recover after a software or hardware failure. |
| **Definition of Recovery Time Objective (RTO)** | The amount of time allowed for the recovery of a business function or resource after a disaster occurs. (See also Recovery Point Objective [RPO].) |
| **Definition of Resilience** | The ability of a system or network to resist failure or to recover quickly from any disruption, usually with minimal recognizable effect |

## D. Process Overview

### Introduction

COBIT defines continuous service assurance as "preventing, mitigating and recovering from disruption" and explains that the following terms can also be used in the same context:

- **Business resumption planning**—the resumption of business operations following an interruption
- **Disaster recovery planning**—the recovery of IT systems and processes needed to support business operations
- **Contingency planning**—alternative processing arrangements to support critical products or services in the event of a disruption to normal operations

The purpose of continuous service assurance is to identify threats to an enterprise, and their probability and impact on the continuity of business processes. This builds the basis for plans to reduce the likelihood of an interruption, minimize the impact of the interruption should one occur and provide priorities for recovery of business processes, so that critical products and services may be recovered within acceptable time limits.

### Relevance

The need for providing continuous business and IT services requires developing, maintaining and testing of both business and IT continuity plans, utilizing offsite backup storage and providing periodic continuity plan training. An effective continuous service assurance process minimizes the likelihood and impact of a major interruption on key business functions and processes.

Continuous service assurance is critical to the ongoing operation of an enterprise after an incident has occurred—after an incident has interrupted the process by impacting one or more of the process dependencies. The continuous service assurance process provides a framework to protect the interests of the enterprise and its key stakeholders by identifying critical business processes and their dependencies, developing recovery strategies, documenting recovery plan tests, testing the effectiveness of those plans, and ensuring that the plans are kept up to date.

### Process Objectives

Key objectives of the continuous service assurance process, and the roles and responsibilities for those objectives, are to:

- Develop a combined business/IT continuity framework.
- Conduct a business impact analysis (BIA) and risk assessment.
- Develop and maintain business and IT continuity plans.
- Identify and categorize information resources based on recovery objectives, including people, information, architecture and applications.
- Define and execute change control procedures to ensure that the continuity plan is current.
- Regularly test the business and IT continuity plans.
- Develop a follow-up action plan from test results.
- Plan and conduct business and IT continuity training.
- Plan business and IT services recovery and resumption.
- Plan and implement backup storage and protection.
- Establish procedures for conducting postresumption review and planning

## E. Risk Management Considerations

### Introduction

This section discusses risk management practices related to the continuous service assurance process. The following points are addressed:
- Risk factors
- Generic threats and vulnerabilities
- Key risk indicators (KRIs)

### Risk Factors

Examples of factors affecting the risk related to the continuous service assurance process are:
- **Enterprise size and complexity**—Size, complexity, geographic location, etc., all affect the ease with which continuous service assurance can be implemented and maintained.
- **Recovery strategies**—In-house, outsourced (vendor-supplied); hot sites, warm sites, cold sites
- **Geographic location**—The hazards inherent to the location
- **Process complexity**—The complexity of the business processes and the amount of resources (time, location, people and technology) required for continuity
- **Use of third-party services**—Outsourcing of IT services places reliance on the service provider's ability to provide continuous services.
- **Time sensitivity of business processes**—Can be impacted by many factors such as regulatory, contractual, financial, work flow peaks, and legal time constraints on completing process activities

> **Note:** Many enterprises do not know their business priorities and interdependencies well enough to create adequate recovery plans.

### Generic Threats and Vulnerabilities

Generic threats and vulnerabilities related to the continuous service assurance process are:
- Continuity practices are not sufficient to support business processes:
  - Recovery plans are not in alignment with the business impact analysis (BIA).
  - Recovery plans are developed in isolation from business requirements.
  - There is no BIA.
- IT continuity services are mismanaged.
- A lack of documentation and/or training results in an increased dependency on key individuals.
- Failure to recover IT systems and services in a timely manner
- Unrealistic timelines for the recovery of critical products or services
- Lack of required recovery resources (hardware, software and personnel)
- Communication to internal and external stakeholders is inadequate.
- Critical business and IT resources are unavailable.
- Increased costs for continuity management
- The prioritization of services recovery is not based on business needs.
- Plans fail to reflect changes to business needs and technology.
- There is a lack of integration of change management procedures with the continuity plan.
- Outdated recovery plans do not reflect the current hardware and software architecture.
- Inadequate recovery steps and processes exist.
- Confidential information in the plans is compromised.
- Plans are not accessible to all required parties.
- Backup data and media are unavailable due to missing documentation in offsite storage.

## KRIs

Examples of KRIs related to continuous services assurance are the:

- Number and location of mission-critical business process
- Redundancy in process and location
- Recovery time objective (RTO) and recovery time capability comparisons
- Percentage of successful tests
- Backup of information to support business operations and recover lost data according to recovery point objectives (RPOs)
- Probability and impact of residual risk
- Training of staff for roles related to business continuity planning

# F. Information Systems Control Design, Monitoring and Maintenance

## Introduction

This section provides an overview of common controls related to continuous service assurance. The following points regarding IS controls are addressed:
- Objectives and activities
- Metrics
- Monitoring practices

## IS Control Objectives and Activities

Key control activities for continuous service assurance are:
- A business and IT continuity framework
- A business and IT continuity plan
- Identification of critical business and IT resources
- Maintenance of the business and IT continuity plan
- Testing the business and IT continuity plan
- Business and IT continuity plan training
- Distribution of the business and IT continuity plan
- Business and IT services recovery and resumption
- Offsite backup storage
- Postresumption review (application of lessons learned)

## Key Control Activities Related to the IT Continuity Framework

Key control activities may include:
- Assign responsibility for, and establish, an enterprisewide business continuity management process. This process should include an IT continuity framework to ensure that a business impact analysis (BIA) is completed and the IT continuity plan supports the business strategy, a prioritized recovery strategy, and necessary operational support based on these strategies and any compliance requirements.
- Identification of continuity plan ownership and steering committee to oversee the operation of the continuity program
- Selection of tools and standards to be used in developing the plan
- Determining budget and resources to be used in creating, executing and testing the plan
- Creation of policy for business continuity program
- The determination of IT recovery priorities based on the priorities for recovery of business products and services
- Ensure that the IT continuity framework addresses the:
  - Organizational structure for IT continuity management as a liaison to organizational continuity management
  - Roles, tasks and responsibilities defined by service level agreements (SLAs) and/or contracts for internal and external service providers
  - Documentation standards and change management procedures for all IT continuity-related procedures and tests
  - Policies for conducting regular tests
  - Frequency and conditions (triggers) for updating the IT continuity plan
  - Results of the risk assessment process

## Key Control Activities Related to the IT Continuity Planning

Key control activities may include:

- Create an IT continuity plan that includes the:
  - Conditions and responsibilities for activating and/or escalating the plan
  - Prioritized recovery strategy, including the necessary sequence of activities
  - Minimum recovery requirements to maintain adequate business operations and service levels with diminished resources
  - Emergency procedures
  - Fallback procedures
  - Temporary operational procedures
  - IT processing resumption procedures
  - Maintenance and test schedule
  - Awareness, education and training activities
  - Responsibilities of individuals
  - Regulatory requirements
  - Critical assets and resources and up-to-date personnel contact information needed to perform emergency, fallback and resumption procedures
  - Alternative processing facilities as determined within the plan
  - Alternative suppliers for critical resources
- Define the underlying assumptions (e.g., level and scope of outage covered by the plan) in the IT continuity plan and which systems (i.e., computer systems, network components and other IT infrastructure) and sites are to be included. Describe alternative processing options for each site.
- Ensure that the IT continuity plan includes a defined checklist of recovery events and a form for event logging.
- Establish and maintain detailed information for every recovery site, including assigned staff and logistics (e.g., transport of media to the recovery site). This information should include the:
  - Processing requirements for each site
  - Location
  - Resources (e.g., systems, staff, support) available at each location
  - Utility companies on which the site depends
- Define the response and recovery team structures, including the reporting requirements for roles and responsibilities, and the requirements for knowledge, skills and experience for all team members. Include the contact details of all team members and ensure that the details are maintained and readily available (e.g., offsite team, backup managing team).
- Define and prioritize communication processes and define responsibility for communication (e.g., public, press, government). Maintain the contact details of relevant stakeholders (e.g., the crisis management team, IT recovery staff, business stakeholders, staff), service providers (e.g., vendors, telecommunications provider) and external parties (e.g., business partners, media, government bodies, public).
- Maintain procedures to protect and restore the affected part of the enterprise, including, where necessary, reconstruction of the affected site or its replacement. This also includes procedures to respond to further disasters while at the backup site.
- Create emergency procedures to ensure the safety of all affected parties, including coverage of occupational health and safety requirements (e.g., counseling services) and coordination with public authorities.

### Key Control Activities Related to Critical IT Resources

Key control activities may include:
- Define priorities for all applications, systems and sites that are in line with business objectives (usually through the execution of a business impact analysis (BIA). Include these priorities in the continuity plan. When defining priorities, consider the:
  - Business risk and IT operational risk
  - Interdependencies
  - Data classification framework
  - SLAs and operational level agreements (OLAs)
  - Costs
- Consider resilience, response and recovery requirements for different tiers, e.g., one to four hours, four to 24 hours, more than 24 hours and critical business operational periods.

### Key Control Activities Related to Maintenance of the IT Continuity Plan

Key control activities may include:
- Maintain a change history of the IT continuity plan. Ensure proper version management of the plan, e.g., through the use of document management systems, and that all distributed copies are the same version.
- Involve the business continuity and IT continuity manager(s) in the change management process to ensure awareness of important changes that would require updates to the IT continuity plan.
- Update the IT continuity plan as described by the IT continuity framework. Triggering events for the update of the plan include:
  - Important architecture changes
  - Important business changes
  - Key staff changes or organizational changes
  - Incidents/disasters and the lessons learned
  - Results from continuity plan tests

### Key Control Activities Related to Testing the IT Continuity Plan

Key control activities may include:
- Schedule IT continuity tests on a regular basis or after major changes in the IT infrastructure or to the business and related applications. Ensure that all new components (e.g., hardware and software updates, new business processes) are included in the schedule.
- Create a detailed test schedule based on established recovery priorities. Ensure that test scenarios are realistic. Tests should include recovery of critical business application processing and should not be limited to recovery of infrastructure. Make sure that testing time is appropriate and will not impact ongoing business operations.
- Establish an independent test task force that keeps track of all events and records all results to be discussed in the debriefing. The members of the task force should not be key personnel defined in the plan. This task force should independently report to senior management and/or the board of directors.
- Perform a debriefing event at which all failures are analyzed and solutions are developed or handed over to task forces. Ensure that all outstanding issues related to continuity planning are analyzed and resolved in an appropriate time frame. Schedule a retesting of the changes using similar or stronger parameters to ensure that the changes resulted in a positive impact on the recovery procedures.
- If testing is not feasible, evaluate alternative means for ensuring resources for business continuity (e.g., a dry run).
- Measure and report the success or failure of the test and, therefore, the continuity and contingency ability for services to the risk management process.

**Key Control Activities Related to IT Continuity Plan Training**

Key control activities may include:
- On a regular basis (at least annually) or upon plan changes, provide training to the required staff members with respect to their roles and responsibilities.
- Assess all needs for training periodically and update all schedules appropriately. While planning the training, take into account the timing and the extent of plan updates and changes, turnover of recovery staff, and recent test results.
- Perform regular IT continuity awareness programs for all levels of employees and IT stakeholders to increase awareness of the need for an IT continuity strategy and their key role within it.
- Measure and document training attendance, training results and coverage.

**Key Control Activities Related to Distribution of the IT Continuity Plan**

Key control activities may include:
- Define a proper distribution list for the IT continuity plan and keep this list up to date. Include people and locations in the list on a need-to-know basis. Ensure that procedures exist with instructions for storage of confidential information.
- Define a distribution process that:
  - Distributes the IT continuity plan in a timely manner to all recipients and locations on the distribution list
  - Collects and destroys obsolete copies of the plan in line with the enterprise's policy for discarding confidential information
- Ensure that all digital and physical copies of the plan are protected in an appropriate manner (e.g., encryption, password protection) and that the document is accessible only by authorized personnel (recovery staff).

**Key Control Activities Related to IT Services Recovery and Resumption**

Key control activities may include:
- Activate the IT continuity plan when conditions require it.
- Maintain an activity and problem log during recovery activities to be used during postresumption review.

**Key Control Activities Related to Offsite Backup Storage**

Key control activities may include:
- Provide protection for the data commensurate with the value and security classification—for the time period starting with the data being taken offsite, in transport to/from the enterprise and located at the storage location.
- Ensure that the backup facilities are not subject to the same risks (e.g., geography, weather, key service provider, utilities, human factors) as the primary site.
- Perform regular testing of:
  - The quality of the backups and media
  - The ability to meet the committed recovery time frame
- Ensure that the backups contain all data, programs and associated resources needed for recovery according to the plan.
- Provide sufficient recovery instructions and adequate labeling of backup media.
- Maintain an inventory of all backups and backup media. Ensure inclusion of all departmental processing, if applicable.

## Key Control Activities Related to Postresumption Review

Key control activities may include:
- Using the problem and activity log of recovery activities
- Identifying the shortcomings of the plan after reestablishing normal processing
- Identifying opportunities for improvement to include in the next update of the IT continuity plan.

## IS Control Metrics for Monitoring

The following tables provide an overview and examples of IS control metrics for monitoring continuous service assurance.

| Monthly Continuous Service Assurance Monitoring | | | | | |
|---|---|---|---|---|---|
| **Attribute** | **Target** | **Current Period** | **Prior Period** | **Prior Period – 1** | **Prior Period – 2** |
| Number of hours lost per user per month due to unplanned outages | | | | | |
| Percent of availability met by SLAs | | | | | |
| Percent of tests that achieve recovery objectives | | | | | |
| Number of service interruptions of critical systems | | | | | |
| Number of service interruptions of critical systems exceeding the tolerance level | | | | | |
| Number of all service interruptions | | | | | |
| Number of all service interruptions exceeding the tolerance level | | | | | |
| Average duration of service interruptions of critical systems | | | | | |
| Average duration of all interruptions | | | | | |

**IS Control Metrics for Monitoring *(cont.)***

| Annual Continuous Service Assurance Monitoring | | | |
|---|---|---|---|
| **Attribute** | **Target** | **Current Period** | **Prior Period** |
| Number of hours lost per user per month due to unplanned outages | | | |
| Percent of availability met by SLAs | | | |
| Number of business-critical processes relying on IT that are not covered by IT continuity plan | | | |
| Percent of tests that achieve recovery objectives | | | |
| Number of service interruptions of critical systems | | | |
| Number of service interruptions of critical systems exceeding the tolerance level | | | |
| Number of all service interruptions | | | |
| Number of all service interruptions exceeding the tolerance level | | | |
| Average duration of service interruptions of critical systems | | | |
| Average duration of all interruptions | | | |
| Elapsed time between tests of any given element of the IT continuity plan | | | |
| Number of IT continuity training hours per year per relevant IT employee | | | |
| Percent of critical infrastructure components with automated availability monitoring | | | |
| Date of last review of the IT continuity plan | | | |
| Applications in IT continuity plan that are not supported by the BIA | | | |

## G. The Practitioner's Perspective

### Introduction

Continuity is not just a technology issue. Some of most significant areas of business risk in the continuous service assurance domain come from inadequate understanding of, and involvement by, the business. Historically, continuity processes have been "owned" by technology departments. In some cases, continuity plans were limited to the resumption and recovery of technology resources. The following section provides an overview of how the five CRISC domains (listed below) relate to the continuous service assurance process in practice:

- Domain 1—Risk Identification, Assessment and Evaluation
- Domain 2—Risk Response
- Domain 3—Risk Monitoring
- Domain 4—Information Systems Control Design and Implementation
- Domain 5—Information Systems Control Monitoring and Maintenance

### Domain 1

The risk practitioner must identify any risk that the enterprise is not prepared for in the case of an adverse event that could impact its ability to meet business requirements. This would include reviewing whether the enterprise has plans in place, whether those plans would be adequate to recover operational capability, whether the leadership is in place to manage a disaster should it happen and whether the staff and plan are up to date with training and according to the current business environment.

Continuous service assurance begins with the business impact analysis (BIA). BIA identifies the risks associated with the business processes, and the supporting IT functions that are required to ensure continuous business processing. Once the BIA identifies the required processing functions, IT can develop a plan that addresses these concerns within the risk tolerances and costs the enterprise can accept.

Outsourced environments add an additional complexity. The service provider is responsible for the IT continuity plan, but the customer is responsible for the BIA, establishing requirements (depending on the outsourcing agreement) and the customer processes.

### Domain 2

The IT disaster recovery plan is the risk response based on the priorities identified in the BIA. How the enterprise responds to known and unforeseen risk will determine the effectiveness of the continuous service assurance. Different plans should be in place depending on the nature of the incident and the priority of the services and systems affected.

Practice drills should be organized.

During the drills, potential situations should be included in the simulation, e.g., individuals who are not available, software that is missing, delays.

### Domain 3

The key risk indicators (KRIs) described previously provide a monitoring process for key issues that would normally be experienced by an IT operations organization.

The risk practitioner should ensure that the plans are tested on a regular basis, staff has been provided adequate training, and the plan is being updated in alignment with changes to the enterprise.

**Domain 4**

The controls identified in the IS controls and objectives provide a resource of best practices for an IT operations unit.

The risk practitioner should ensure that the recovery alternative (hot site, warm site, cold site) chosen will meet business expectations.

The offsite storage requirement incorporates data management processes affecting data restoration.

**Domain 5**

Monthly and annual continuous service assurance monitoring provides a monitoring benchmark. This report should be provided to IT management and business stakeholders on a monthly basis and summarized on an annual basis.

Timely reporting using the KRIs defined previously should be distributed to stakeholders within IT and the business units affected.

Annual evaluation of the maturity of controls provides a barometer of controls in their current state, comparison to previous periods, and the target maturity level. Target maturity levels should be agreed on by stakeholders and evaluated by senior management as part of the annual review.

Maturity assessment can be performed as a self-assessment, with oversight by an objective third-party, peer review, or independent assessment (internal audit or external provider).

Issue monitoring related to continuous service assurance should be reported and evaluated routinely.

## H. Suggested Resources for Further Study

**Suggested Resources for Further Study**

In addition to the resources cited throughout this manual, the following resources are suggested for further study:

- ISACA:
  - COBIT 4.1, 2007
    **Note:** The COBIT 4.1 framework is available at no charge from ISACA and can be downloaded at *www.isaca.org/cobit*. The new COBIT 5 framework will be available in 2012.
  - *The Risk IT Framework*, 2009
  - *The Risk IT Practitioner Guide*, 2009
  - *Enterprise Value: Governance of IT Investments, The Val IT Framework 2.0*, 2008
  - *Implementing and Continually Improving IT Governance*, 2010
- British Standards Institution (BSI), BS 25999, *A Code of Practice for Business Continuity Management*, UK, 2007
- Business Continuity Institute (BCI), *Good Practice Guidelines 2010*, UK, 2010

**Page intentionally left blank**

# 6. Information Security Management

## A. Chapter Overview

### Introduction

This chapter provides an overview of the information security management process and:
- Explains its importance to achieving business objectives
- Outlines a high-level process overview
- Introduces related key concepts
- Presents examples of common risk
- Lists selected key risk indicators (KRIs)
- Provides examples of common IS controls supporting the process
- Describes the practitioner's perspective
- Offers suggested reading materials and references

### Learning Objectives

Information security management is a major risk element. The CRISC candidate should have a comprehensive understanding of information security management processes and how information security management interrelates with risk management.

### Contents

This chapter contains the following topics:

## B. Related Knowledge Statements

**Contents**

The following table lists the applicable knowledge statements from the CRISC job practice.

| | Knowledge Statement (KS) |
|---|---|
| **No.** | **Knowledge of:** |
| KS1.10 | Information security concepts |
| KS4.6 | Controls related to information security |
| KS5.9 | Control objectives, activities and metrics related to information security |

## *C. Key Terms and Principles*

**Introduction**

This section introduces terms and principles related to information security management as well as terms that help relate the process to other key business processes.

---

**Definition of Access Controls**

The rules, procedures, practices and devices intended to preclude unauthorized entry or right of use, whether physical or logical

---

**Definition of Authentication**

The act of verifying the identity of an entity, such as a user, system, network node

---

**Definition of Accountability**

The ability to map a given activity or event back to the responsible party

---

**Definition of Confidentiality**

The protection of sensitive or private information from unauthorized disclosure

---

**Definition of Data Classification**

The assignment of a level of sensitivity to data (or information) that results in the specification of controls for each level of classification. Levels of sensitivity of data are assigned according to predefined categories as data are created, amended, enhanced, stored or transmitted. The classification level is an indication of the value or importance of the data to the enterprise.

---

**Definition of Integrity**

The accuracy, completeness and validity of information

## *D. Process Overview*

### Introduction

This section introduces the information security management process and its importance to the achievement of business objectives. The risk practitioner is expected to influence and be active in the design and monitoring of information systems controls.

### Relevance

The need to maintain the integrity, confidentiality and availability of information and the protection of IT assets requires a security management process. An information security management process is almost always based on risk management and the need to implement risk-based, cost-effective controls that will protect information systems from adverse events. The information security management process includes establishing and maintaining information security roles and responsibilities, policies, standards and procedures. Security management also includes performing security monitoring, periodic testing and implementing corrective actions for identified security weaknesses or incidents. Effective security management protects all IT assets to minimize the business impact of security vulnerabilities and incidents.

### Process Objectives

Key objectives of the information security management process are to:

- Define and maintain an information security plan.
- Define, establish and operate an identity (account) management process.
- Monitor potential and actual security incidents.
- Periodically review and validate user access rights and privileges.
- Establish and maintain procedures to safeguard cryptographic keys.
- Implement and maintain technical and procedural controls to protect information in storage, process and transmission as it flows between systems and across networks.
- Conduct regular vulnerability assessments.

# E. Risk Management Considerations

## Introduction

This section discusses risk management practices related to information security management. The following points are addressed:
- Risk factors
- Generic threats and vulnerabilities
- Key risk indicators (KRIs)

## Relevance

An information security management process can only be effective and efficient if it is properly managed and based on a valid risk assessment.
Otherwise, the likelihood is very high that security investments are made in the wrong areas, inappropriate controls are implemented or that chosen solutions do not meet regulatory requirements. The risk practitioner must consider the following when assessing the risk factors related to information security management:
- Emerging risk factors
- New technology
- Changes in business strategy or culture
- Incident handling
- Response to previously identified risk
- Accountability and ownership for security management and risk

## Risk Factors

Examples of factors affecting information security management are:
- **Enterprise size and complexity**—Size, complexity, geographic range, number of users to be managed, types of devices to be secured, etc., all affect the ease with which information security management can be implemented and maintained.
- **Operational model:**
  - Infrastructure: insourced, facilities managed, outsourced—specific facility, outsourced—cloud
  - Organizational structure: in-sourced, outsourced (consultants), hybrid
  - Control environment: mature, immature, resilient
  - Applications source: in-house, purchased off the shelf, customized
  - Use or lack of use of an identity management system (i.e., single sign-on technology)
- **Risk management capabilities**—The enterprise's risk culture and integration of information security functions with risk management

## Generic Threats and Vulnerabilities

Generic threats and vulnerabilities related to information security management are:
- Management of information security:
  - A lack of information security governance
  - IT and business objectives are misaligned.
  - Unprotected data and information assets
- Information security plan:
  - The information security plan is not aligned with business requirements.
  - The information security plan is not cost effective.
  - The business is exposed to threats not covered in the security strategy.
  - Gaps exist between known risk and planned and implemented information security measures.
  - Users are not aware of the information security plan.
  - Security measures are compromised by stakeholders and users.
- Identity management:
  - Unauthorized changes are made to hardware and software.
  - Access management fails business requirements and compromises the security of business-critical systems.
  - Security requirements are not specified for all systems.
  - Segregation-of-duty violations
  - System information is compromised.

## Generic Threats and Vulnerabilities *(cont.)*

- User account management:
  - Security breaches exist.
  - Users fail to comply with security policy.
  - Incidents are not solved in a timely manner.
  - There is a failure to terminate unused accounts in a timely manner, thus impacting corporate security.
- Security, testing, surveillance and monitoring:
  - User accounts are misused or set up incorrectly, which compromises organizational security.
  - Undetected security breaches exist.
  - Security logs are unreliable or are not being monitored.
- Security incident definition:
  - Undetected security breaches exist.
  - No prescribed incident management procedure
  - Lack of information for performing counterattacks
  - Classification of security breaches is missing.
- Protection of security technology:
  - Keys are misused.
  - Nonverified users are registered, thus compromising system security.
  - Unauthorized access is available to cryptographic keys.
- Malicious software prevention, detection and correction:
  - Information is exposed.
  - Violations of legal and regulatory requirements occur.
  - Systems and data are prone to virus attacks.
  - Ineffective countermeasures exist.
- Network security:
  - Firewall rules do not reflect the enterprise's security policy.
  - Undetected unauthorized modifications to firewall rules are made.
  - The overall security architecture is compromised.
  - Security breaches are not detected in a timely manner.
- Exchange of sensitive data:
  - Sensitive information is exposed.
  - Inadequate physical security measures exist.
  - Unauthorized external connections to remote sites
  - Corporate assets are disclosed and sensitive information is accessible to unauthorized parties.

## KRIs

Examples of KRIs include:

- Excessive numbers of security incidents
- A vacant or underfunded chief information security officer (CISO) position
- A high turnover ratio or unfilled positions among information security staff
- Excessive security projects that are delayed or cancelled
- Actual information security expenditure significantly over or under budget
- Excessive user violations
- Numerous user access changes after quarterly managerial user privileges review
- Excessive time to investigate and close security violations
- Open security vulnerabilities identified by scans
- Malware attacks affecting operations
- Malware attacks requiring resources to clean up from events
- Excessive access rights that are authorized, revoked, reset or changed

# F. Information Systems Control Design, Monitoring and Maintenance

## Introduction

This section provides an overview of common controls related to information security management. The following points regarding IS control are addressed:
- Objectives and activities
- Metrics
- Monitoring practices

## IS Control Objectives and Activities

Key control activities are provided for the following aspects of information security management:
- Management of information security
- The information security plan
- Personnel management
- Physical security of IT assets
- Identity management
- User account management
- Security testing, surveillance and monitoring
- Security incident definition
- Protection of security technology
- Cryptographic key management
- Prevention, detection and correction of malware
- Network security
- Exchange of sensitive data
- Disaster recovery plans

## Key Control Activities Related to Management of Information Security

Key control activities may include:
- Define a charter for information security;
  - Define the security management function:
  - Scope and objectives
  - Responsibilities
  - Drivers (e.g., compliance, risk, performance)
- Confirm that the board, executive management and line management direct, and sign off on, the policy development process to ensure that the information security policy reflects the requirements of the business.
- Set up an adequate organizational structure and reporting line for information security, ensuring that the security management and administration functions have sufficient authority. Define the interaction with enterprise functions, particularly the control functions such as risk management, compliance, physical security, business continuity management and audit.
- Implement an information security management reporting mechanism, regularly informing the board, and business and IT management of the status of information security so that appropriate management actions can be taken:
  - Determine whether a security steering committee exists, with representation from key functional areas, including internal audit, human resources (HR), operations, information security and legal.
  - Determine whether a process exists to prioritize proposed security initiatives, including required levels of policies, standards and procedures.
  - Inquire whether, and confirm that, an information security charter exists.
  - Review and analyze the charter to verify that it refers to the organizational risk appetite relative to information security, and that the charter clearly includes:
    - Scope and objectives of the security management function
    - Responsibilities of the security management function
    - Compliance and risk drivers

**Key Control Activities Related to Management of Information Security *(cont.)***

- Inquire whether, and confirm that, the information security policy covers the responsibilities of board, executive management, line management, staff members and all users of the enterprise IT infrastructure and that it is supported by, and refers to, detailed security standards and procedures.
- Inquire whether, and confirm that, detailed security policy, standards and procedures exist. Examples of policies, standards and procedures include:
  - A security compliance policy
  - Management risk acceptance (acknowledgment of security noncompliance)
  - An external communications security policy
  - A firewall policy
  - An e-mail security policy
  - An agreement to comply with IS policies
  - A laptop/desktop computer security policy
  - An Internet usage policy
  - A mobile device (i.e., universal serial bus [USB], external hard drives) policy
- Inquire whether, and confirm that, an adequate organizational structure and reporting line for information security, and security incidents, exist, and assess whether the security management and administration functions have sufficient authority.
- Inquire whether and confirm that a security management reporting mechanism exists that informs the board and business and IT management of the status of information security.
- Ensure that the information security department supports compliance and audit functions.
- Ensure that the information technology projects are aligned with business needs.
- Ensure that information security is considered in all business processes and IT initiatives.

**Key Control Activities Related to the Information Security Plan**

Key control activities may include:

- Define and maintain an overall information security plan that includes:
  - A complete set of security policies and standards in line with the established information security policy framework
  - Procedures to implement, monitor and enforce the policies and standards
  - Ongoing risk evaluation
  - Roles and responsibilities
  - Staffing requirements
  - Security awareness and training
  - Enforcement practices
  - Investments in required security resources
- Collect information security requirements from IT tactical operations and projects for integration into the overall information security plan.
- Translate the overall information security plan into enterprise information security baselines for all major platforms and integrate it into the configuration baseline.
- Provide information security requirements and implementation advice to other processes, including the development of automated solution requirements for service level agreements (SLAs) and operating level agreements (OLAs), and the development of application software and IT infrastructure components.
- Communicate to all stakeholders and users in a timely and regular fashion updates of the information security strategy, plans, policies and procedures.
- Determine the effectiveness of the collection and integration of information security requirements into an overall information security plan that is responsive to the changing needs of the enterprise.
- Verify that the information security plan considers IT tactical plans, data classification, technology standards, security and control policies, risk management, and external compliance requirements.

## Key Control Activities Related to the Information Security Plan

- Ensure that there is an up-to-date and complete asset inventory of all IT systems, equipment, applications, configurations and documentation.
- Determine whether a process exists to periodically update the information security plan, and if the process requires appropriate levels of management review and approval of changes.
- Determine whether enterprise information security baselines for all major platforms are commensurate with the overall information security plan, if the baselines have been recorded in the configuration baseline (COBIT 4.1 DS9) central repository, and if a process exists to periodically update the baselines based on changes in the plan.

  **Note:** The COBIT 4.1 framework is available at no charge from ISACA and can be downloaded at *www.isaca.org/cobit*. The new COBIT 5 framework will be available in 2012.
- Determine whether a process exists to integrate information security requirements and implementation advice from the information security plan into other processes, including the development of automated solution requirements for SLAs and OLAs, and the development of application software and IT infrastructure components.

## Key Control Activities Related to Personnel Management

Key control activities may include:
- Ensure that all employees, contractors and anyone else accessing the systems of the enterprise are informed of security policies.
- Ensure that all staff attends periodic security awareness programs.
- Ensure that staff is trained in security according to job responsibilities.
- Ensure that incidents of misuse are investigated and reported to management for appropriate action.
- Ensure that hiring and termination practices include security practices such as nondisclosure agreements (NDAs), noncompete agreements and tracking of all equipment provided to the employee.

## Key Control Activities Related to Physical Security of IT Assets

Key control activities may include:
- Ensure that access is restricted to all data processing centers and areas containing sensitive information.
- Have visitor logs and controls over access to buildings and nonpublic areas.
- Have access controls, locks and tracing software on all equipment.

## Key Control Activities Related to Identity Management

Key control activities may include:

- Establish and communicate policies and procedures to uniquely identify, authenticate and authorize access mechanisms and access rights for all users on a need-to-know/need-to-have/least privilege basis, based on predetermined and preapproved roles. Clearly state the accountability of any user for any action on any of the systems and/or applications involved.
- Ensure that roles and access authorization criteria for assigning user access rights take into account:
  - Sensitivity of information and applications involved (data classification)
  - Policies for information protection and dissemination (legal, regulatory, internal policies and contractual requirements)
  - Roles and responsibilities as defined within the enterprise
  - The need-to-have access rights associated with the function
  - Standard, but individual, user access profiles for common job roles in the enterprise
  - Requirements to implement appropriate segregation of duties
- Establish a method for authenticating and authorizing users to establish responsibility and enforce access rights in line with sensitivity of information and functional application requirements and infrastructure components, and in compliance with applicable laws, regulations, internal policies and contractual agreements.
- Define and implement a procedure for identifying new users and recording, approving and maintaining access rights. This needs to be requested by user management, approved by the system and information owners and implemented by the responsible security person.
- Ensure that a timely information flow is in place that reports changes in jobs (i.e., people in, people out, job responsibility changes). Grant, revoke and adapt user access rights in coordination with HR and departments for users who are new, who have left the enterprise, or who have changed roles or jobs.

## Key Control Activities Related to User Account Management

Key control activities may include:

- Ensure that access control procedures include but are not limited to:
  - Using unique user IDs to enable users to be linked to, and held accountable for, their actions
  - Awareness that the use of group IDs results in the loss of individual accountability and is permitted only when justified for business or operational reasons and compensated by mitigating controls. Use of group IDs must be approved, monitored and documented.
  - Checking that the user has authorization from the system owner for the use of the information system or service, and the level of access granted is appropriate to the business purpose and consistent with the organizational security policy
  - A procedure to require users to understand and acknowledge their access rights and the conditions of such access
  - Ensuring that internal and external service providers do not provide access until authorization procedures have been completed
  - Maintaining a formal record, including access levels, of all persons registered to use the service
  - A timely and regular review of user IDs and access rights
- Ensure that management reviews or reallocates user access rights at regular intervals using a formal process. User access rights should be reviewed or reallocated after any job changes, such as transfer, promotion, demotion or termination of employment. Authorizations for special privileged access rights should be reviewed independently at more frequent intervals.

**Key Control Activities Related to Security Testing, Surveillance and Monitoring**

Key control activities may include:
- Implement monitoring, testing, review and other controls to:
  - Promptly prevent/detect errors in the results of processing
  - Promptly identify attempted, successful and unsuccessful security breaches and incidents
  - Detect security events and thereby prevent security incidents by using detection and prevention technologies
  - Determine whether the actions taken to resolve a breach of security are effective
- Conduct effective and efficient security testing procedures at regular intervals to:
  - Verify that identity management procedures are effective
  - Verify that user account management is effective
  - Validate that security-relevant system parameter settings are defined correctly and are in compliance with the information security baseline
  - Validate that network security controls/settings are configured properly and are in compliance with the information security baseline
  - Validate that security monitoring procedures are working properly
  - Consider, where necessary, obtaining expert reviews of the security perimeter

**Key Control Activities Related to Security Incident Definition**

Key control activities may include:
- Describe what a security incident entails. Document within the characteristics a limited number of impact levels to allow a commensurate response. Communicate and distribute information related to security incidents, or relevant parts, to previously identified people who need to be notified.
- Establish security incident response processes and incident response teams. Ensure that all team members have been training appropriately.
- Ensure that security incidents and appropriate follow-up actions, including root cause analysis, follow the existing incident and problem management processes.
- Define measures to protect confidentiality of information related to security incidents.

**Key Control Activities Related to Protection of Security Technology**

Protection of security technology activities may include:
- Ensure that all hardware, software and facilities related to the security function and controls, e.g., security tokens and encryptions, are tamperproof.
- Secure security documentation and specifications to prevent unauthorized access. However, do not make security of systems reliant solely on secrecy of security specifications.
- Make the security design of dedicated security technology (e.g., encryption algorithms) strong enough to resist exposure, even if the security design is made available to unauthorized individuals.
- Evaluate the protection mechanisms on a regular basis (at least annually) and perform updates to the protection of the security technology, if necessary.

### Key Control Activities Related to Cryptographic Key Management

Key control activities may include:

- Ensure that there are appropriate procedures and practices in place for the generation, storage and renewal of the root key, including dual custody and observation by witnesses.
- Make sure that procedures are in place to determine when any cryptographic key (especially the root key) renewal is required (e.g., the root key is compromised or expired).
- Create and maintain a written certification practice statement that describes the practices that have been implemented in the certification authority, registration authority and directory when using a public-key-based encryption system.
- Create cryptographic keys in a secure manner. When possible, enable only individuals not involved with the operational use of the keys to create the keys. Verify the credentials of key requestors (e.g., registration authority).
- Ensure that cryptographic keys are distributed in a secure manner (e.g., offline mechanisms) and stored securely:
  - In an encrypted form, regardless of the storage media used (e.g., write-once disk with encryption)
  - With adequate physical protection (e.g., sealed, dual custody vault), if stored on paper
- Create a process that identifies and revokes compromised keys. Notify all stakeholders as soon as possible of the compromised key.
- Verify the authenticity of the counterparty before establishing a trusted path.

### Key Control Activities Related to Prevention, Detection and Correction of Malware

Key control activities may include:

- Establish, document, communicate and enforce a malicious software prevention policy in the enterprise. Ensure that people in the enterprise are aware of the need for protection against malicious software, and their responsibilities.
- Install and activate malicious software protection tools on all processing facilities, with malicious software definition files that are updated as required (automatically or semi-automatically).
- Use both network-based and host-based malware tools.
- Distribute all protection software centrally (version and patch-level) using centralized configuration and change management.
- Regularly review and evaluate information on new potential threats.
- Filter incoming traffic, such as e-mail and downloads, to protect against unsolicited information (e.g., spyware, phishing e-mails).
- Ensure that suspicious activity is restricted to a sandbox or other quarantine area.

## Key Control Activities Related to Network Security

Key control activities may include:
- Establish, maintain, communicate and enforce a network security policy (e.g., provided services, allowed traffic, types of connections permitted) that is reviewed and updated on a regular basis (at least annually).
- Establish and regularly update the standards and procedures for administering all networking components (e.g., core routers, demilitarized zone [DMZ], virtual private network [VPN] switches, wireless).
- Properly secure network devices with special mechanisms and tools (e.g., authentication for device management, secure communications, strong authentication mechanisms). Implement active monitoring and pattern recognition to protect devices from attack.
- Configure operating systems with minimal features enabled (e.g., features that are necessary for functionality and are hardened for security applications). Remove all unnecessary services, functionalities and interfaces (e.g., graphical user interface [GUI]). Apply all relevant security patches and major updates to the system in a timely manner.
- Plan the network security architecture (e.g., DMZ architectures, internal and external network segmentation, intrusion prevention and detection systems [IPS/IDS] placement, and wireless equipment) to address processing and security requirements. Ensure that documentation contains information on how traffic is exchanged through systems and how the structure of the enterprise's internal network is hidden from the outside world (i.e., network address translation [NAT]).
- Subject devices to reviews by experts who are independent of the implementation or maintenance of the devices.

## Key Control Activities Related to Exchange of Sensitive Data

Key control activities may include:
- Determine, by using the established information classification scheme, how the data should be protected when exchanged.
- Apply appropriate application controls to protect the data exchange.
- Apply appropriate infrastructure controls, based on information classification and technology in use, to protect the data exchange.

## Key Control Activities Related to Disaster Recovery Plans

Key control activities may include:
- Ensure that disaster recovery plans have been created for all critical systems and networks.
- Review training, maintenance and distribution procedures for all disaster recovery plans to ensure that plans are up-to-date and will work effectively in a crisis.
- Ensure that backups of all data, configurations, documentation and applications are being conducted and tested for completeness.

## IS Control Metrics for Monitoring Information Security Management

The following table provides an overview and examples of IS control metrics for monitoring information security management.

| Monthly Systems Security Management Monitoring | | | | | |
|---|---|---|---|---|---|
| **Attribute** | **Target** | **Current Period** | **Prior Period** | **Prior Period − 1** | **Prior Period − 2** |
| Security incidents by type: | | | | | |
| User | | | | | |
| Application/operating system | | | | | |
| Vendor-related incidents | | | | | |
| Malware infections | | | | | |
| Management | | | | | |
| Network security | | | | | |
| Information security personnel headcount/turnover | | | | | |
| Percent of projects delayed or cancelled | | | | | |
| Percent of budget vs. expenditure | | | | | |
| Number of user access violations | | | | | |
| Number of accounts not modified after user transfer | | | | | |
| Number of accounts not disabled after employee termination | | | | | |
| Number of contractor accounts not disabled after contract closure | | | | | |
| Number of security incidents involving sensitive data | | | | | |
| Excessive time to investigate and close security incidents | | | | | |
| Number of security incidents identified by vulnerability testing | | | | | |
| Number of malware attacks | | | | | |
| Hours required to remediate malware attacks | | | | | |

## G. The Practitioner's Perspective

### Introduction

The following section provides an overview of how the five CRISC domains (listed below) relate to information security management in practice:

- Domain 1—Risk Identification, Assessment and Evaluation
- Domain 2—Risk Response
- Domain 3—Risk Monitoring
- Domain 4—Information Systems Control Design and Implementation
- Domain 5—Information Systems Control Monitoring and Maintenance

### Domain 1

Identification is the process of transforming uncertainties and issues related to how well an enterprise's assets are being protected into distinct (tangible) areas of risk. The objective of this activity is to anticipate risk before it becomes a problem and to incorporate this information into the enterprise's information security risk management process.

Risk identification, assessment and evaluation are ongoing processes. Initially, risk identification establishes the baseline of information security requirements, prioritization of activities, and an execution plan. Information security management is all about risk. Risk changes as technology changes and new vulnerabilities are identified. A structured approach to risk identification is required. This process should include IT operations, the business risk management and other stakeholders.

Risk assessment should be initiated as conditions change; if no change is identified, risk assessment should be reviewed annually.

### Domain 2

The risk response will be dependent on the risk identified.

The enterprise should develop risk mitigation strategies—including the development of managerial controls (such as policy), technical controls (such as access control) and physical controls (such as fire prevention) where appropriate in order to reduce the level of risk to an acceptable level.

An emergency response team should be established to address security incidents and report them through the problem/incident management system. This system should have a specific category for information security issues.

Anticipated issues should have pre-established responses, with practice drills, documentation of the event, and review of the process after the event.

### Domain 3

The key risk indicators (KRIs) described previously are an example of metrics that can identify trends and identify new risk affecting the enterprise. KRIs should be analyzed and reported to management as part of an ongoing monitoring process. In addition, where significant risks are identified, the risk assessment process should be initiated.

### Domain 4

The suggested controls described in the IS control objectives and activities provides a best practice resource for designing and implementing effective systems security controls.

**Domain 5**

Control monitoring should be integrated into the information security management and the incident reporting systems.

Timely reporting using KRIs defined previously should be distributed to stakeholders within IT and the business units affected.

Annual evaluation of the maturity of controls provides a barometer of controls in their current state, comparison to previous periods, and the target maturity level. Target maturity levels should be agreed on by stakeholders and evaluated by senior IT and business operations management as part of the IT annual review.

## H. Suggested Resources for Further Study

### Suggested Resources for Further Study

In addition to the resources cited throughout this manual, the following resources are suggested for further study:

- ISACA:
  - COBIT 4.1, 2007
    **Note:** The COBIT 4.1 framework is available at no charge from ISACA and can be downloaded at *www.isaca.org/cobit*. The new COBIT 5 framework will be available in 2012.
  - *The Risk IT Framework*, 2009
  - *The Risk IT Practitioner Guide*, 2009
  - *Enterprise Value: Governance of IT Investments, The Val IT Framework 2.0*, 2008
  - *Implementing and Continually Improving IT Governance*, 2010
- Alberts, Christopher; Audrey Dorofee; *Managing Information Security Risks: The OCTAVE*$^{SM}$ *Approach*, Addison-Wesley Professional, USA, 2002
- ISO, ISO/IEC 27002:2005, *Code of Practice for Information Security Management*, Switzerland, 2005

Page intentionally left blank

# 7. Configuration Management

## A. Chapter Overview

### Introduction

This chapter provides an overview of configuration management and:
- Explains its importance to achieving business objectives
- Outlines a high-level process overview
- Introduces related key concepts
- Provides examples of common risk
- Lists selected key risk indicators (KRIs)
- Provides examples of common IS controls supporting the process
- Describes the practitioner's perspective
- Offers suggested reading materials and references

### Learning Objectives

The CRISC candidate should have a general understanding of configuration management processes and how configuration management interrelates with risk management.

### Contents

This chapter contains the following topics:

## *B. Related Knowledge Statements*

### Contents

The following table lists the applicable knowledge statements from the CRISC job practice.

| No. | Knowledge Statement (KS)<br>Knowledge of: |
|---|---|
| KS1.9 | Information systems architecture (e.g., platforms, networks, applications, databases and operating systems) |
| KS4.5 | Information systems architecture (e.g., platforms, networks, applications, databases and operating systems) |
| KS5.8 | Control objectives, activities and metrics related to information systems architecture (e.g., platforms, networks, applications, databases and operating systems) |

## C. Key Terms and Principles

### Introduction

This section introduces terms and principles related to configuration management as well as terms that help relate the process to other key business processes.

### Definition of Change Management

Effective change management—within the IT service domain—helps ensure that changes to the IT infrastructure are applied using standardized methods and procedures to minimize the number and impact of any related incidents upon service. Change management comprises that aspect of the overall IT governance framework which is enterprise sensitive and deals with IT configuration, release and problem management issues.

> **Note:** Changes in the IT infrastructure may arise as part of project, program or service improvement initiatives, based on problems related to IT services or in response to changing legislative requirements.

### Exhibit 7.1: Relationship Among Change Management, Configuration Management, Release Management and Problem Management

**Exhibit 7.1** describes the relationship among change management, configuration management, release management and problem management.

**Exhibit 7.1: Relationship Among Change Management, Configuration Management, Release Management and Problem Management**

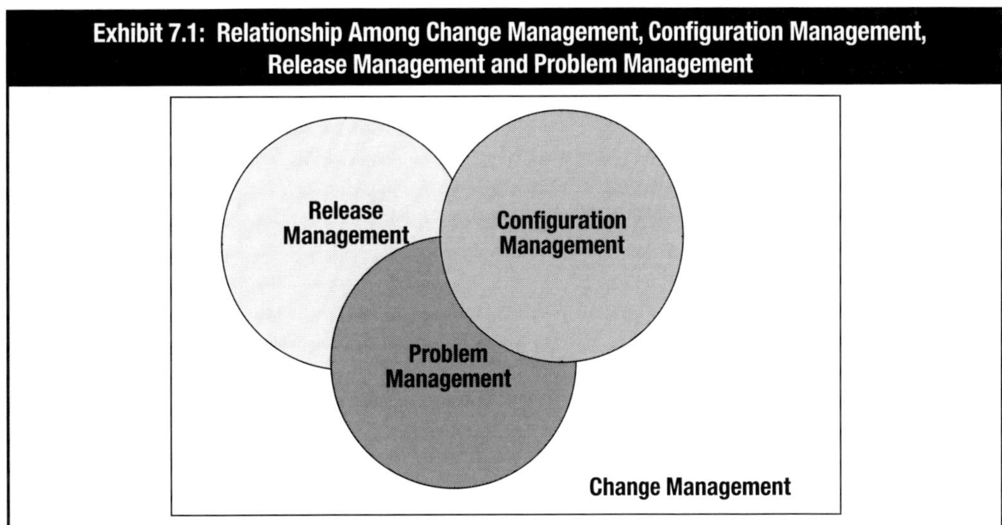


Change management should be identified as the governance framework of processes and controls within which configuration management (baseline definition and maintenance control), release management (software and hardware release controls) and problem management (issue resolution) take place.

### Definition of Release Management

Release management focuses on the software release life cycle processes, ranging from the initial development of a piece of software to its eventual release, and creating updated versions of the released version to help improve the software or to fix existing software bug(s). This terminology can also be applied to the hardware acquisition, deployment, upgrade and retirement processes.

### Definition of Configuration Management

The control of changes to a set of configuration items over a system life cycle

Configuration management focuses on the processes relating to the establishment and maintenance of hardware and software baselines, hardware versions and models, software versions (commercial off-the shelf software [COTS]) and release baselines (changes to internally developed systems) and related service agreements.

| | |
|---|---|
| **Definition of Problem Management** | Systematic assessment and resolution through a defined set of IT change and control activities, of technical and procedural issues arising from the use of IT systems in a production environment |
| **Definition of Configuration Item (CI)** | Component of an infrastructure—or an item, such as a request for change, associated with an infrastructure—which is (or is to be) under the control of configuration management<br><br>**Note:** CIs may vary widely in complexity, size and type, from an entire system (including all hardware, software and documentation) to a single module or a minor hardware component. |

## D. Process Overview

### Introduction

This section introduces the configuration management process and its importance to the achievement of business objectives.

### Relevance

Ensuring the integrity of hardware and software configurations requires the establishment and maintenance of an accurate and complete inventory of all information technology assets (i.e., software, databases, routers, switches, firewalls, operating systems, applications and communications devices) and a configuration repository (configuration management database [CMDB]). This process includes collecting initial configuration information, establishing baselines, verifying and auditing configuration information, and updating the configuration repository as needed. Effective configuration management facilitates greater system availability, minimizes production issues and resolves issues more quickly.

Uncontrolled changes to configurations may result in unauthorized access to systems, networks or data, errors in processing, unavailability of systems and loss of data integrity.

For many enterprises, configuration management also includes patch and vulnerability management and change management.

### Process Objectives

Key steps of the configuration management process are:
- Develop a configuration management policy.
- Ensure that access to administration and configuration management of systems is carefully restricted and monitored.
- Develop configuration management planning procedures.
- Collect initial configuration information and establish configuration baselines.
- Verify and audit configuration information (including detection of unauthorized hardware and software and unauthorized changes to hardware and software configurations).
- Update the configuration repository.

# E. Risk Management Considerations

## Introduction

This section discusses risk management practices related to the configuration management process. The following points will be addressed:
- Risk factors
- Generic threats and vulnerabilities
- Key risk indicators (KRIs)

## Risk Factors

Examples of factors affecting the configuration management process are:
- **Enterprise size and complexity**—Size, complexity, geographic range, international borders, industry, regulatory requirements, etc., all affect the ease with which configuration management can be implemented and maintained.
- **Operational model:**
  - Infrastructure: insourced, facilities managed, outsourced—specific facility, outsourced—cloud
  - Organizational structure: insourced, outsourced (consultants), hybrid
  - Infrastructure design: mainframe, distributed, virtualized
  - Control environment: mature, immature, resilient
  - Application source: in-house, purchased off the shelf, customized
  - Application portfolio: managed, *ad hoc*
  - Hardware/software purchasing: centralized, decentralized
  - Budgetary responsibility: centralized, decentralized
  - Hardware/software acquisition: centralized, decentralized
- **Risk management capabilities:**
  - The enterprise risk management (ERM) function extends to the IT function or the IT function operates in its own silo
  - The absence of ERM, the maturity of IT risk management processes
  - Risk tolerance of IT and business
  - IT risk levels aligned with business risks
- **Portfolio management capability**—The alignment of the application portfolio with the hardware and systems software architecture
- **Staff expertise**—The skill and knowledge levels of staff responsible for administration of devices

## Generic Threats and Vulnerabilities

Generic threats and vulnerabilities related to the configuration management process are:
- Uncontrolled change management of configuration, causing business disruptions
- Inability to accurately account for assets
- Failure of changes to comply with the overall technology architecture
- Unauthorized changes to hardware and software that are not discovered, which could result in security breaches
- Assets that are not properly protected
- Documented information that fails to reflect the current architecture
- Inability to fall back to the baseline configuration
- Inability to assess the impact of a change because of inaccurate information
- Failure to identify business-critical components
- Misused assets
- Increased costs for problem solving

## KRIs

Examples of KRIs are the:
- Number of business compliance issues caused by improper configuration of assets
- Number of deviations identified between the configuration repository and actual asset configurations
- Percent of licenses purchased and not accounted for in the repository
- Average time period (lag) between identifying a discrepancy and rectifying it
- Number of discrepancies relating to incomplete or missing configuration information
- Percent of configuration items in line with service levels for performance, security and availability
- Excessive number of unique workstation images
- Number of configuration deviations or policy overrides to approved architecture
- Time between the release of vendor patches and deployment across the enterprise
- Number of staff that have administrator level access to devices

# F. Information Systems Control Design, Monitoring and Maintenance

## Introduction

This section provides an overview of common controls related to configuration management. The following points regarding IS control are addressed:
- Objectives and activities
- Metrics
- Monitoring practices

## IS Control Objectives and Activities

Key control activities for configuration management are:

Policies and Procedures:
- Define and implement a policy requiring that all configuration items and their attributes and versions be identified and maintained.
- Establish a configuration policy and baseline for each operating platform and operating function (database server, file server, web server, etc.). Define a policy that integrates incident, change and problem management procedures with the maintenance of the configuration repository.

Asset Inventory:
- Record all assets—including new hardware and software, procured or internally developed—within the configuration management data repository.
- Tag physical assets according to a defined policy. Consider using an automated mechanism, such as barcodes.
- Define a process to identify critical configuration items in relationship to business functions (component failure impact analysis).
- Provide a unique identifier to a configuration item so the item can be easily tracked and related to physical asset tags and financial records.
- Implement a configuration repository to capture and maintain configuration management items.
- Implement a tool to enable the effective logging of configuration management information within a repository.

Configuration Baseline:
- Define and document configuration baselines for components across development, test and production environments, to enable identification of system configuration at specific points in time (past, present and planned).
- Establish an approved configuration baseline for each operating platform and operating function.
- Establish a process to revert to the baseline configuration in the event of problems, if determined appropriate after an initial investigation.

Change Monitoring Management:
- Install mechanisms to monitor changes against the defined repository and baseline. Provide management reports for exceptions, reconciliation and decision making.
- Define a process to record new, modified and deleted configuration items and their relative attributes and versions. Identify and maintain the relationships between configuration items in the configuration repository.
- Establish a process to maintain an audit trail for all changes to configuration items.
- To validate the integrity of configuration data, implement a process to ensure that configuration items are monitored. Compare recorded data against actual physical existence, and ensure that errors and deviations are reported and corrected.
- Using automated discovery tools where appropriate, reconcile actual installed software and hardware periodically against the configuration database, license records and physical tags.

## IS Control Objectives and Activities *(cont.)*

Compliance:
- Define and implement a process to ensure that valid licenses are in place to prevent the inclusion of unauthorized software.
- Periodically review (against the policy for software usage) the existence of any software in violation or in excess of current policies and license agreements. Report deviations for correction.

## IS Control Metrics for Monitoring

The following tables provide an overview and examples of IS controls metrics for monitoring configuration management.

| Monthly Configuration Monitoring | | | | | |
|---|---|---|---|---|---|
| **Attribute** | **Target** | **Current Period** | **Prior Period** | **Prior Period −1** | **Prior Period −2** |
| Total incident/problem reports for the period | | | | | |
| Total incident/problem reports identified as configuration issues for the period | | | | | |
| Total incidents/problems identified as both configuration and high-priority issues and/or requiring off-hours remediation | | | | | |
| Average time (hours) between identification and remediation of configuration issues | | | | | |
| Total down time due to configuration issues | | | | | |
| Total incidents affecting compliance | | | | | |
| Number of service level agreements (SLAs) not achieved due to configuration issues | | | | | |
| Number of high-priority SLAs not achieved due to configuration issues | | | | | |
| Percent of work stations monitored during the period for unauthorized or unlicensed software | | | | | |
| Percent of work stations discovered to have unauthorized or unlicensed software during the period | | | | | |

**IS Control Metrics for Monitoring *(cont.)***

| Annual Configuration Monitoring | | | |
|---|---|---|---|
| **Attribute** | **Target** | **Current Period** | **Prior Period** |
| Percent of total asset value inventoried | | | |
| Percent of asset value requiring adjustment | | | |
| Percent of servers (mainframe, mid-range, and dedicated servers) audited during the year for baseline compliance | | | |
| Percent of servers not in baseline compliance | | | |
| Percent of configuration changes supported and authorized by an appropriate change request form | | | |
| Percent of configuration changes supported by problem/incident tickets | | | |
| Number of high-priority SLAs not achieved due to configuration issues | | | |

## G. The Practitioner's Perspective

### Introduction

The following section provides an overview of how the five CRISC domains (listed below) relate to configuration management in practice:
- Domain 1—Risk Identification, Assessment and Evaluation
- Domain 2—Risk Response
- Domain 3—Risk Monitoring
- Domain 4—Information Systems Control Design and Implementation
- Domain 5—Information Systems Control Monitoring and Maintenance

### Domain 1

Configuration management focuses on:
- Achieving processing service level agreements (SLAs)
- Secure computing
- Achieving compliance
- Responding to events to minimize operational disruption
- Tracking assets

The risk identification, assessment and evaluation process addresses those issues that would preclude the achievement of the above focus areas.

The initial identification of risk should be integrated into the systems architecture, hardware/software acquisition, and systems development processes. Assessment and evaluation can be performed as part of the initial process; however, configuration risk is also associated with subsequent reconfiguration of the assets, requiring risk monitoring (Domain 3) as well as reassessment of Domain 1. The risk identification process should include an annual review of previous assessments and evaluations for material changes affecting the risk profile. Any changes that have been made to a system without proper documentation and approval should be thoroughly investigated and reported to management.

The practitioner should focus attention on new operating systems introduced into the environment, new operating platforms (i.e., Windows Servers dedicated to e-mail, web services, database management), virtualization, network design and modification (firewalls, intrusion detection, etc.), and system software patches, updates or upgrades.

## Domain 2

The risk response will be based on the risk identification, assessment and evaluation. Configuration management has an inherent risk in the dynamic nature of the controls. Many configuration-based controls can be circumvented intentionally or accidentally in the course of normal maintenance. Since these issues may be a result of human error or software "bugs," a pre-planned response process for high-risk and/or high-probability issues should be established, documented and appropriate practice drills implemented.

Systems should be hardened against threats by disabling unneeded services. An error in the configuration of a system or device may present an opportunity for a successful attack. Therefore, all changes must be reviewed and gaps in the configuration management process remedied as quickly as possible.

As vendors become aware of threats, they will release patches to close any vulnerability that the threat could exploit. For this reason, it is critically important for an enterprise to test and deploy patches through an approved process as quickly as possible.

Risk responses require a formal approach to issues, opportunities and events to ensure that solutions are in alignment with the business objectives and are cost effective. The following should be considered:

- When preparing the risk response, identify the risks in business terms: loss of productivity, disclosure of confidential information, lost opportunity costs, etc.
- Understand the business risk appetite, acceptable service interruptions, confidentiality of data, compliance requirements, etc.
- Keep the business stakeholders apprised of identified risks and how IT will respond to these risks. In defining the response, be specific and describe it in nontechnical terms. Transparency is key.
- The risk response requiring major changes to system configuration should be documented to describe the justification for the alternative selected, and how the changes support the business objective.
- The impact of the configuration change on the business environment and the need to have a maintenance window to allow for the deployment of the new configuration
- When building the response, consider how the response will be measured (leading to Domain 3: Risk Monitoring).

## Domain 3

Risk monitoring provides the communication between the affected stakeholders. Most configuration incidents do not suddenly occur. There are symptoms, sometimes ignored. The function of risk monitoring is to minimize this occurrence through effective key risk indicators (KRIs). The KRIs cited in this document are generic. The identification of meaningful KRIs is essential to transparency and effective communication with stakeholders and requires attention:

- The KRI should describe the risk issue and provide a trend line to identify performance patterns and measure remediation processes.
- An issue monitoring process should be implemented to provide IT and the stakeholders with an understanding of open issues, estimated issue closure dates, a means to prioritize and track issue resolution, and to provide status reports to interested parties.

**Domain 4**

Risk frameworks require effectively designed controls that reduce inherent risk to a residual risk level that is within levels and costs acceptable to the stakeholders:

- Control design must be a part of the systems development, acquisition and maintenance processes.
- Configuration requirements should be defined during the system development life cycle (SDLC) requirements definition. The costs associated with processes required to achieve the requirements should be defined and accepted by the business. This may include detection systems or services, additional monitoring and change management processes, etc.
- Automated tools should be designed to report when configuration changes are implemented and these changes should be matched to configuration change requests.
- The user acceptance tests of new or modified systems should include configuration tests to ensure that the settings and approved processes (services, started tasks and other utilities) are according to design.
- KRIs designed to monitor the key configuration processes should be included in the control design.

**Domain 5**

Effective monitoring of configuration management, operations and changes should be built into an effective monitoring system:

- Timely reporting using KRIs defined previously should be distributed to stakeholders within IT and the business units affected.
- Annual evaluation of the maturity of controls provides a barometer of controls in their current state, comparison to previous periods, and the target maturity level. Target maturity levels should be agreed on by stakeholders and evaluated by senior management as part of the IT annual review.
- Maturity assessment can be performed as a self-assessment, with oversight by an objective third party, peer review or independent assessment (internal audit or external provider).
- Configuration-related issue monitoring should be reported and evaluated routinely.

## H. Suggested Resources for Further Study

### Suggested Resources for Further Study

In addition to the resources cited throughout this manual, the following resources are suggested for further study:

- ISACA:
  - COBIT 4.1, 2007
    **Note:** The COBIT 4.1 framework is available at no charge from ISACA and can be downloaded at *www.isaca.org/cobit*. The new COBIT 5 framework will be available in 2012.
  - *The Risk IT Framework*, 2009
  - *The Risk IT Practitioner Guide*, 2009
  - *Enterprise Value: Governance of IT Investments, The Val IT Framework 2.0*, 2008
  - *Implementing and Continually Improving IT Governance*, 2010

# 8. Problem Management

## A. Chapter Overview

### Introduction

This chapter provides an overview of problem management and:
- Explains its importance to achieving business objectives
- Outlines a high-level process overview
- Introduces related key concepts
- Presents examples of common risk
- Lists selected key risk indicators (KRIs)
- Provides examples of common IS controls supporting the process
- Describes the practitioner's perspective
- Offers suggested reading materials and references

### Learning Objectives

The CRISC candidate should have a general understanding of the problem management process and how problem management interrelates with risk management.

### Contents

This chapter contains the following topics:

## B. Related Knowledge Statements

### Contents

The following table lists the applicable knowledge statements from the CRISC job practice.

| No. | Knowledge Statement (KS)<br>Knowledge of: |
|---|---|
| KS5.6 | Control objectives, activities and metrics related to incident and problem management |

## *C. Key Terms and Principles*

**Introduction**

This section introduces terms and principles related to problem management as well as terms that help relate the process to other key business processes.

**Definition of Problem**

In IT, the unknown underlying cause of one or more incidents

**Definition of Problem Escalation Procedure**

The process of escalating a problem up from junior to senior support staff, and ultimately to higher levels of management

**Definition of Problem Management**

Problem management includes the activities required to diagnose the root cause of incidents and to determine the resolution to underlying problems.

> **Note:** It is also responsible for ensuring that the resolution is implemented through the appropriate control procedures, especially change management and release management.

## D. Process Overview

### Introduction

This section introduces the problem management process and its importance to the achievement of business objectives.

### Relevance

Effective problem management requires:
- Identification and classification of problems
- Root cause analysis
- Determining an approved solution
- Assignment of resolution procedures through case management
- Monitoring of the resolution process
- Escalation to appropriate functions, where necessary
- Closure of the case
- Analysis of cases for process improvement

An effective problem management process maximizes system availability, improves service levels, reduces costs, and improves customer convenience and satisfaction.

### Process Phases

Key phases of the problem management process are:
1. Identify and classify problems.
2. Perform root cause analysis.
3. Determine an approved solution.
4. Assign resolution to a specific team or individual.
5. Monitor the status of resolution.
6. Escalate resolution issues, where necessary.
7. Complete remediation and close the case.
8. Maintain problem records.
9. Review problems for trends, process reengineering and training opportunities.

## E. Risk Management Considerations

### Introduction

This section discusses risk management practices related to the problem management process. The following points are addressed:
- Risk factors
- Generic IT risks and their potential root causes
- Key risk indicators (KRIs)

### Relevance to Risk Management

Every enterprise will encounter problems and incidents that affect normal business operations and the ability of the enterprise to meet its goals and objectives. Problem management ensures that the enterprise is able to identify the root cause of incidents, prepared to address problems through an incident management process, and then also able to detect and react to any incident in an effective manner. This should work to contain the incident and prevent the degree of damage from expanding to a point where the very existence of the enterprise may be in jeopardy. As can be quoted from the United States White House Executive Order 13231, Critical Infrastructure Protection in the Information Age (October 2001), "any disruptions that occur are infrequent, of minimal duration and manageable, and cause the least damage possible."

The controls chosen, monitored, and reported on, by the risk practitioner, including the problem management process, are based on the likelihood and potential impact that a problem could have on the enterprise.

### Risk Factors

Examples of factors affecting the problem management process are:
- **Enterprise size and complexity**—Size, complexity, geographic range, etc., all affect the ease with which problem management can be implemented and maintained.
- **Operational model:**
  - Infrastructure: insourced, facilities managed, outsourced—specific facility, outsourced—cloud
  - Organizational structure: insourced, outsourced (consultants), hybrid
  - Infrastructure design: mainframe, distributed, virtualized
  - Control environment: mature, immature, resilient
  - Applications source: in-house, purchased off the shelf, customized
  - Application portfolio: managed, *ad hoc*
  - Hardware/software purchasing: centralized, decentralized
  - Budgetary responsibility: centralized, decentralized
  - Hardware/software acquisition: centralized, decentralized
- **Risk management capabilities:**
  - The enterprise risk management (ERM) function extends to the IT function or the IT function operates in its own silo.
  - In the absence of ERM, the maturity of IT risk management processes
  - Risk tolerance of IT and the business
  - IT risk levels aligned with business risks
- **Portfolio management capability**—The diligence with which business cases are assessed
- **Risk culture**—The risk-related patterns of behaviors, assumptions and beliefs that exist throughout an enterprise, together with the geographic location, political and economic climate, etc.

## Generic Risks

Generic risks and potential root causes related to the problem management process are:
- Financial loss
- Impact on employee morale
- Disruption of IT and business services
- Loss of information
- Increased likelihood of, and actual, problem recurrence
- Problems and incidents not solved in a timely manner, not satisfying service level agreements (SLAs) or regulatory requirements
- Lack of audit trails of problems, incidents and their solutions for proactive problem and incident management
- Dissatisfaction and/or decreased confidence with/in IT services
- Increasing number of problems
- Unplanned expenses

## KRIs

Examples of KRIs are the:
- Excessive time to detect and react to a problem
- High number of recurring problems categorized as critical with an impact on the business (customers, core functions)
- High number of problems originating from a single functional area (applications development, scheduling, end user, technical support, information security, third party)
- Repeated problems originating from the same root cause
- High number of problems not resolved according to the SLA
- High number of problems requiring escalation
- High number of emergency changes that were initiated to fix a problem introduced directly after a change

# F. Information Systems Control Design, Monitoring and Maintenance

## Introduction

This section provides an overview of common controls related to problem management. The following points regarding IS controls are addressed:
- Objectives and activities
- Metrics
- Monitoring practices

## IS Control Objectives and Activities

Key control activities for problem management are:
- Policy
- Process
- Monitoring

## Policy

Key control activities include:
- Establish and maintain a single problem management system to register and report problems identified and to establish audit trails of the problem management processes, including the status of each problem (i.e., open, reopened, in progress, closed). The system should register each problem, including the:
  - Needed information to understand the problem
  - Relevant documentation of the problem
  - Contact persons
  - Time the problem was identified
  - Method by which the problem was identified/detected
  - Known consequences/impact
  - Actual problem owner (department)
  - Workaround performed, if any
  - How and when of the solutions implemented
  - Identification of the root cause
  - Lessons learned
  - Status of problem
- Identify and initiate sustainable solutions (permanent fix) addressing the root cause, and initiate change requests via the established change management process.
- Define priority levels through consultation with the business to ensure that problem identification and root cause analysis are handled in a timely manner according to the agreed-on service level agreements (SLAs). Base priority levels on business impact and urgency.
- Define appropriate support groups to assist with problem identification, root cause analysis and solution determination to support problem management. Determine the support groups based on predefined categories such as hardware, network, software, applications and support software.

## Process

Key control activities include:

- Define and implement a problem-handling process that has access to all relevant data, including information from the change management system and IT configuration/asset and incident details, to effectively address the root cause(s).Identify problems through the correlation of incident reports, error logs and other problem identification resources. Determine priority levels and categorization to address problems in a timely manner.
- Define problem ownership and the team responsible for problem management and reporting. Team members may include:
  - A senior management representative
  - Information technology experts
  - Representatives of the business units
  - Communications and public relations
  - Human resources (HR)
  - Legal
  - Internal audit
  - Health and safety
  - Outside experts

> **Note:** Not all team members will be contacted for each incident, but the people required are identified and can be deployed to the problem location as quickly as possible. This requires a communication system that can reach each team member as needed. Team members should be chosen based on the skills, knowledge or authority that they will provide the team. All team members should be aware of the need to maintain confidentiality of the incident.

- Ensure that process owners and managers from problem, change and configuration management meet regularly to discuss future planned changes.
- Inform the service desk (so users and customers can be informed) of the schedule of problem closure—e.g., the schedule for fixing the known errors, the possible workaround or the fact that the problem will remain until the change is implemented—and the consequences of the approach taken.
- Develop and implement a process to capture problem information related to IT changes and communicate it to key stakeholders. This communication could take the form of reports to and periodic meetings among problem, change and configuration management process owners to consider recent problems and potential corrective actions.
- Define and implement a process to close problem records either after confirmation of successful elimination of the known error or after agreement with the business on how to alternatively handle the problem.
- To maximize resources and reduce turnaround, define and implement problem management procedures for the tracking of problem trends.
- Produce reports to monitor the problem resolution against the business and customer SLAs. Ensure the proper escalation of problems—e.g., escalation to a higher management level according to agreed-on criteria, contacting external vendors or referring to the change advisory board—to increase the priority of an urgent request for change to implement a temporary workaround.
- Identify and implement a process for problems to be assigned and analyzed in a timely manner to determine the root cause. Identify problems by comparing incident data with the database of known and suspected errors (e.g., those communicated by external vendors). Upon successful root cause identification, classify problems as known errors. Associate the affected configuration items to the established/known error.

## Monitoring

Key control activities include:
- To determine the overall improvement of availability of IT services, monitor changes resulting from the problem management process. Monitor how the results of the problem management process decrease repeat incidents and reactive support requirements.
- Produce reports to communicate the progress in resolving problems and to monitor the continuing impact of problems not solved. Monitor the status of the problem handling process throughout its life cycle, including input from change and configuration management.
- To enable the enterprise to monitor the total costs of problems, develop and implement a process to capture change efforts resulting from problem management process activities (e.g., fixes to problems and known errors) and report on them.
- Report the status of identified problems to the service desk so customers and IT management can be kept informed.

## IS Control Metrics for Measuring Performance

The following IS control metrics measure the performance of problem management:
- Number of recurring problems categorized as critical with an impact on the business (customers, core functions)
- Number of problems by functional responsibility (applications, scheduling, end user, technical support, information security, third party)
- Number of open/new/closed problems, by severity
- Percent of problems that recur (within a time period), by severity
- Percent of problems resolved according to the SLA
- Percent of problems not resolved according to the SLA
- Percent of problems requiring escalation
- Aging of open items, problems not yet closed off
- Average and standard deviation of time lag between problem identification and resolution
- Average and standard deviation of time lag between problem resolution and closure
- Average duration between the logging of a problem and the identification of the root cause
- Number of emergency changes that were invoked to fix a problem
- Number of changes required due to previous incidents that are not yet completed

## IS Control Metrics for Monitoring

The following table provides an overview and examples of IS control metrics for monitoring problem management.

| Monthly Problem Management Monitoring | | | | | |
|---|---|---|---|---|---|
| **Attribute** | **Target** | **Current Period** | **Prior Period** | **Prior Period − 1** | **Prior Period − 2** |
| Total incident/problem reports for period | | | | | |
| Number of recurring problems categorized as critical with an impact on the business (customers, core functions) | | | | | |
| Number of problems by functional responsibility (applications, scheduling, end user, technical support, information security, third party) | | | | | |
| Number of open/new/closed problems, by severity | | | | | |
| Percent of problems that recur (within a time period), by severity | | | | | |
| Percent of problems resolved according to the SLA | | | | | |
| Percent of problems not resolved according to the SLA | | | | | |
| Percent of problems requiring escalation | | | | | |
| Aging of open items | | | | | |
| Average and standard deviation of time lag between problem identification and resolution | | | | | |
| Average and standard deviation of time lag between problem resolution and closure | | | | | |
| Average duration between the logging of a problem and the identification of the root cause | | | | | |
| Number of emergency changes that were invoked to fix a problem | | | | | |

## G. The Practitioner's Perspective

### Introduction

Problem management is a key process in the mitigation of risk. The problem management process identifies, categorizes and follows an issue from its origination to closure. Because problems may originate from several sources within the enterprise, a central clearinghouse of all incidents is critical to the assurance that all issues have been remediated in a timely manner.

Outsourcing adds a level of complexity. Communication between the customer and servicer requires documentation, monitoring and management oversight. In addition, escalation processes need to be established where normal processes are not effective. The following section provides an overview of how the five CRISC domains (listed below) relate to problem management in practice:

- Domain 1—Risk Identification, Assessment and Evaluation
- Domain 2—Risk Response
- Domain 3—Risk Monitoring
- Domain 4—Information Systems Control Design and Implementation
- Domain 5—Information Systems Control Monitoring and Maintenance

### Domain 1

Problem management focuses on:

- Identifying issues that could affect the ability of the enterprise to perform operational processes
- Prioritizing problems and determining remediation actions
- Ensuring the monitoring of the status of remediation and closure activities

Once a problem has been identified, the risk identification, assessment and evaluation function is the determinant used to establish the prioritization of the issue, the solution to be implemented and the process to be initiated for remediation. Domain 1 is the key element of problem management.

### Domain 2

The risk response will be based on the risk identification, assessment and evaluation. Problem management has an inherent risk in the dynamic nature of the controls. Most problems are a result of inadequate testing, ineffective internal program quality controls, poor internal systems design, software "bugs" introduced by outside parties, human error, and other unexpected incidents.

Risk responses require a formal approach to issues, opportunities and events to ensure that solutions are in alignment with the business objectives and are cost effective. The following should be considered:

- When preparing the risk response, identify the risks in business terms: loss of productivity, disclosure of confidential information, lost opportunity costs, etc.
- Understand the business risk appetite, acceptable extent of service interruptions, confidentiality of data, compliance requirements, etc.
- Keep the business stakeholders apprised of identified risks and how IT will respond to these risks. In defining the response, be specific and describe in nontechnical terms. Transparency is key.
- Risk response requiring an investment in assets or resources should be supported by a well-thought-out business case that justifies the expenditure, outlines alternatives, describes the justification for the alternative selected, and supports the business objective.
- When building the response, consider how the response will be measured (leading to Domain 3: Risk Monitoring).

**Domain 3**

Risk monitoring provides the communication between the affected stakeholders. Most problems occur with little warning. The key risk indicators (KRIs) are a tool to identify trends and initiate action based on the trends before the issue becomes a defined "problem." The KRIs cited in this document are generic, but provide a good basis for monitoring most problems. The problem management process should utilize the KRI-identified trends to initiate a problem management assessment and remediation process.

**Domain 4**

Risk frameworks require effectively designed controls that reduce inherent risk to a residual risk level that is within levels and costs acceptable to the stakeholders.
Problem management policies should be established, defining:

- A "problem"
- A system which records the:
  - Problem
  - Risk rating
  - Remediation priority
  - Assignment of remediation responsibility
  - Cost estimate to remediate
  - Target completion date
  - Escalation details
  - Closure date
  - Actual remediation costs
- Target time from problem identification to closure by risk category
- Escalation policy
- How to detect the problem earlier (new controls, closer monitoring)

Automated tools should be designed to report when problems are identified, track all data elements in the policy, and maintain a status of the problem as it progresses through the remediation cycle.

**Domain 5**

Effective monitoring of problem management, configuration management, operations, and changes should be built into an effective monitoring system:

- KRIs defined previously should be distributed to stakeholders within IT and the business units affected on a monthly basis, describing current and past period activity.
- Annual evaluation of the maturity of controls provides a barometer of controls in their current state, comparison to previous periods, and the target maturity level. Target maturity levels should be agreed on by stakeholders and evaluated by senior management as part of the IT annual review.
- Maturity assessment can be performed as a self-assessment, with oversight by an objective third party, peer review, or independent assessment (internal audit or external provider).

Problems requiring a preset investment, affecting critical processes, and/or exceeding defined resolution duration should be reported and evaluated according to a defined policy.

## H. Suggested Resources for Further Study

### Suggested Resources for Further Study

In addition to the resources cited throughout this manual, the following resources are suggested for further study:

- ISACA:
  - COBIT 4.1, 2007
    **Note:** The COBIT 4.1 framework is available at no charge from ISACA and can be downloaded at *www.isaca.org/cobit*. The new COBIT 5 framework will be available in 2012.
  - *The Risk IT Framework*, 2009
  - *The Risk IT Practitioner Guide*, 2009
  - *Enterprise Value: Governance of IT Investments, The Val IT Framework 2.0*, 2008
  - *Implementing and Continually Improving IT Governance*, 2010
  - *Information Security Governance: Guidance for Boards of Directors and Executive Management, 2nd Edition*, 2006
- Blokdijk, Gerard; Claire Engle; Jackie Brewster; *IT Risk Management Guide—Risk Management Implementation Guide: Presentations, Blueprints, Templates; Complete Risk Management Toolkit Guide for Information Technology Processes and Systems*, Emereo Publishing, Australia, 2008
- Kouns, Jake; Daniel Minoli; *Information Technology Risk Management in Enterprise Environments: A Review of Industry Practices and a Practical Guide to Risk Management Teams*, Wiley-Interscience, USA, 2010
- Westerman, George; Richard Hunter; *IT Risk: Turning Business Threats Into Competitive Advantage*, Harvard Business School Press, USA, 2007

**Page intentionally left blank**

# 9. Data Management

## A. Chapter Overview

### Introduction

This chapter provides an overview of the data management process and:
- Explains its importance to achieving business objectives
- Outlines a high-level process overview
- Introduces related key concepts
- Presents examples of common risk
- Lists selected key risk indicators
- Provides examples of common IS controls supporting the process
- Describes the practitioner's perspective
- Offers suggested reading materials and references

### Learning Objectives

The CRISC candidate should have a general understanding of the data management processes and how data management interrelates with risk management.

### Contents

This chapter contains the following topics:

## *B. Related Knowledge Statements*

### Contents

The following table lists the applicable knowledge statements from the CRISC job practice.

| No. | Knowledge Statement (KS)<br>Knowledge of: |
|---|---|
| KS1.12 | Threats and vulnerabilities related to data management |
| KS4.8 | Controls related to data management |
| KS5.11 | Control objectives, activities and metrics related to data management |

## C. Key Terms and Principles

| | |
|---|---|
| **Introduction** | This section introduces terms and principles related to data management as well as terms that may help relate the process to other key business processes. |
| **Definition of Availability** | Information that is accessible when required by the business process, now and in the future |
| **Definition of Confidentiality** | The protection of sensitive or private information from unauthorized disclosure |
| **Definition of Data Classification** | The assignment of a level of sensitivity to data (or information) that results in the specification of controls for each level of classification. Levels of sensitivity of data are assigned according to predefined categories as data are created, amended, enhanced, stored or transmitted. The classification level is an indication of the value or importance of the data to the enterprise. |
| **Definition of Data Classification Scheme** | An enterprise scheme for classifying data by factors such as criticality, sensitivity and ownership |
| **Definition of Data Custodian** | The individuals and departments responsible for the storage and safeguarding of computerized data |
| **Definition of Data Owner** | The individuals—normally managers or directors—who have responsibility .for the integrity, accurate reporting and use of computerized data |
| **Definition of Database Management System (DBMS)** | A software system that controls the organization, storage and retrieval of data in a database |
| **Definition of Integrity** | The accuracy, completeness and validity of information. Also see Glossary: Availability, Confidentiality. |

## D. Process Overview

**Introduction**

This section introduces the data management process and its importance to the achievement of business objectives.

**Relevance**

Data are some of the most valuable assets possessed by an enterprise. Each enterprise needs to manage its data throughout the information life cycle—from the initial receipt/creation of new data, through data processing, storage, transmission, reporting and deletion. Effective data management identifies and maintains data protection and handling requirements, including data ownership; the controls needed to maintain the confidentiality, integrity and availability of the data; and the regulations that may affect how the enterprise protects, uses, shares, retains and deletes data.

The data management process also includes the establishment of effective procedures to manage the media library, backup and recovery of data, and proper disposal of media that haves held sensitive data. Effective data management helps ensure the quality, timeliness and availability of business data.

**Process Phases**

Key phases of the data management process methodology are:
1. Identify data ownership and data classification scheme.
2. Classify data according to their sensitivity or criticality to business operations.
3. Label data electronically and physically to indicate their classification level.
4. Translate data storage and retention requirements into procedures (business requirements).
5. Define, maintain and implement procedures to manage the media library (media management).
6. Define, maintain and implement procedures for secure disposal of media and equipment (data disposal).
7. Back up data according to scheme; and define, maintain and implement procedures for data restoration (data retention).

# *E. Risk Management Considerations*

## Introduction

This section discusses risk management practices related to the data management process. The following points will be addressed:
- Risk factors
- Generic threats and vulnerabilities
- Key risk indicators (KRIs)

## Risk Factors

Examples of factors affecting the data management process are:
- **Enterprise size and complexity**—Size, complexity, geographic range, international borders, industry, regulatory requirements, etc., all affect the ease with which data management can be implemented and maintained.
- **Operational model:**
  - Infrastructure: insourced, facilities managed, outsourced—specific facility, outsourced—cloud
  - Organizational structure: insourced, outsourced (consultants), hybrid
  - Infrastructure design: mainframe, distributed, virtualized
  - Control environment: mature, immature, resilient
  - Database management models, data storage models and data warehousing
- **Risk management capabilities**—The enterprise risk management (ERM) function extends to and is practiced by the IT function. In the absence of an ERM, the maturity of IT risk management processes would determine the level of risk. IT risk levels should be aligned with business risk.

## Generic Threats and Vulnerabilities

Generic threats and vulnerabilities related to the data management process are:
- Data management fails to support business requirements.
- Security breaches exist.
- Business, legal and regulatory data storage and retention requirements are not met.
- Media integrity is compromised.
- Backup media are unavailable when needed.
- Unauthorized access to data sources exists.
- Backups are destroyed.
- The location of backup media cannot be determined.
- Corporate information is disclosed.
- The integrity of sensitive data is compromised.
- Backup data cannot be recovered when needed.
- Recovery procedures fail to meet business requirements.
- The time required for performing backups in inappropriate.
- Sensitive data are misused or destroyed.
- Data are altered by unauthorized users.

## KRIs

Examples of KRIs are:

- There is no routine review of data management process to ensure alignment with business requirements.
- There are security breaches related to data management.
- Media is not inventoried on a routine basis.
- Media inventory discrepancies exist.
- Unusual media read errors occur during test or production data restoration activities.
- There are times when sensitive data are retrieved after media are disposed.
- Media are reported lost and data are not sanitized.
- Disposed media are located and found not to be sanitized.
- Media are lost in transit.
- There are many data restoration requests.
- There are times when data critical to the business process cannot be recovered.
- The number of hours between the request for and completion of data restoration exceeds the service level agreement (SLA).
- The enterprise is cited for noncompliance with data storage laws.
- Downtime in hours or data integrity incidents is/are caused by insufficient storage capacity.
- Data required for restoration are unavailable.
- Restoration capabilities are not routinely tested.
- Third parties retain access to and responsibility for data management, backup and restoration.

## F. Information Systems Control Design, Monitoring and Maintenance

### Introduction

This section provides an overview of common controls related to data management. The following points regarding IS control are addressed:
- Objectives and activities
- Metrics
- Monitoring practices

### IS Control Objectives and Activities

Key control activities for data management focus on:
- Business alignment
- Data security
- Media management
- Data disposal
- Data retention

### Key Control Activities Related to Business Alignment

Key control activities may include:
- Define the business requirements for the management of data by IT.
- Periodically identify critical data that affect business operations, in alignment with the risk management model and IT service as well as the business continuity plan.

### Key Control Activities Related to Data Security

Key control activities may include:
- Define and implement a process to clearly identify sensitive data. Consider the business need for confidentiality of the data, and applicable laws and regulations. Communicate and agree on the classification of data with the business process owners.
- Define and implement a policy to protect sensitive data and messages from unauthorized access and incorrect transmission and transport, including, but not limited to:
  - Encryption
  - Message authentication codes
  - Hash totals
  - Bonded couriers
  - Tamper-resistant packaging for physical transport
- Ensure that appropriate programs are instituted to create and maintain awareness of security in the handling and processing of sensitive data.
- Ensure that sensitive information processing facilities are within secure physical locations. These should be protected by defined security perimeters coupled with appropriate surveillance, security barriers and entry controls. Consider the design of the physical infrastructure to prevent losses from fire, interference, external attack or unauthorized access. Consider secure output drop-off points for sensitive outputs or transfer of data to third parties.
- As appropriate, and in accordance with defined security policies, communicate to management the security breaches during any operational phase of data receipt, processing and transmission.

## Key Control Activities Related to Media Management

Key control activities may include:
- Assign responsibilities within the IT function for the development and maintenance of policies and procedures for media library management.
- Analyze media types of stored and archived data to define environmental requirements, e.g., humidity and temperature. Monitor and review the physical storage environment.
- Ensure that the media library management system specifies security and access rights.
- Maintain an inventory list of archived media to limit the opportunity for data loss.
- Review on a regular basis the media inventoried against the list. Investigate and correct any discrepancies and missing media, and report to management.
- Periodically review the integrity and usability (age) of magnetic media. Periodically report and track disk errors. Investigate trends to ensure that media can still be used. Replace media susceptible to degradation, such as tape and DVD-ROMs.
- Define a procedure to remove active media from the media inventory list upon disposal.
- Consider the impact of current and future changes in hardware and software standards on retrieval and processing of archived data.

## Key Control Activities Related to Data Disposal

Key control activities may include:
- Clearly define responsibility for the development and communication of policies on disposal.
- Institute policies and procedures for retention of data received from the business and their subsequent destruction according to the sensitivity of the data.
- Sanitize equipment and media containing sensitive information prior to reuse or disposal. Such processes should ensure that data marked as "deleted" or "to be disposed" cannot be retrieved (e.g., media containing highly sensitive data should be physically destroyed).
- Transport nonsanitized equipment and media in a secure way throughout the disposal process.
- To maintain an audit trail, log the disposal of equipment or media containing sensitive information.
- Require disposal contractors to have the necessary physical security and procedures to store and handle the equipment and media before and during disposal.

## Key Control Activities Related to Data Retention

Key control activities related to backup, restoration and retention may include:

- Agree on and communicate with the business or IT process owner the time frame required for restoration.
- Define requirements for onsite and offsite storage of backup data that meet the business requirements. Consider the accessibility required to back up data.
- Prioritize data recovery based on business requirements and IT service continuity procedures.
- Establish storage and retention procedures that address the enterprise's security policy and change management procedures, including encryption and authentication. Consider the data and the keys and certificates used for encryption and authentication.
- Institute policies and procedures for retention of data received from the business and their subsequent destruction according to the sensitivity of the data.
- Establish storage and retention arrangements to satisfy legal, regulatory and business requirements for documents, data, archives, programs, reports and messages (incoming and outgoing).
- Define policies and procedures for the backup of systems, applications, data and documentation that consider factors including:
  - Type of backup (e.g., full vs. incremental)
  - Type of media
  - Automated online backups
  - Data types (e.g., voice, optical)
  - Creation of logs
  - Critical end-user computing data (e.g., spreadsheets)
  - Physical and logical location of data sources
  - Security and access rights
  - Encryption
  - Frequency of backup (e.g., disk mirroring for real-time backups vs. DVD-ROM for long-term retention)
- Assign responsibilities for taking and monitoring backups.
- Ensure that systems, applications, data and documentation maintained or processed by third parties are adequately backed up or otherwise secured. Consider requiring return of backups from third parties. Consider escrow or deposit arrangements.
- Schedule, take and log backups in accordance with established policies and procedures.
- Periodically perform sufficient restoration tests to ensure that all components of backups can be effectively restored.

## IS Control Metrics for Monitoring

The following table provides an overview and examples of IS control metrics for monitoring data management.

| Monthly Data Management Monitoring Report | | | | | |
|---|---|---|---|---|---|
| **Attribute** | **Target** | **Current Period** | **Prior Period** | **Prior Period − 1** | **Prior Period − 2** |
| Number of security breaches related to data management per month | | | | | |
| Number of security breaches related to data management, by priority | | | | | |
| Average period to resolve security breaches related to data management | | | | | |
| Asset value lost due to security breaches related to data management during the period | | | | | |
| Number of media inventories performed during the period (assumes more than one data media library) | | | | | |
| Number of missing media identified due to inventory being taken during the period | | | | | |
| Number of incidents where sensitive data were retrieved after media were disposed (incident explanation attached) | | | | | |

**IS Control Metrics for Monitoring *(cont.)***

| Monthly Data Management Monitoring Report *(cont.)* | | | | | |
|---|---|---|---|---|---|
| **Attribute** | **Target** | **Current Period** | **Prior Period** | **Prior Period −1** | **Prior Period −2** |
| Number of incidents where media were reported lost (incident explanation attached) | | | | | |
| Number of requests for data restoration (user error and system error) | | | | | |
| Number of incidents where data could not be restored (explanation attached) | | | | | |
| Number of occurrences where time between request and completion of data restoration exceeds the SLA | | | | | |
| Number of citations received from regulators for noncompliance with data storage and related laws/ covenants/requirements | | | | | |
| Number of restoration tests scheduled for the period | | | | | |
| Number of problems reported from restoration tests | | | | | |
| Number of third-party data management incidents reported during the period | | | | | |
| Number of third-party incidents reported during the period classified as critical or serious (explanation of critical incidents attached.) | | | | | |
| Average time between identification and resolution | | | | | |

# G. The Practitioner's Perspective

## Introduction

The practitioner's perspective for data management is organized first by the following topics:
- Business alignment
- Media management
- Data disposal
- Data security

Then an overview is provided of how the five CRISC domains (listed below) relate to data management in practice, if applicable:
- Domain 1—Risk Identification, Assessment and Evaluation
- Domain 2—Risk Response
- Domain 3—Risk Monitoring
- Domain 4—Information Systems Control Design and Implementation
- Domain 5—Information Systems Control Monitoring and Maintenance

## Business Alignment for Domain 1

The risk identification, assessment, and evaluation process addresses those issues that would affect the integrity, availability and security of corporate data.

The initial identification of risk should be integrated into the systems architecture definition, database management selection, software acquisition, and systems development processes. The assessment and evaluation can be performed as part of the initial process; however, data management risk requires routine monitoring. The risk identification process should include a periodic assessment of critical data that could affect business operations, and evaluate its alignment with the risk management model.

The practitioner should focus attention on changes to the data management environment. Special focus should be directed to data managed and/or stored with third parties.

The risk practitioner will have to ensure that the management of data is compliant with organizational policy and external regulations. This may require implementation of monitoring controls that capture a record of activity related to data access. The risk of mishandling data—whether accidentally or through a breach—may be severe and result in significant financial, reputational and legal impact on the enterprise.

## Business Alignment for Domain 2

The risk response should address new data management requirements resulting from an enterprise risk assessment, strategic changes in the business requirements, systems development activities and other management initiatives.

Data protection starts with the identification of what data are sensitive and/or critical, and the determination of the ownership of the data. The data access controls must do exactly that—control access—which means that access is granted to authorized individuals or processes, but denied to unauthorized personnel. Even the granting of access must have a response to the risk of improper disclosure, modification or destruction of data by enforcing principles such as separation of duties, least privilege, mutual exclusivity, dual control and need to know.

The challenge from the perspective of risk response is to align the response to the risk of unauthorized access to data with the need for the business to operate efficiently and allow authorized access according to business requirements.

## Business Alignment for Domain 3

Business alignment is a management function. IT's participation in strategic meetings and board and strategy committee meeting minutes provides a good source of management direction.

Risk monitoring for data protection focuses on the logging and reporting of all access requests—whether permitted or denied—to data and providing those reports to management or regulatory agencies as required to provide for proper governance of data.

## Media Management for Domain 1

Media management concerns the maintenance of the media used to store or transport data. IT management should focus on the following:
- Where media are stored—temperature, humidity, sunlight, etc.
- How media are maintained and protected—age, number of uses, labeling
- Extending the definition of the media library to other storage forms in addition to tapes—controlled access, encryption of sensitive data
- Desktop and laptop local disk drives—loss protection, encryption
- Universal serial bus (USB) memory sticks (thumb drives)
- Smartphones

Recognize the limitations of media storage technology and ensure that data continue to be readable (degradation of media).

The risk identification, assessment and evaluation process should be routinely performed as business storage requirements change, new technologies are introduced or limitations to existing technology are identified.

The cost of media management should be assessed against the risk to determine whether these costs are appropriate or can be reduced through efficiencies or newer technology.

## Media Management for Domain 2

Risk responses require a formal approach to issues, opportunities and events to ensure that solutions are in alignment with the business objectives and are cost effective.

Media management is perceived as an operational issue. A robust incident reporting and resolution system will demonstrate IT management focus and promote transparency with the business community.

Focused procedures relating to the event of media residing on portable equipment being lost or stolen.

## Media Management for Domain 3

Metrics that will assist in the monitoring of media management issues are the:
- Number of media inventories performed during the period (assumes more than one data media library)
- Number of incidents of unauthorized access to media libraries
- Number of missing media identified during periodic inventory
- Number of incidents where media were lost

## Media Management for Domain 4

Media management controls address:
- Media protection from degradation
- Media inventory control
- Media access control
- Viability of media storage as technology changes
- The establishment of data access controls—labeling, encryption
- Hardware architecture design
- Data center design— secure storage areas
- IT operations

## Media Management for Domain 5

Effective monitoring of media operations, incident reports and issue monitoring should be part of IT operations and included in reports to management.

Timely reporting using key risk indicators (KRIs) defined previously should be distributed to stakeholders within IT and the business units affected.

Annual evaluation of the maturity of controls provides a barometer of controls in their current state, comparison to previous periods, and the target maturity level. Target maturity levels should be agreed on by stakeholders and evaluated by senior management as part of the IT annual review.

Maturity assessment can be performed as a self-assessment, with oversight by an objective third party, peer review or independent assessment (internal audit or external provider).

## Data Disposal for Domain 1

The risk of improper disposal of data is a serious matter for any enterprise that handles sensitive data. The types of storage requiring disposal takes many forms:
- Deleted tapes
- Obsolete computer equipment
- Disk drive failures
- CDs, DVDs and memory sticks containing sensitive data for retirement
- Loss of smartphones
- Paper

The risk identification, assessment and evaluation process should be routinely performed since the methods used for disposal of media need to be adapted for new technologies.

## Data Disposal for Domain 2

Disposal of data is considered an operational issue and should be addressed in the IT operations function. Risk response should focus on nonoperational data disposal (loss or theft).

The method of disposal will depend on the level of risk associated with the data. Media that have contained very sensitive (classified) data may need to be destroyed in a multistep process including overwriting, degaussing and even physical destruction. Other media that have contained less sensitive data may only need to be overwritten or degaussed.

The proliferation of data on small portable devices is a serious risk. Smartphones, USB drives, DVDs and portable hard drives can contain large amounts of sensitive data and may be shared with unauthorized users, may not be adequately tracked, and may be lost or disposed of inappropriately. The risk practitioner should encourage the development of policy, procedures, encryption, labeling and other controls to mitigate this risk.

**Data Disposal for Domain 3**

Metrics that will assist in the monitoring of data disposal issues are the:
- Records of secure disposal of media
- Reviews of disposal process to ensure compliance with policy and procedures
- Number of incidents where sensitive data were retrieved after media were disposed
- Number of incidents where media were reported lost

**Data Disposal for Domain 4**

Data disposal controls address:
- Disposal policies
- Management oversight to enforce policies
- Identification of data loss
- Equipment available for secure disposal (i.e., shredders, degaussing equipment, incinerators)

Controls need to be established regarding:
- IT operations for data within the custody of the data center
- Data maintained in user departments
- Recycle bins for data, followed by data sanitization
- Awareness programs
- Strong data disposal security awareness for end users to address smartphone, memory stick and disk drive concerns
- Contractual arrangements and oversight with third-party vendors maintaining and destroying company data

**Data Disposal for Domain 5**

Effective monitoring of data disposal processes includes:
- Reports on data destruction from vendors
- Incident reporting and issue management processes to identify operational critical risk issues
- Filtering refuse for data media
- Timely reporting of previously defined KRIs to stakeholders within IT and affected business units
- Annual evaluation of the control maturity provides a barometer of controls in their current state, comparison to previous periods, and the target maturity level. Target maturity levels should be agreed on by stakeholders and evaluated by senior management as part of the IT annual review.
- Maturity assessment can be performed as a self-assessment, with oversight by an objective third party, peer review or independent assessment (internal audit or external provider).

**Data Security for Domain 1**

The primary issues within data security are:
- Balancing the risk of data disclosure or unauthorized access with the ability to access data to accomplish the operational tasks
- Identifying and classifying data risk
- Determining data ownership
- Maintaining the risk classification of data as applications, access methods (network and user hardware) and infrastructure change
- Conducting awareness programs for all staff regarding data protection requirements and procedures

**Data Security for Domain 2**

Risk response focuses on the actions taken after a data security incident has been identified.

The response includes triaging incidents:
- Identify the incident.
- Determine the potential for loss, the effect on the enterprise, and its stakeholders, compliance, assets and reputation.
- Determine pre-established actions based on known risk.
- Perform drills based on predefined incident scenarios.

---

**Data Security for Domain 3**

Metrics include the:
- Number of attempts to gain unauthorized access to data
- Number of security breaches related to data management per month
- Number of security breaches related to data management by priority
- Average period to resolve security breaches related to data management
- Asset value lost due to security breaches related to data management during the period

---

**Data Security for Domain 4**

Data security includes the following:
- It is a component of the system development life cycle (SDLC).
- Control design should include identified risks in initial phases.
- Controls should align with business requirements and be efficient.

---

**Data Security for Domain 5**

Data security includes the following:
- Control monitoring should be integrated into the information security management and the incident reporting systems.
- Timely reporting using the KRIs defined previously should be distributed to stakeholders within IT and the business units affected.
- Annual evaluation of the maturity of controls provides a barometer of controls in their current state, comparison to previous periods, and the target maturity level. Target maturity levels should be agreed on by stakeholders and evaluated by senior management as part of the IT annual review.
- Maturity assessment can be performed as a self-assessment, with oversight by an objective third party, peer review or independent assessment (internal audit or external provider).

## H. Suggested Resources for Further Study

**Suggested Resources for Further Study**

In addition to the resources cited throughout this manual, the following resources are suggested for further study:

- ISACA:
  - COBIT 4.1, 2007
    **Note:** The COBIT 4.1 framework is available at no charge from ISACA and can be downloaded at *www.isaca.org/cobit*. The new COBIT 5 framework will be available in 2012.
  - *The Risk IT Framework*, 2009
  - *The Risk IT Practitioner Guide*, 2009
  - *Enterprise Value: Governance of IT Investments, The Val IT Framework 2.0*, 2008
  - *Implementing and Continually Improving IT Governance*, 2010

**Page intentionally left blank**

# 10. Physical Environment Management

## A. Chapter Overview

### Introduction

This chapter provides an overview of the physical environment management process and:
- Explains its importance to achieving business objectives
- Outlines a high-level process overview
- Introduces related key concepts
- Presents examples of common risk
- Lists selected key risk indicators (KRIs)
- Provides examples of common IS controls supporting the process
- Describes the practitioner's perspective
- Offers suggested reading materials and references

### Learning Objectives

The CRISC candidate should have a general understanding of the physical environment management processes and how physical environment management interrelates with risk management.

### Contents

This chapter contains the following topics:

## B. Related Knowledge Statements

### Contents

The following table lists the applicable knowledge statements from the CRISC job practice.

| No. | Knowledge Statement (KS)<br>Knowledge of: |
|---|---|
| KS1.16 | Threats and vulnerabilities related to management of IT operations |
| KS4.12 | Controls related to management of IT operations |
| KS5.5 | Control objectives, activities and metrics related to IT operations and business processes and initiatives |

## C. Key Terms and Principles

| | |
|---|---|
| **Introduction** | This section introduces terms and principles related to determination of physical environment management as well as terms that help relate the process to other key business processes—particularly data classification, which provides input into the physical environment management process. |
| **Definition of Data Classification** | The assignment of a level of sensitivity to data (or information) that results in the specification of controls for each level of classification. Levels of sensitivity of data are assigned according to predefined categories as data are created, amended, enhanced, stored or transmitted. The classification level is an indication of the value or importance of the data to the enterprise. |
| **Definition of Data Classification Scheme** | An enterprise scheme for classifying data by factors such as criticality, sensitivity and ownership |
| **Definition of Data Custodian** | The individuals and departments responsible for the storage and safeguarding of computerized data |

## D. Process Overview

### Introduction

This section introduces the physical environment management process and its importance to the achievement of business objectives.

### Relevance

Protection for computer equipment and personnel requires well-designed and well-managed physical facilities. The process of managing the physical environment includes defining the physical site requirements, selecting appropriate facilities, and designing effective processes for monitoring environmental factors and managing physical access. Effective management of the physical environment prevents or minimizes harm to personnel and computer equipment and reduces business interruptions.

### Process Phases

Key phases of managing the physical environment are:

1. Define the required level of physical protection.
2. Select and commission the site (data center, office, etc.).
3. Implement physical environment measures.
4. Manage the physical environment (maintaining, monitoring and reporting included).
5. Define and implement procedures for physical access authorization and maintenance.

## *E. Risk Management Considerations*

### Introduction

This section discusses risk management practices related to the physical environment management process. The following points are addressed:
- Risk factors
- Generic threats and vulnerabilities
- Key risk indicators (KRIs)

### Risk Factors

Examples of factors affecting the physical environment management process are:
- **Enterprise size and complexity**—Number and size of installations, complexity of installations, geographic range, including international borders, etc., all affect the ease with which physical environment management can be implemented and maintained.
- **Operational model:**
  - Infrastructure: insourced, facilities managed, outsourced—specific facility, outsourced—cloud
  - Organizational structure: insourced, outsourced (consultants), hybrid
  - Control environment: mature, immature, resilient
- **Physical location of facilities**—Violent weather and earthquake zones, political instability or high crime areas, power and communications quality and resiliency, building structure, heating, cooling and humidity requirements, proximity to neighboring hazards (old buildings, military, airports, railways)
- **Risk management capabilities:**
  - The enterprise risk management (ERM) function extends to the IT function or the IT function operates in its own silo.
  - In absence of ERM, maturity of IT risk management processes
  - Risk tolerance of IT and business
  - IT risk levels are aligned with business risks

### Generic Threats and Vulnerabilities

Generic threats and vulnerabilities related to the physical environment management process are:
- Physical attack on the IT site
- Hardware theft
- Confidential information being accessed by devices configured to read the radiation emitted by the equipment or cabling
- Visitors gaining unauthorized access to IT equipment or information
- Unauthorized entry to secure areas
- Facilities exposed to environmental impacts (heat, cooling, humidity)
- Noncompliance with health and safety regulations
- Accidents to staff members
- Fire, flooding
- Inadequate environmental threat detection
- Inadequate measures for environmental threat protection
- Threats to physical security not identified
- Increased vulnerability to security risks, resulting from site location and/or layout, damage to building (roof collapse, window breakage)
- IT systems failure due to improper protection from power outages and other facility-related risk (broken pipes, air conditioning failure)

## KRIs

Examples of KRIs are the:

- Frequency of training of personnel in safety, security and facilities measures
- Level of facility/campus resident awareness of their physical security oversight responsibilities
- Completeness of tracking of personnel with access to facility (unsecured and secured areas)
- Frequency and thoroughness of period access privilege review and follow-up
- Completeness and thoroughness of background checks for personnel with access privileges
- Control over visitors, contractors and maintenance personnel (escort, identification, logging of access)
- Access control system up-time reliability
- Adequacy of control design and operational effectiveness
- Amount of downtime arising from physical environment incidents
- Number of injuries caused by the physical environment
- Number of security exposures arising from physical environment incidents
- Number of incidents due to physical security breaches or failures
- Number of incidents of unauthorized access to computer facilities
- Percent of personnel trained in safety, security and facilities measures
- Number of risk mitigation tests conducted over the last year
- Frequency of physical risk assessment and reviews
- Number of tests of physical security controls (fire evacuation, fire suppression, backup power, etc.)

[1] The issues listed may not be appropriate for outsourced environments because they are the responsibility of the servicer. The customer is ultimately responsible for the achievement of service level agreements (SLAs) and security of processes.

# F. Information Systems Control Design, Monitoring and Maintenance

## Introduction

This section provides an overview of common controls related to physical environment management. The following points regarding IS control are addressed:
- Objectives and activities
- Metrics
- Monitoring practices

## IS Control Objectives and Activities

Key control activities for physical environment management[1] are:
- Physical protection risk assessment
- Site selection and layout
- Physical security measures
- Physical access
- Protection against environmental factors
- Facilities management

| Key control activities related to... | May be to... |
|---|---|
| Physical protection risk assessment | • Define a process that identifies the potential risks and threats related to the physical environment management.<br>• Align the risks identified with cost and effect with the enterprise's ability to maintain normal business activities. |
| Site selection and layout | • Determine the technology strategy; select a site for IT equipment that meets business requirements and the security policy.<br>• Define a process that identifies potential risks and threats to the enterprise's IT sites and assesses the business impact on an ongoing basis.<br>• Ensure that the selection and design of the site take into account relevant laws and regulations, such as building codes and environmental, fire, electrical engineering, and occupational health and safety regulations. |
| Physical security measures | • Define and implement a policy for physical security and access control measures to be followed for IT sites. Regularly review the policy to ensure that it remains relevant and up to date.<br>• Design physical security measures that:<br>–Limit the access to information about sensitive IT sites and the design plans.<br>–Provide alarm systems, building hardening, armored cabling protection, secure partitioning, and other measures commensurate with the risks associated with the business and operation.<br>–Periodically test and document the preventive, detective and corrective physical security measures to verify design implementation and effectiveness.<br>• Establish procedures addressing:<br>– Secure removal of IT equipment, supported by the appropriate authorization.<br>– Controlled receiving and shipping areas for IT equipment.<br>– Secure transportation and storage of equipment, backups and documents.<br>• Define a process for recording, monitoring, managing, reporting and resolving physical security incidents, in line with the overall IT incident management process.<br>• Label all equipment and provide security measures to prevent theft.<br>• Ensure the secure disposal of media that contain sensitive information.<br>• Ensure the secure storage of equipment and media.<br>• Label all media indicating retention time, data sensitivity levels and classification, age and ownership. |

[1] The issues listed may not be appropriate for outsourced environments because they are the responsibility of the service provider (facility manager). The customer is ultimately responsible for the achievement of service level agreements (SLAs) and security of processes.

## IS Control Objectives and Activities *(cont.)*

| Key control activities related to ... | May be to ... |
|---|---|
| Physical access | • Define and implement a process that governs the requesting and granting of access to the computing facilities as defined by the site access policy.<br>• Define a process to log and monitor all entry points to IT sites, including visitor registration.<br>• Define, implement and monitor a policy addressing identity card issuance and mandatory usage.<br>• Define and implement a policy requiring visitors to be escorted at all times while onsite by a member of the IT operations group.<br>• Develop policy and procedures to prevent multiple users from entering together (tailgating or piggybacking) through access control points.<br>• Restrict access to sensitive IT sites by establishing perimeter restrictions, such as fences, walls and security devices on interior and exterior doors.<br>• Define a process to conduct regular physical security awareness training. |
| Protection against environmental factors | • Establish and maintain a process to identify natural and man-made disasters that may occur in the area within which the IT facilities are located. Assess the potential effect on the IT facilities.<br>• Define and implement a policy that identifies how IT equipment, including mobile and offsite equipment, is protected against environmental threats.<br>• Define and implement a process to regularly monitor and maintain devices that proactively detect environmental threats (e.g., fire, water, smoke, humidity). Establish and maintain an incident response and monitoring process to respond to environmental alarms.<br>• Compare measures and contingency plans against insurance policy requirements, report results, and maintain an issue monitoring of open points.<br>• Ensure that IT sites are built and designed to minimize the impact of environmental risks. |
| Facilities management | • Keep the IT sites and server rooms clean and in a safe condition.<br>• Define and implement a process to examine the IT facility requirement for protection against environmental conditions as well as power fluctuations and outages, in conjunction with other business continuity planning requirements.<br>• Regularly test the uninterruptible power supply mechanisms and ensure that power can be switched to the supply without any significant effect on business operations and address requirements for high-availability systems.<br>• Ensure that the facilities housing the IT systems have more than one source for dependent utilities.<br>• Define and implement a process to ensure that IT sites and facilities are in compliance with relevant health and safety laws, regulations, guidelines and vendor specifications.<br>• Define and implement a process to record, monitor, manage and resolve facilities incidents in line with the IT incident management process. Make available reports on facilities incidents where disclosure is required in terms of laws and regulations.<br>• Define a process to ensure that IT sites and equipment are maintained as per the supplier's recommended service intervals and specifications by authorized personnel. |
| Measuring performance | • Measure performance of access control systems, monitoring devices and alarms. |

## IS Control Metrics for Monitoring

The following table provides an overview and examples of IS control metrics monitoring physical environment management:

| Monthly Physical Security Monitoring Report | | | | | |
|---|---|---|---|---|---|
| **Attribute** | **Target** | **Current Period** | **Prior Period** | **Prior Period −1** | **Prior Period −2** |
| **Incident management** | | | | | |
| Number of physical incidents opened/closed | | | | | |
| Number of physical incidents classified as critical | | | | | |
| Average interval between opening and closing an incident | | | | | |
| Number of physical incidents escalated to the information security function | | | | | |
| Number of physical incidents escalated to internal or external enforcement | | | | | |
| **Physical access** | | | | | |
| Number of incidents involving identity cards with physical access to the IT environment | | | | | |
| Number of identity cards issued with access to the IT environment | | | | | |
| Number of identity cards terminated or employees transferred to lesser access | | | | | |
| Number of identity cards revoked | | | | | |
| Number of identity cards lost | | | | | |
| Number of physical incidents involving unauthorized entry | | | | | |
| **Environmental response** | | | | | |
| Number of environmental incidents | | | | | |
| Number of false alarms | | | | | |
| Number of incidents requiring action | | | | | |

## *G. The Practitioner's Perspective*

### Introduction

The following section provides an overview of how the five CRISC domains (listed below) relate to physical environment management in practice:
- Domain 1—Risk Identification, Assessment and Evaluation
- Domain 2—Risk Response
- Domain 3—Risk Monitoring
- Domain 4—Information Systems Control Design and Implementation
- Domain 5—Information Systems Control Monitoring and Maintenance

### Domain 1

Risk identification traditionally focused on the data center. Today's data center definition is more diverse and includes:
- Hardware room (old computer room)
- Server rooms
- Outsourced environments

Risk identification and assessment should be integrated into the site identification process. In most enterprises, data center placement is an infrequent activity. However, with the expansion of dedicated and enterprisewide computer equipment in central and distributed locations, site expansion or modification is a more frequent occurrence. Server rooms, telephone systems, network switch/router rooms, etc., have proliferated and are often added due to an immediate need without due diligence, and consideration of nonevident factors. The risk assessment and evaluation should be a collaborative effort between the IT architecture, operations and information security teams.

A higher risk profile exists for outsourced environments. The various types of outsourcing provide the customer with different levels of influence. A facilities management agreement within the customer's former data center provides the most influence. Once the processing environment is wholly within the servicer's environment, the influence is primarily through service level agreements (SLAs) and compliance. Cloud computing provides a higher risk because the location of the processing can move to different sites through load balancing. Third-party reports and compliance requirements are the front-line in locating services. Outsourcing contracts should specifically describe requirements for minimizing risks associated with data processing facilities.

### Domain 2

Risk response will vary depending on the ownership and location of the processing resources. For corporate owned facilities, internally defined risk assessment processes should identify and document risks and mitigation processes. In an outsourced environment, the risk response will include contractual remedies and alternate processing solutions.

### Domain 3

Risk monitoring is part of the incident management system. The quality of risk monitoring is dependent on the timeliness and accuracy of the incident reporting system. The incident reporting system should have triggers built into the notification system, to direct physical environmental issues to the facilities/physical environment department. In an outsourced environment, the risk level is higher and risk monitoring processes require cooperative measures between customer and servicer. The retained enterprise has the primary responsibility of monitoring risks from the servicer.

**Domain 4**

As previously noted in the risk monitoring section, the IS control design and implementation consists of measured solutions to defined risks. Control design is an integral part of the systems development, hardware/software acquisition, and business process life cycles. The practitioner should consider the control design addressing the risk mitigation, cost/benefit of the control design, ongoing operating maintenance and cost, and the overall effectiveness in reducing residual risk.

**Domain 5**

The focus of physical environment management is incident management, physical access, and environmental response. Incident management is an enterprise system of monitoring problems from various sources, and directing the response by the responsible organization. This process must be effective, provide alerts to the appropriate job functions with responsibility to implement corrective action, and monitor the closure of open incidents. The issue monitoring process closes the loop between identified incidents and closure of issues. Physical access combines automated solutions (badge readers, surveillance and automated detection system) with human interaction (human resources policy, supervisory monitoring, guard observation, etc.) and policy. Environmental monitoring also combines automated and human interaction, but relies heavily on the former.

The metrics defined previously measure the effectiveness of control. Reporting these metrics on a regular basis provides trend lines for more effective management and enforcement.

## H. Suggested Resources for Further Study

**Suggested Resources for Further Study**

In addition to the resources cited throughout this manual, the following resources are suggested for further study:

- ISACA:
  - COBIT 4.1, 2007
    **Note:** The COBIT 4.1 framework is available at no charge from ISACA and can be downloaded at *www.isaca.org/cobit*. The new COBIT 5 framework will be available in 2012.
  - *The Risk IT Framework*, 2009
  - *The Risk IT Practitioner Guide*, 2009
  - *Enterprise Value: Governance of IT Investments, The Val IT Framework 2.0*, 2008
  - *Implementing and Continually Improving IT Governance*, 2010

# 11. IT Operations Management

## A. Chapter Overview

### Introduction

This chapter provides an overview of IT operations management and:
- Explains its importance to achieving business objectives
- Outlines a high-level process overview
- Introduces related key concepts
- Presents examples of common risk
- Lists selected key risk indicators (KRIs)
- Provides examples of common IS controls supporting the process
- Describes the practitioner's perspective
- Offers suggested reading materials and references

### Learning Objectives

The CRISC candidate should have a general understanding of IT operations management processes and how IT operations management interrelates with risk management.

### Contents

This chapter contains the following topics:

## B. Related Knowledge Statements

**Contents** The following table lists the applicable knowledge statements from the CRISC job practice.

| No. | Knowledge Statement (KS)<br>Knowledge of: |
|---|---|
| KS1.16 | Threats and vulnerabilities related to management of IT operations |
| KS4.12 | Controls related to management of IT operations |
| KS5.5 | Control objectives, activities and metrics related to IT operations and business processes and initiatives |

## *C. Key Terms and Principles*

**Introduction**

This section introduces terms and principles related to IT operations management as well as terms that may help relate the process to other key business processes.

---

**Definition of Service Level Agreement (SLA)**

An agreement, preferably documented, between a service provider and the customer(s)/user(s) that defines minimum performance targets for a service and how they will be measured

---

**Definition of Operational Level Agreement (OLA)**

An internal agreement covering the delivery of services that support the IT organization in its delivery of services

## D. Process Overview

### Introduction

This section introduces the IT operations management process and its importance to the achievement of business objectives.

### Relevance

Complete and accurate processing of data requires effective management of data processing procedures, job scheduling and monitoring, and diligent maintenance of hardware and software. This process includes defining operating policies and procedures for effective management of scheduled processing, protecting sensitive output, monitoring infrastructure and application performance and ensuring preventive maintenance of hardware and software. Effective IT operations management helps maintain data integrity and reduces business delays and IT operating costs.

### Process Phases

Key phases of the IT operations management process are to:

1. Create/modify operations procedures (including manuals, checklists, shift planning, handover documentation, escalation procedures, etc.).
2. Schedule workload and batch jobs.
3. Monitor infrastructure and processing, document and resolve problems.
4. Manage and secure physical output (e.g., paper, media).
5. Apply fixes or changes to the schedule and infrastructure, apply emergency fixes.
6. Implement authorized changes according to the change management process.
7. Implement/establish a process for safeguarding authentication devices against interference, loss and theft.
8. Schedule and perform preventive maintenance of hardware and software.

### Structure

Key segments of a generic operations management structure for IT operations may include:

- Production control
- Job scheduling
- Computer room operations
- Media handling
- Output distribution

# E. Risk Management Considerations

## Introduction

This section discusses risk management practices related to the IT operations management process. The following points are addressed:
- Risk factors
- Generic threats and vulnerabilities
- Key risk indicators (KRIs)

## Risk Factors

Examples of factors affecting the operations management process are:
- **Enterprise size and complexity**—Size, complexity, geographic range, etc., all affect the ease with which operations management can be implemented and maintained.
- **Operational model**—Insourced, outsourced (consultants), mainframe, distributed computing, etc.
- **Maturity of operations and change control procedures**—poorly defined, not consistently followed, lack of oversight/enforcement

## Generic Threats and Vulnerabilities

Generic threats and vulnerabilities related to the IT operations management process are:
- **Operations procedures:**
  - Errors and rework due to misunderstanding of procedures
  - Inefficiencies due to unclear and/or nonstandard procedures
  - The inability to deal quickly with operational problems, new staff and operational changes
  - Misuse of elevated privileges by operators and administrators
  - Use of equipment or software for personal use
  - Illegal copying of software
  - Unauthorized installation of equipment or programs
- **Job scheduling:**
  - Resource utilization peaks
  - Problems with scheduling of *ad hoc* jobs
  - Reruns or restarts of jobs
- **IT infrastructure monitoring:**
  - Undetected infrastructure problems and occurrence of incidents
  - Infrastructure problems causing greater operational and business impact than if they had been prevented or detected earlier
  - Poorly utilized and deployed infrastructure resources
- **Sensitive documents and output devices:**
  - Misuse of sensitive IT assets, leading to financial losses and other business impacts
- **Preventive maintenance for hardware:**
  - The inability to account for all sensitive IT assets
  - Infrastructure problems that could have been avoided or prevented
  - Warranties violated due to noncompliance with maintenance requirements

## KRIs

Examples of KRIs are a/an:
- Excessive number of operational incidents
- Excessive number of hours of unplanned downtime resulting from operational incidents
- Presence of unauthorized equipment or software
- Excessive number of instances of service level agreements (SLAs) exceeding thresholds
- High average downtime due to operational incidents
- High number of hours lost due to late completion of scheduled work
- Excessive average time to research and remediate operations incidents

# F. Information Systems Control Design, Monitoring and Maintenance

## Introduction

This section provides an overview of common controls related to IT operations management. The following points regarding IS control are addressed:
- Objectives and activities
- Metrics
- Monitoring practices

## IS Control Objectives and Activities

Key control activities for IT operations management are:
- Operations processes and instructions
- Job scheduling
- IT infrastructure monitoring
- Sensitive documents and output devices
- Preventive maintenance for hardware

## Key Control Activities Related to Operations Processes and Instructions

Key control activities may include:
- Develop, implement and maintain standard IT operational procedures covering the definition of roles and responsibilities, including those of external service providers.
- Train support personnel in operational procedures and related tasks for which they are responsible.
- Define procedures and responsibilities for formal handover of duties (e.g., for shift change, planned or unplanned absence).
- Define procedures for exception handling in line with the incident management and change management procedures and to address security aspects.
- Ensure that segregation of duties is in line with the associated risk, security and audit requirements.

## Key Control Activities Related to Job Scheduling

Key control activities may include:
- Use formal procedures for planning and scheduling processing activities. Gain authorization for the initial schedules and changes to these schedules.
- Schedule authorized changes to systems, applications, configurations, networks and procedures with consideration of maintenance windows, minimizing impact on business operations, urgency of change.
- Ensure that the scheduling of batch jobs takes into consideration business requirements, priorities, conflicts between jobs and workload balancing. Put procedures in place to identify, investigate and approve departures from standard job schedules.

## Key Control Activities Related to IT Infrastructure Monitoring

Key control activities may include:
- Define and implement a procedure to resolve and correct job failures, including balancing or abnormal condition controls.
- Implement automated tools and processes to immediately detect, notify and rectify critical processing failures.
- Define and implement a process for event logging that identifies the level of information to be recorded based on a consideration of risk and performance.
- Identify and maintain a list of infrastructure assets that need to be monitored based on service criticality, and the relationship between configuration items and services that depend on them.
- Define and implement rules that identify and record threshold breaches and event conditions. Find a balance between generating spurious minor events and significant events so event logs are not overloaded with unnecessary information.
- Produce event logs and retain them for an appropriate period to assist in future investigations.
- Ensure that incident tickets are created in a timely manner when monitoring activities identify deviations.

## Key Control Activities Related to Sensitive Documents and Output Devices

Sensitive documents and output devices activities may include:
- Establish procedures for monitoring event logs and conduct regular reviews.
- Establish procedures to govern the receipt, storage, use, removal and disposal of special forms and output devices into, within and outside of the enterprise.
- Assign access privileges to sensitive documents and output devices based on the least privilege principle, balancing risk and business requirements.
- Establish an inventory of sensitive documents and output devices, and conduct regular reconciliations.
- Establish appropriate physical safeguards over special forms and sensitive devices.
- Ensure that sensitive documents and media containing sensitive information are labeled, stored and transported securely.
- Define and implement a process for destroying sensitive information and output devices (e.g., degaussing of electronic media, physical destruction of memory devices, making shredders or locked paper baskets available to destroy special forms and other confidential papers).

## Key Control Activities Related to Preventive Maintenance for Hardware

Preventive maintenance for hardware activities may include:
- Establish a preventive maintenance plan for all hardware and software, considering cost/benefit analysis, vendor recommendations and support, risk of outage, qualified personnel and other relevant factors.
- Review all activity logs on a regular basis to identify critical hardware and software components that require preventive maintenance, and update the maintenance plan accordingly.
- Establish maintenance agreements involving third-party access to the enterprise's IT facilities for onsite and offsite activities (e.g., outsourcing). Establish formal service contracts containing or referring to all necessary security conditions, including access authorization procedures, to ensure compliance with the enterprise's security policies and standards.
- In a timely manner, communicate to affected customers and users the expected impact (e.g., performance restrictions or system unavailability) of maintenance activities.
- Ensure that ports, services, user profiles or other means used for maintenance or diagnosis are active only when required.
- Incorporate planned downtime in an overall production schedule, and schedule the maintenance activities to minimize the adverse impact on business processes.

## IS Control Metrics for Monitoring

The following table provides an overview and examples of IS control metrics for monitoring IT operations management.

| Monthly Operations Management Monitoring | | | | | |
|---|---|---|---|---|---|
| **Attribute** | **Target** | **Current Period** | **Prior Period** | **Prior Period − 1** | **Prior Period − 2** |
| Number of operational incidents | | | | | |
| Total hours of unplanned downtime resulting from operational incidents | | | | | |
| Average downtime due to operational incidents | | | | | |
| Percent of operational SLAs exceeding established thresholds | | | | | |
| Percent of unplanned downtime incidents by cause: | | | | | |
| Equipment malfunction | | | | | |
| Procedural malfunction | | | | | |
| Documentation inefficiency | | | | | |
| Nonoperational incidents (programming, network, configuration, etc.) | | | | | |
| Number of hours lost due to late completion of scheduled work | | | | | |
| Average time to research and remediate operations incidents | | | | | |

## G. The Practitioner's Perspective

### Introduction

IT operations is the production arm of IT. Its primary role is the execution and delivery of scheduled and unscheduled IT services that support business processes.

The following section provides an overview of how the five CRISC domains (listed below) relate to IT operations management in practice:
- Domain 1—Risk Identification, Assessment and Evaluation
- Domain 2—Risk Response
- Domain 3—Risk Monitoring
- Domain 4—Information Systems Control Design and Implementation
- Domain 5—Information Systems Control Monitoring and Maintenance

### Domain 1

During systems development and/or acquisition, the risk identified with the delivery of processes should be included as part of the overall selection process. Once this has been established, the primary risk identification is focused on those incidents or issues that could affect the delivery of IT production services. Distributed systems, in which many operational activities are performed by the business unit, create additional risk since they are generally not managed by IT professionals, and their job function includes other operational duties.

Additional risk is introduced where applications or operations are outsourced.

### Domain 2

Risk response is based on the effective mitigation, resolution and notification of system or equipment failures and ensuring that all failures are rapidly identified, analyzed and contained.

The problem/incident tracking system should provide the notification and tracking of the response. Most operations risk is predictable and responses should be included in the operations procedures.

### Domain 3

The key risk indicators (KRIs) described previously provide a monitoring process for key issues that would normally be experienced by an IT operations organization. They provide a good view of trends and will highlight problems requiring management attention.

### Domain 4

The controls identified in the IS controls and objectives provides a resource of best practices for an IT operations unit. The risk practitioner will be concerned with the operations of IS controls being conducted with effective selection, configuration, implementation and maintenance of IS controls.

### Domain 5

IT operations management monitoring activities include:
- Monthly operations management monitoring described previously provides a monitoring benchmark. This report should be provided to IT management and business stakeholders on a monthly basis and summarized on an annual basis.
- Timely reporting using KRIs defined previously should be distributed to stakeholders within IT and the business units affected.
- Annual evaluation of the maturity of controls provides a barometer of controls in their current state, comparison to previous periods, and the target maturity level. Target maturity levels should be agreed on by stakeholders and evaluated by senior management as part of the IT annual review.
- Maturity assessment can be performed as a self-assessment, with oversight by an objective third party, peer review, or independent assessment (internal audit or external provider).
- Monitoring related to IT operations should be reported and evaluated routinely.

## H. Suggested Resources for Further Study

### Suggested Resources for Further Study

In addition to the resources cited throughout this manual, the following resources are suggested for further study:

- ISACA:
  - COBIT 4.1, 2007

    **Note:** The COBIT 4.1 framework is available at no charge from ISACA and can be downloaded at *www.isaca.org/cobit*. The new COBIT 5 framework will be available in 2012.
  - *The Risk IT Framework*, 2009
  - *The Risk IT Practitioner Guide*, 2009
  - *Enterprise Value: Governance of IT Investments, The Val IT Framework 2.0*, 2008
  - *Implementing and Continually Improving IT Governance*, 2010

# Study Questions, Answers and Explanations

**Note:** For more practice questions, you may also want to obtain a copy of the *CRISC® Review Questions, Answers & Explanations Manual 2011*, which consists of 100 multiple-choice study questions, answers and explanations, and the *CRISC® Review Questions, Answers & Explanations Manual 2012 Supplement*, which consists of 100 new multiple-choice study questions, answers and explanations.

## Study Questions

**Introduction**

Study questions are grouped by domain. Answers and explanations are provided following the 25 questions.

### Domain 1—Risk Identification, Assessment and Evaluation

**Question 1**

The **MOST** significant drawback of using quantitative analysis instead of qualitative risk analysis is the:

A. higher cost.
B. lower objectivity.
C. higher reliance on skilled personnel.
D. lower management buy-in.

**Question 2**

Which of the following business requirements **MOST** relates to the need for resilient business and information systems processes?

A. Effectiveness
B. Confidentiality
C. Integrity
D. Availability

**Question 3**

An enterprise that chooses not to engage in e-commerce is demonstrating a form of:

A. risk avoidance.
B. risk transfer.
C. risk treatment.
D. risk acceptance.

**Question 4**

The risk to an information system that supports a critical business process is owned by:

A. the IT director.
B. senior management.
C. the risk management department.
D. the system users.

**Question 5**

The **FIRST** step in the risk assessment process is the identification of:

A. assets.
B. threats.
C. vulnerabilities.
D. threat sources.

## Domain 2—Risk Response

**Question 6**

The preparation of a risk register begins in which risk management process?

A. Risk response planning
B. Risk monitoring and control
C. Risk identification
D. Risk management strategy planning

**Question 7**

To address the risk of the failure of operations staff to perform the daily backup, management requires that the systems administrator sign off on the daily backup. This is an example of:

A. risk transference.
B. risk avoidance.
C. risk mitigation.
D. risk acceptance.

**Question 8**

When a risk cannot be sufficiently mitigated through manual or automatic controls, which of the following options will **BEST** protect the enterprise from the potential financial impact of the risk?

A. Insuring against the risk
B. Updating the IT risk registry
C. Improving staff training in the risk area
D. Outsourcing the related business process to a third party

**Question 9**

When responding to an identified risk event, the **MOST** important stakeholders involved in reviewing risk response options to an IT risk are the:

A. information security managers.
B. internal auditors.
C. incident response team members.
D. business managers.

**Question 10**

What risk elements **MUST** be known in order to accurately calculate residual risk?

A. Threats and vulnerabilities
B. Inherent risk and control risk
C. Compliance risk and reputation
D. Risk governance and risk response

## Domain 3—Risk Monitoring

**Question 11**

The **MOST** important reason to maintain key risk indicators (KRIs) is because:

A. complex metrics require fine-tuning.
B. threats and vulnerabilities change over time.
C. risk reports need to be timely.
D. they help to avoid risk.

**Question 12**

To be effective, risk mitigation **MUST**:

A. minimize the residual risk.
B. minimize the inherent risk.
C. reduce the frequency of a threat.
D. reduce the impact of a threat.

**Question 13**

Which of the following is the **BEST** measure of the operational effectiveness of risk management capabilities?

A. Key performance indicators (KPIs)
B. Key risk indicators (KRIs)
C. Capability maturity models (CMMs)
D. Metric thresholds

**Question 14**

During a data extraction process the total number of transactions per year was forecasted by multiplying the monthly average by twelve. This is considered:

A. a controls total.
B. simplistic and ineffective.
C. a duplicates test.
D. a reasonableness test.

**Question 15**

The **PRIMARY** objective difference between an internal and an external risk management assessment would be the reviewer's:

A. professionalism.
B. quality of work.
C. independence.
D. ease of access.

## Domain 4—Information Systems Control Design and Implementation

**Question 16**

The **BEST** control to prevent unauthorized access to an enterprise's information is user:

A. accountability.
B. authentication.
C. identification.
D. authorization.

**Question 17**

Which of the following should be considered **FIRST** when designing information systems controls?

A. The organizational strategic plan
B. The existing IT environment
C. The present IT budget
D. The IT strategic plan

**Question 18**

Which of the following controls **BEST** protects an enterprise from unauthorized individuals gaining access to sensitive information?

A. Using a challenge response system
B. Forcing periodic password changes
C. Monitoring and recording unsuccessful logon attempts
D. Providing access on a need-to-know basis

**Question 19**

A poor choice of passwords and transmission over unprotected communications lines are examples of:

A. vulnerabilities.
B. threats.
C. probabilities.
D. impacts.

**Question 20**

Which of the following is the **BEST** defense against successful phishing attacks?

A. An intrusion detection system (IDS)
B. Spam filters
C. End-user awareness
D. Application hardening

## Domain 5—Information Systems Monitoring and Maintenance

**Question 21**

An enterprise has implemented a tool that correlates information from multiple sources. This is an example of a monitoring tool that focuses on:

A. transaction data.
B. configuration settings.
C. system changes.
D. process integrity.

**Question 22**

The **BEST** test for confirming the effectiveness of the system access management process is to map:

A. access requests to user accounts.
B. user accounts to access requests.
C. user accounts to human resources (HR) records.
D. the vendor database to user accounts.

**Question 23**

Which of the following provides the **BEST** assurance that a firewall is configured in compliance with an enterprise's security policy?

A. Review the actual procedures.
B. Interview the firewall administrator.
C. Review the parameter settings.
D. Review the device's log file for recent attacks.

**Question 24**

One way to verify control effectiveness is by determining:

A. its reliability.
B. whether it is preventive or detective.
C. the capability of providing notification of failure.
D. the test results of intended objectives.

**Question 25**

Which of the following is the **MOST** effective way to ensure that outsourced service providers comply with the enterprise's information security policy?

A. Periodic audits
B. Security awareness training
C. Penetration testing
D. Service level monitoring

# Study Questions, Answers and Explanations

## Answers and Explanations

**Introduction**

Answers and explanations are provided for the 25 study questions.

**Answer Key**

The following table:
- Lists and explains the correct answers to the study questions
- Explains why the other answer choices are incorrect *ISACA JOURNAL* VOLUME 1, 2012

| Domain 1—Risk Identification, Assessment and Evaluation | | | |
|---|---|---|---|
| Answers and Explanations | | | |
| **Question No.** | **Correct Answer** | **Incorrect Choice** | **Explanation** |
| 1 | **A** | | Quantitative risk analysis is generally more complex and thus more costly than qualitative risk analysis. |
| | | B | Neither of the two risk analysis methods is fully objective; while the qualitative method subjectively assigns High, Medium and Low frequency and impact categories to a specific risk, subjectivity within the quantitative method is often expressed in mathematical “weights.” |
| | | C | To be effective, both processes require personnel who have a good understanding of the business. |
| | | D | Quantitative analysis generally has a better buy-in than qualitative analysis to the point where it can cause overreliance on the results. |
| 2 | | A | Confidentiality deals with the protection of sensitive information from unauthorized disclosure. While the lack of system resilience can in some cases affect data confidentiality, resilience is more closely linked to the business information requirement of availability. |
| | | B | Integrity relates to the accuracy and completeness of information as well as to its validity in accordance with business values and expectations. While the lack of system resilience can in some cases affect data integrity, resilience is more closely linked to the business information requirement of availability. |
| | | C | Effectiveness deals with information being relevant and pertinent to the business process as well as being delivered in a timely, correct, consistent and usable manner. While the lack of system resilience can in some cases affect effectiveness, resilience is more closely linked to the business information requirement of availability. |
| | **D** | | Availability relates to information being available when required by the business process—now and in the future. Resilience is the ability to provide and maintain an acceptable level of service during disasters or when facing operational challenges. |

| Domain 1—Risk Identification, Assessment and Evaluation *(cont.)* | | | |
|---|---|---|---|
| **Answers and Explanations** | | | |
| **Question No.** | **Correct Answer** | **Incorrect Choice** | **Explanation** |
| 3 | **A** | | Each business process involves inherent risk. Not engaging in any activity avoids the inherent risk associated with the activity. |
| | | B | Risk transfer/sharing means reducing either risk frequency or impact by transferring or otherwise sharing a portion of the risk. Common techniques include insurance and outsourcing. These techniques do not relieve an enterprise of a risk, but can involve the skills of another party in managing the risk and reducing the financial consequence if an adverse event occurs. |
| | | C | Risk treatment means that action is taken to reduce the frequency and/or impact of a risk. |
| | | D | Acceptance means that no action is taken relative to a particular risk, and loss is accepted when/if it occurs. This is different from being ignorant of risk; accepting risk assumes that the risk is known, i.e., an informed decision has been made by management to accept it as such. |
| 4 | | A | The IT director manages the IT systems on behalf of the business owners. |
| | **B** | | Senior management is responsible for the acceptance and mitigation of all risk. |
| | | C | The risk management department determines and reports on level of risk, but does not own the risk. |
| | | D | The system users are responsible for utilizing the system properly and following procedures, but they do not own the risk. |
| 5 | **A** | | Asset identification is the most crucial and first step in the risk assessment process. Risk identification, assessment and evaluation (analysis) should always be clearly aligned to assets. Assets can be people, processes, infrastructure, information or applications. |
| | | B | While threats tie into the risk assessment process, they are not relevant unless they can be related to specific assets. |
| | | C | While vulnerabilities tie into the risk assessment process, they are not relevant unless they can be related to specific assets. |
| | | D | While threat sources tie into the risk assessment process, they are not relevant unless they can be related to specific assets. |

| Domain 2—Risk Response | | | |
|---|---|---|---|
| Answers and Explanations | | | |
| **Question No.** | **Correct Answer** | **Incorrect Choice** | **Explanation** |
| 6 | | A | In the risk response planning process, appropriate responses are chosen, agreed on, and included in the risk register. |
| | | B | Risk monitoring and control often requires identification of new risks and reassessment of risks. Outcomes of risk reassessments, risk audits and periodic risk reviews trigger updates to the risk register. |
| | **C** | | While the risk register details all identified risks, including description, category, cause, probability of occurring, impact(s) on objectives, proposed responses, owners, and current status, the primary outputs from risk identification are the initial entries into the risk register. |
| | | D | Risk management strategy planning describes how risk management will be structured and performed. |
| 7 | | A | The stem does not describe the sharing of risk. Transference is the strategy that provides for sharing risk with partners or taking insurance coverage. |
| | | B | The stem does not describe risk avoidance. Avoidance is a strategy that provides for not implementing certain activities or processes that would incur risk. |
| | **C** | | Mitigation is the strategy that provides for the definition and implementation of controls to address the risk described. |
| | | D | The stem does not describe risk acceptance. Acceptance is a strategy that provides for formal acknowledgment of the existence of a risk and the monitoring of that risk. |
| 8 | **A** | | An insurance policy can compensate the enterprise up to 100 percent by transferring the risk to another company. |
| | | B | Updating the risk registry (with lower values for impact and probability) will not actually change the risk, only management's perception of it. |
| | | C | Staff capacity to detect or mitigate the risk may potentially reduce the financial impact, but insurance allows for the risk to be completely mitigated. |
| | | D | Outsourcing the process containing the risk does not necessarily remove or change the risk. |
| 9 | | A | Information security managers may best understand the technical tactical situation, but business managers are accountable for managing the associated risk and will determine what actions to take based on the information provided by others, which includes collaboration with, and support from, IT security managers. |
| | | B | This is not internal audit's function. Business managers set priorities, possibly consulting with other parties, which may include internal audit. |
| | | C | The incident response team must ensure open communication to management and stakeholders to ensure that business managers/leaders understand the associated risk and are provided enough information to make informed risk-based decisions. |
| | **D** | | Business managers are accountable for managing the associated risk and will determine what actions to take based on the information provided by others. |

| Domain 2—Risk Response *(cont.)* | | | |
|---|---|---|---|
| **Answers and Explanations** | | | |
| **Question No.** | **Correct Answer** | **Incorrect Choice** | **Explanation** |
| 10 | | A | Threats and vulnerabilities are elements of inherent risk. They do not accurately calculate residual risk. |
| | **B** | | Inherent risk (threats × vulnerabilities) multiplied by control risk is the formula to calculate residual risk. |
| | | C | Compliance risk is the current and prospective risk to earnings or capital arising from violations of, or nonconformance with, laws, rules, regulations, prescribed practices, internal policies and procedures, or ethical standards. Compliance risk can lead to reputational damage. |
| | | D | Risk governance and risk response are risk domains, not risk elements, to calculate residual risk. |

| Domain 3—Risk Monitoring | | | |
|---|---|---|---|
| **Answers and Explanations** | | | |
| **Question No.** | **Correct Answer** | **Incorrect Choice** | **Explanation** |
| 11 | | A | While most key risk indicator (KRI) metrics need to be optimized in respect to their sensitivity, the most important objective of KRI maintenance is to ensure that KRIs continue to effectively capture the changes in threats and vulnerabilities over time. |
| | **B** | | Threats and vulnerabilities change over time and KRI maintenance ensures that KRIs continue to effectively capture these changes. |
| | | C | Risk reporting timeliness is a business requirement, but is not a driver for KRI maintenance. |
| | | D | Risk avoidance is one possible risk response. Risk responses are based on KRI reporting. |
| 12 | | A | The objective of risk reduction is to reduce the residual risk to levels below the enterprise's risk tolerance level. |
| | **B** | | The inherent risk of a process is a given and cannot be affected by risk reduction/risk mitigation efforts. |
| | | C | Risk reduction efforts can focus on either avoiding the frequency of the risk or reducing the impact of a risk. |
| | | D | Risk reduction efforts can focus on either avoiding the frequency of the risk or reducing the impact of a risk. |
| 13 | **A** | | Key performance indicators (KPIs) provide insights into the operational effectiveness of the concept or capability that they monitor. |
| | | B | Key risk Indicators (KRIs) only provide insights into potential risks that may exist or be realized within a concept or capability that they monitor. |
| | | C | Capability maturity models (CMMs) assess the maturity of a concept or capability and do not provide insights into operational effectiveness. |
| | | D | Metric thresholds are decision or action points that are enacted when a KPI or KRI reports a specific value or set of values. |

## Domain 3—Risk Monitoring *(cont.)*

### Answers and Explanations

| Question No. | Correct Answer | Incorrect Choice | Explanation |
|---|---|---|---|
| 14 | | A | The described test does not ensure that all transactions have been extracted. |
| | | B | While simplistic, the reasonableness test is a valid foundation for more elaborate data validation tests. |
| | | C | The described test does not identify duplicate transactions. |
| | **D** | | Reasonableness tests make certain assumptions about the information as the basis for more elaborate data validation tests. |
| 15 | | A | This choice can vary subjectively. |
| | | B | This choice can vary subjectively. |
| | **C** | | Independence is the freedom from conflict of interest and undue influence. By the mere fact that the external auditors are from a different entity, their independence level is higher than if the reviewer were from inside the entity for which they are performing a review. Independence is directly linked to objectivity. |
| | | D | This choice can vary subjectively. |

## Domain 4—Information Systems Control Design and Implementation

### Answers and Explanations

| Question No. | Correct Answer | Incorrect Choice | Explanation |
|---|---|---|---|
| 16 | | A | User accountability does not grant access. |
| | **B** | | Authentication verifies the user's identity and the right to access information according to the access rules. |
| | | C | User identification without authentication does not grant access. |
| | | D | User authorization without authentication does not grant access. |
| 17 | **A** | | Review of the enterprise's strategic plan is the first step in designing effective IS controls that would fit the enterprise's long-term plans. |
| | | B | Review of the existing IT environment, although useful and necessary, is not the first task that needs to be undertaken. |
| | | C | The present IT budget is just one of the components of the strategic plan. |
| | | | The IT strategic plan exists to support the enterprise's strategic plan. |
| 18 | | A | Verifying the user's identification through a challenge response does not completely address the issue of access risk if access was not appropriately designed in the first place. |
| | | B | Forcing users to change their passwords does not guarantee that access control is appropriately assigned. |
| | | C | Logon and monitoring unsuccessful access attempts does not address the risk of appropriate access rights. |
| | **D** | | Physical or logical system access should be assigned on a need-to-know basis, where there is a legitimate business requirement based on least privilege and segregation of duties. |

| Domain 4—Information Systems Control Design and Implementation *(cont.)* | | | |
|---|---|---|---|
| **Answers and Explanations** | | | |
| **Question No.** | **Correct Answer** | **Incorrect Choice** | **Explanation** |
| 19 | **A** | | Vulnerabilities represent characteristics of information resources that may be exploited by a threat. The stem describes such a situation. |
| | | B | Threats are circumstances or events with the potential to cause harm to information resources. The stem does not describe a threat. |
| | | C | Probabilities represent the likelihood of the occurrence of a threat. The stem does not describe a probability. |
| | | D | Impacts represent the outcome or result of a threat exploiting a vulnerability. The stem does not describe an impact. |
| 20 | | A | An intrusion detection system (IDS) does not protect against phishing attacks since phishing attacks usually do not have the same patterns or unique signatures. |
| | | B | While certain highly specialized spam filters can reduce the number of phishing e-mails that reach their addressees' "in" boxes, they are not as effective in addressing phishing attacks as end-user awareness. |
| | **C** | | Phishing attacks are a type of to social engineering attack and are best defended by end-user awareness training. |
| | | D | Application hardening does not protect against phishing attacks since phishing attacks generally use e-mail as the attack vector, with the end-user as the vulnerable point, not the application. |

| Domain 5—Information Systems Control Monitoring and Maintenance | | | |
|---|---|---|---|
| **Answers and Explanations** | | | |
| **Question No.** | **Correct Answer** | **Incorrect Choice** | **Explanation** |
| 21 | **A** | | Monitoring tools focusing on transaction data generally correlate information from one system to another, such as employee data from the human resources (HR) system with spending information from the expense system or the payroll system. |
| | | B | Configuration settings are generally compared against predefined values and not based on the correlation between systems. |
| | | C | System changes are compared from a previous state to the current state. |
| | | D | Process integrity is confirmed within the system. |
| 22 | | A | Tying access requests to user accounts confirms that all access requests have been processed; however, the test does not consider user accounts that have been established without the supporting access request. |
| | **B** | | Tying user accounts to access requests confirms that all existing accounts have been approved. |
| | | C | Tying user accounts to human resources (HR) records confirms whether user accounts are uniquely tied to employees. |
| | | D | Tying vendor records to user accounts may confirm valid accounts on an e-commerce application, but is similarly flawed as choices A and C since it does not consider user accounts that have been established without the supporting access request. |

| Domain 5—Information Systems Control Monitoring and Maintenance *(cont.)* | | | |
|---|---|---|---|
| **Answers and Explanations** | | | |
| **Question No.** | **Correct Answer** | **Incorrect Choice** | **Explanation** |
| 23 | | A | While procedures may provide a good understanding of how the firewall is supposed to be managed, they do not reliably confirm that the firewall configuration complies with the enterprise's security policy. |
| | | B | While interviewing the firewall administrator may provide a good process overview, it does not reliably confirm that the firewall configuration complies with the enterprise's security policy. |
| | **C** | | A review of the parameter settings will provide a good basis for comparison of the actual configuration to the security policy and will provide reliable audit evidence documentation. |
| | | D | While reviewing the device's log file for recent attacks may provide indirect evidence about the fact that logging is enabled, it does not reliably confirm that the firewall configuration complies with the enterprise's security policy. |
| 24 | | A | Reliability is not an indication of control strength; weak controls can be highly reliable, even if they do not meet the control objective. |
| | | B | The type of control (preventive or detective) does not help determine control effectiveness. |
| | | C | Notification of failure does not determine control strength. |
| | **D** | | Control effectiveness requires a process to verify that the control process worked as intended and meets the intended control objectives. |
| 25 | **A** | | Regular audits can spot gaps in information security compliance. |
| | | B | Training can increase user awareness of the information security policy, but is not more effective than auditing. |
| | | C | Penetration testing can identify security vulnerability, but cannot ensure information compliance. |
| | | D | Service level monitoring can only pinpoint operational issues in the enterprise's operational environment. |

# Glossary

| Term | CRISC Definition |
|---|---|
| **Access control** | The processes, rules and deployment mechanisms which control access to information systems, resources and physical access to premises |
| **Access rights** | The permission or privileges granted to users, programs or workstations to create, change, delete or view data and files within a system, as defined by rules established by data owners and the information security policy |
| **Application control** | The policies, procedures and activities designed to provide reasonable assurance that objectives relevant to a given automated solution (application) are achieved |
| **Asset** | Something of either tangible or intangible value worth protecting, including people, information, infrastructure, finances and reputation |
| **Authentication** | 1. The act of verifying identity, i.e., user, system<br>2. Can also refer to the verification of the correctness of a piece of data |
| **Availability** | Information that is accessible when required by the business process now and in the future |
| **Balanced scorecard (BSC)** | Developed by Robert S. Kaplan and David P. Norton<br><br>A coherent set of performance measures organized into four categories that includes traditional financial measures, but adds customer, internal business process, and learning and growth perspectives |
| **Business case** | Documentation of the rationale for making a business investment, used both to support a business decision on whether to proceed with the investment and as an operational tool to support management of the investment through its full economic life cycle |
| **Business continuity plan (BCP)** | A plan used by an organization to respond to disruption of critical business processes<br><br>Depends on the contingency plan for restoration of critical systems |
| **Business goal** | The translation of the enterprise's mission from a statement of intention into performance targets and results |
| **Business impact** | The net effect, positive or negative, on the achievement of business objectives |
| **Business impact analysis/ assessment (BIA)** | Evaluating the criticality and sensitivity of information assets<br><br>An exercise that determines the impact of losing the support of any resource to an organization, establishes the escalation of that loss over time, identifies the minimum resources needed to recover, and prioritizes the recovery of processes and supporting system<br><br>**Scope note:**<br>This process also includes addressing:<br>• Income loss<br>• Unexpected expense<br>• Legal issues (regulatory compliance or contractual)<br>• Interdependent processes<br>• Loss of public reputation or public confidence |
| **Business objective** | A further development of the business goals into tactical targets and desired results and outcomes |
| **Business process owner** | The individual responsible for identifying process requirements, approving process design and managing process performance<br><br>**Scope note:**<br>A business process owner must be at an appropriately high level in the enterprise and have authority to commit resources to process-specific risk management activities. |

| Term | CRISC Definition |
|---|---|
| **Business risk** | A probable situation with uncertain frequency and magnitude of loss (or gain) |
| **Capability** | An aptitude, competency or resource that an enterprise may possess or require at an enterprise, business function or individual level that has the potential or is required to contribute to a business outcome and to create value |
| **Capability maturity model (CMM)** | 1. Contains the essential elements of effective processes for one or more disciplines<br><br>It also describes an evolutionary improvement path from *ad hoc*, immature processes to disciplined, mature processes with improved quality and effectiveness.<br><br>2. CMM for software, from the Software Engineering Institute (SEI), is a model used by many organizations to identify best practices useful in helping them assess and increase the maturity of their software development processes.<br><br>**Scope note:**<br>The CMM ranks software development organizations according to a hierarchy of five process maturity levels. Each level ranks the development environment according to its capability of producing quality software. A set of standards is associated with each of the five levels. The standards for level one describe the most immature or chaotic processes and the standards for level five describe the most mature or quality processes.<br><br>A maturity model that indicates the degree of reliability or dependency the business can place on a process achieving the desired goals or objectives<br><br>A collection of instructions an organization can follow to gain better control over its software development process |
| **Compensating control** | An internal control that reduces the risk of an existing or potential control weakness resulting in errors and omissions |
| **Computer emergency response team (CERT)** | A group of people integrated at the organization with clear lines of reporting and responsibilities for standby support in case of an information systems emergency<br><br>This group will act as an efficient corrective control, and should also act as a single point of contact for all incidents and issues related to information systems. |
| **Confidentiality** | The protection of sensitive or private information from unauthorized disclosure |
| **Data custodian** | The individuals and departments responsible for the storage and safeguarding of computerized data |
| **Data owner** | The individuals, normally managers or directors, who have responsibility for the integrity, accurate reporting and use of computerized data |
| **Detective controls** | Controls that exist to detect and report when errors, omissions and unauthorized uses or entries occur |
| **Disaster recovery plan (DRP)** | A set of human, physical, technical and procedural resources to recover, within a defined time and cost, an activity interrupted by an emergency or disaster |
| **Enterprise resource planning (ERP) system** | An integrated system containing multiple business subsystems |
| **Enterprise risk management (ERM)** | The discipline by which an enterprise in any industry assesses, controls, exploits, finances and monitors risks from all sources for the purpose of increasing the enterprise's short- and long-term value to its stakeholders |
| **Event** | Something that happens at a specific place and/or time |

| Term | CRISC Definition |
|---|---|
| **Event type** | For the purpose of IT risk management, one of three possible sorts of events: threat event, loss event and vulnerability event.<br><br>**Scope note:**<br>Being able to consistently and effectively differentiate the different types of events that contribute to risk is a critical element in developing good risk-related metrics and well-informed decisions. Unless these categorical differences are recognized and applied, any resulting metrics lose meaning and, as a result, decisions based on those metrics are far more likely to be flawed. |
| **Evidence** | 1. Information that proves or disproves a stated issue<br>2. Information an auditor gathers in the course of performing an IS audit; relevant if it pertains to the audit objectives and has a logical relationship to the findings and conclusions it is used to support |
| **Fallback procedures** | A plan of action or set of procedures to be performed if a system implementation, upgrade or modification does not work as intended<br><br>**Scope note:**<br>Fallback procedures may involve restoring the system to its state prior to the implementation or change. Fallback procedures are needed to ensure that normal business processes continue in the event of failure and should always be considered in system migration or implementation. |
| **Feasibility study** | A phase of a system development life cycle (SDLC) methodology that researches the feasibility and adequacy of resources for the development or acquisition of a system solution to a user need |
| **Frequency** | A measure of the rate by which events occur over a certain period of time |
| **Governance** | The oversight, direction and high-level monitoring and control of an enterprise to ensure the achievement of defined and approved objectives |
| **Impact analysis** | A study to prioritize the criticality of information resources for the organization based on costs (or consequences) of adverse events<br><br>In an impact analysis, threats to assets are identified and potential business losses determined for different time periods. This assessment is used to justify the extent of safeguards that are required and recovery time frames. This analysis is the basis for establishing the recovery strategy. |
| **Information systems (IS)** | The combination of strategic, managerial and operational activities involved in the gathering, processing, storing, distributing and use of information, and its related technologies.<br><br>**Scope note:**<br>Information systems are distinct from information technology (IT) in that an information system has an IT component that interacts with the process components. |
| **Inherent risk** | 1. The risk level or exposure without taking into account the actions that management has taken or might take (e.g., implementing controls)<br>2. The risk that a material error could occur, assuming that there are no related internal controls to prevent or detect the error<br><br>**Scope note:**<br>Audit perspective |
| **Integrity** | The accuracy, completeness and validity of information |
| **Internal control** | The policies, procedures, practices and organizational structures designed to provide reasonable assurance that the business objectives will be achieved and undesired events will be prevented or detected |
| **IT architecture** | Description of the fundamental underlying design of the IT components of the business, the relationships among them, and the manner in which they support the organization's objectives |

| Term | CRISC Definition |
|---|---|
| **IT infrastructure** | The set of hardware, software and facilities that integrates an organizations' IT assets<br><br>**Scope note:**<br>Specifically, the equipment (including servers, routers, switches, and ca!bling), software, services and products used in storing, processing, transmitting and displaying all forms of information for the organization's users |
| **IT-related incident** | An IT-related event that causes an operational, developmental and/or strategic business impact |
| **IT risk** | The business risk associated with the use, ownership, operation, involvement, influence and adoption of IT within an enterprise |
| **IT risk issue** | 1. An instance of an IT risk<br>2. A combination of control, value and threat conditions that impose a noteworthy level of IT risk |
| **IT risk profile** | A description of the overall (identified) IT risk to which the enterprise is exposed |
| **IT risk register** | A repository of the key attributes of potential and known IT risk issues<br><br>Attributes may include name, description, owner, expected/actual frequency, potential/actual magnitude, potential/actual business impact and disposition. |
| **IT risk scenario** | The description of an IT-related event that can lead to a business impact |
| **Key performance indicator (KPI)** | A measure that determines how well the process is performing in enabling the goal to be reached<br><br>**Scope note:**<br>A KPI is a lead indicator of whether a goal will likely be reached, and a good indicator of capability, practices and skills. It measures an activity goal, which is an action the process owner must take to achieve effective process performance. |
| **Key risk indicator (KRI)** | A subset of risk indicators that are highly relevant and possess a high probability of predicting or indicating important risk<br><br>**Scope note:**<br>See *risk indicator.* |
| **Loss event** | Any event where a threat event results in loss |
| **Magnitude** | A measure of the potential severity of loss or the potential gain from realized events/scenarios |
| **Objectivity** | The ability to exercise judgment, express opinions and present recommendations with impartiality |
| **Preventive control** | An internal control that is used to avoid undesirable events, errors and other occurrences that an organization has determined could have a negative material effect on a process or end product |
| **Project portfolio** | The set of projects owned by a company<br><br>**Scope note:**<br>It usually includes the main guidelines relative to each project, including objectives, costs, timelines and other information specific to the project. |
| **Recovery point objective (RPO)** | The RPO is determined based on the acceptable data loss in case of a disruption of operations. It indicates the earliest point in time to which it is acceptable to recover the data. The RPO effectively quantifies the permissible amount of data loss in case of interruption. |
| **Recovery time objective (RTO)** | The amount of time allowed for the recovery of a business function or resource after a disaster occurs |
| **Residual risk** | The remaining risk after management has implemented risk response |
| **Resilience** | The ability of a system or network to resist failure or to recover quickly from any disruption, usually with minimal recognizable effect |
| **Risk aggregation** | The process of integrating risk assessments at a corporate level to obtain a complete view on the overall risk for the enterprise |

| Term | CRISC Definition |
|---|---|
| **Risk analysis** | 1. A process by which frequency and magnitude of IT risk scenarios are estimated<br>2. The initial steps of risk management: analyzing the value of assets to the business, identifying threats to those assets and evaluating how vulnerable each asset is to those threats<br><br>**Scope note:**<br>It often involves an evaluation of the probable frequency of a particular event, as well as the probable impact of that event. |
| **Risk appetite** | The amount of risk, on a broad level, that an entity is willing to accept in pursuit of its mission |
| **Risk culture** | The set of shared values and beliefs that governs attitudes toward risk-taking, care and integrity, and determines how openly risks and losses are reported and discussed |
| **Risk factor** | A condition that can influence the frequency and/or magnitude and, ultimately, the business impact of IT-related events/scenarios |
| **Risk indicator** | A metric capable of showing that the enterprise is subject to, or has a high probability of being subject to, a risk that exceeds the defined risk appetite |
| **Risk management** | The coordinated activities to direct and control an organization with regard to risk.<br><br>In the International Standard, the term "control" is used as a synonym for "measure" (ISO/IEC Guide 73:2002). |
| **Risk map** | A (graphic) tool for ranking and displaying risks by defined ranges for frequency and magnitude |
| **Risk portfolio view** | 1. A method to identify interdependencies and interconnections among risks, as well as the effect of risk responses on multiple risks<br>2. A method to estimate the aggregate impact of multiple risks (e.g., cascading and coincidental threat types/scenarios, risk concentration/correlation across silos) and the potential effect of risk response across multiple risks |
| **Risk tolerance** | The acceptable level of variation that management is willing to allow for any particular risk as it pursues objectives |
| **System development life cycle (SDLC)** | The phases deployed in the development or acquisition of a software system<br><br>**Scope note:**<br>An approach used to plan, design, develop, test and implement an application system or a major modification to an application system. Typical phases of SDLC include the feasibility study, requirements study, requirements definition, detailed design, programming, testing, installation and postimplementation review, but not the service delivery or benefits realization activities. |
| **Threat** | Anything (e.g., object, substance, human) that is capable of acting against an asset in a manner that can result in harm<br><br>**Scope note:**<br>A potential cause of an unwanted incident (ISO/IEC 13335) |
| **Threat event** | Any event where a threat element/actor acts against an asset in a manner that has the potential to directly result in harm |
| **Vulnerability** | A weakness in the design, implementation, operation or internal control of a process that could expose the system to adverse threats from threat events |
| **Vulnerability event** | Any event where a material increase in vulnerability results.<br><br>Note that this increase in vulnerability can result from changes in control conditions or from changes in threat capability/force. |

**Page intentionally left blank**

# Suggested Resources for Further Study

As candidates read through this manual and encounter topics that are new to them or ones for which they feel their knowledge and experience are limited, additional references should be sought. Suggested resources for further study are provided in each of the chapters in Parts I and II of this manual. Also presented, below, is a comprehensive alphabetical list that includes all of the references provided in Parts I and II. Publications in **boldface** are available through the ISACA Bookstore.

Alberts, Christopher; Audrey Dorofee; *Managing Information Security Risks: The OCTAVE℠ Approach*, Addison-Wesley Professional, USA, 2002

Barnier, Brian; *The Operational Risk Handbook for Financial Companies: A Guide to the New World of Performance-oriented Operational Risk*, Harriman House Ltd, UK, 2011

Blokdijk, Gerard; Claire Engle; Jackie Brewster; *IT Risk Management Guide—Risk Management Implementation Guide: Presentations, Blueprints, Templates; Complete Risk Management Toolkit Guide for Information Technology Processes and Systems*, Emereo Publishing, Australia, 2008

British Standards Institution (BSI), BS 25999, *A Code of Practice for Business Continuity Management*, UK, 2007

Business Continuity Institute (BCI), *Good Practice Guidelines 2010*, UK, 2010

Committee of Sponsoring Organizations of the Treadway Commission (COSO), *Internal Control—Integrated Framework: Guidance on Monitoring Internal Control Systems*, USA, 2009

IEEE Computer Society, Standard 1074-2006 for Developing a Software Project Lifecycle Process, USA, 2006, *www.ieee.org*

International Organization for Standardization (ISO), ISO 27001, *Information security management—Specification with guidance for use*, Switzerland, 2005. This is the replacement for BS7799-2. It is intended to provide the foundation for third-party audit and is harmonized with other management standards, such as ISO/IEC 9001 and 14001.

ISO, ISO/IEC 27001:2005, *Information technology—Security techniques—Information security management systems—Requirements*, Switzerland, 2005

ISO, ISO/IEC 27002:2005, *Code of practice for information security management*, Appendix 1, Switzerland, 2005

ISO, ISO/IEC 27004:2009, *Information technology—Security techniques—Information security management—Measurement*, Switzerland, 2009

ISO/IEC 27005:2009

ISO, ISO 27005:2011, *Information technology—Security techniques—Information security risk management*, Switzerland, 2011

International Project Management Association (IPMA), *IPMA Competence Baseline (ICB), Version 3.0*, The Netherlands, 2006

## ISACA

- **Frameworks and related publications:**
  - **COBIT 4.1, 2007**
    **Note:** The COBIT 4.1 framework is available at no charge from ISACA and can be downloaded at *www.isaca.org/cobit*. The new COBIT 5 framework will be available in 2012.
  - ***COBIT and Application Controls: A Management Guide*, 2009**
  - ***Implementing and Continually Improving IT Governance*, 2010**
  - ***IT Assurance Guide: Using COBIT*, 2007**
  - ***The Risk IT Framework*, 2009**
  - ***The Risk IT Practitioner Guide*, 2009**
  - ***Enterprise Value: Governance of IT Investments, The Val IT Framework 2.0*, 2008**

***Note: Publications in bold are stocked in the ISACA Bookstore.***

- **Frameworks and related publications: *(cont.)***
  - ***Value Management Guidance for Assurance Professionals: Using Val IT 2.0, 2010***
  - ***ITAF: A Professional Practices Framework for IT Assurance*, 2008**
  - ***The Business Model for Information Security (BMIS)*, 2010**
- **Executive and management guidance:**
  - ***Information Security Governance: Guidance for Boards of Directors and Executive Management, 2nd Edition*, 2006**
- **Practitioner guidance:**
  - ***Change Management Audit/Assurance Program*, 2009**
  - ***Monitoring Internal Control Systems and IT: A Primer for Business Executives, Managers and Auditors on How to Embrace and Advance Best Practices*, 2010**
  - ***Systems Development and Project Management Audit/Assurance Program*, 2009**
- **Certification publications:**
  - ***CISM Review Manual 2012*, 2011**

Jones, J., *An Introduction to Factor Analysis of Information Risk (FAIR)*, Risk Management Insight LLC, USA, November 2006

**Kouns, Jake; Daniel Minoli; *Information Technology Risk Management in Enterprise Environments: A Review of Industry Practices and a Practical Guide to Risk Management Teams*, Wiley-Interscience, USA, 2010**

Kovacich, Gerald L.; Edward Halibozek; *The Manager's Handbook for Corporate Security: Establishing and Managing a Successful Assets Protection Program*, Butterworth-Heinemann, USA, 2003

Krutz, Ronald L.; Russell Dean Vines; *The CISM Prep Guide: Mastering the Five Domains of Information Security Management*, Wiley, USA, 2003

Lientz, Bennet P.; Lee Larssen; *Risk Management for IT Projects: How to Deal With Over 150 Issues and Risks*, Butterworth-Heinemann, USA, 2006

**Maizlish, Bryan; Robert Handler; *IT Portfolio Management Step-by-Step: Unlocking the Business Value of Technology*, John Wiley & Sons, USA, 2005**

National Institute of Technology and Standards (NIST), *Guide for Assessing the Security Controls in Federal Information Systems and Organizations, Building Effective Security Assessment Plans*, Special Publication (SP) 800-53A, Revision 1, USA, 2010

National Institute of Technology and Standards (NIST), *Risk Management Guide for Information Technology Systems*, Special Publication (SP) 800-30, *www.csrc.nist.gov*

National Institute of Technology and Standards (NIST), *Security Controls in External Environments*, Special Publication (SP) 800-53, Revision 3, Section 2.4, Appendix 2, USA, 2009

National Institute of Technology and Standards (NIST), Special Publication (SP) 800-53, Revision 3, On-line Database, USA, *http://web.nvd.nist.gov/view/800-53/home*

Office of Government Commerce (OGC), *ITIL: IT Service Management*, Version 3, UK, 2009, *http://itil.osiatis.es/ITIL_course/it_service_management/change_management/overview_change_management/overview_change_management.php*

OGC, *Projects in Controlled Environments 2 (PRINCE2): Directing Successful Projects With PRINCE2*, UK, 2009

OGC, *Projects in Controlled Environments 2 (PRINCE2): Managing Successful Projects With PRINCE2*, UK, 2009

Peltier, Thomas R.; *Information Security Risk Analysis, 3rd Edition*, Auerbach Publications, USA, 2010

Project Management Institute (PMI), *A Guide to the Project Management Body of Knowledge (PMBOK), 4th Edition*, USA, 2008

Sherwood, John; Andrew Clark; David Lynas; *Enterprise Security Architecture: A Business-Driven Approach*; CMP Books, USA, 2005

**Westerman, George; Richard Hunter; *IT Risk: Turning Business Threats Into Competitive Advantage*, Harvard Business School Press, USA, 2007**


***Note: Publications in bold are stocked in the ISACA Bookstore.***

# General CRISC Information

## Requirements for Certification

To earn the CRISC designation, the following requirements must be met:
1. Pass the CRISC exam
2. Submit an application (within five years of the passing date) with verified evidence of a minimum of at least three years of cumulative work experience performing the tasks of a CRISC professional across at least three CRISC domains. There will be no substitutions or experience waivers. A processing fee of $50 must accompany all applications.
3. Adhere to the ISACA Code of Professional Ethics
4. Agree to comply with the CRISC continuing education policy

Please note that certification application decisions are not final as there is an appeal process for certification application denials. Inquiries regarding denials of certification can be sent to *certification@isaca.org*.

## Work Experience

Work experience must be gained within the 10-year period preceding the application for certification or within five years from the date of initially passing the exam. An application for certification must be submitted within five years from the passing date of the CRISC exam. All experience must be verified independently with employers.

**Note:** A CRISC candidate may choose to take the CRISC exam prior to meeting the experience requirements.

## Description of the Exam

The CRISC Certification Committee oversees the development of the exam and ensures the currency of its content.

The exam consists of 200 multiple-choice questions that cover the CRISC job practice domains.

The job practice was developed and validated using prominent industry leaders, subject matter experts and industry practitioners.

## Exam Schedule

The CRISC exam will be administered twice in 2012:
- Saturday, 9 June 2012
- Saturday, 8 December 2012

Any changes to this schedule will be specified in the *CRISC Bulletin of Information* (*www.isaca.org/criscboi*).

## Exam Registration

Refer to the *CRISC™ Bulletin of Information* at *www.isaca.org/criscboi* for specific registration deadlines and registration forms. Registration for the exam can be completed online at *www.isaca.org/examreg*.

The *Candidate's Guide to the CRISC™ Exam and Certification* will be sent to candidates upon receipt and recording of their exam registration and payment.

## Exam Administration

ISACA has contracted with an internationally recognized testing agency. This not-for-profit corporation engages in the development and administration of credentialing exams for certification and licensing purposes. It assists ISACA in the construction, administration and scoring of the CRISC exam.

## Exam Admission Requirements

All of the following are required to be admitted to take the exam:

**Timely arrival**—Report to the testing site at the report time indicated on the admission ticket.

NO CANDIDATE WILL BE ADMITTED TO THE TEST CENTER ONCE THE CHIEF EXAMINER BEGINS READING THE ORAL INSTRUCTIONS.

> **Note:** Candidates who arrive after the oral instructions have begun will not be allowed to sit for the exam and will forfeit their registration fee.

**Admission ticket**—To be admitted into the test site, candidates must bring the e-mail printout OR hard copy admission ticket. Candidates can use their admission tickets only at the designated test center on the admission ticket.

**Government issued-identification**—To be admitted to the test site, candidates must bring an acceptable form of photo identification (ID) such as a driver's license, passport or other government ID. This ID must be a current and original government-issued identification that is not handwritten and that contains both the candidate's name as it appears on the admission ticket and the candidate's photograph.

> **Note:** Candidates who do not provide an acceptable form of identification will not be allowed to sit for the exam and will forfeit their registration fee.

## Exam Guidelines: Logistics

DO:
- Become familiar with the exact location of, and the best travel route to, the exam site prior to the date of the exam.
- Arrive at the exam testing site at least 30 minutes before the exam instructions are read, allowing for time to locate a seat and get acclimated.

## Exam Guidelines: Tools at the Exam Center

DO:
- Bring several no. 2 pencils since pencils will not be provided at the exam site.

DO NOT bring any of the following to the exam site:
- Study materials (including notes, paper, books or study guides)
- Scratch paper
- Notepads
- Communication devices (e.g., cellular phone, PDA, BlackBerry®, etc.)
- Reference materials, including language dictionaries
- Calculators
- Food or beverages (without advanced authorization)

> **Note: Candidates are not allowed to bring any type of communication device (e.g., cellular phone, PDA, BlackBerry, etc.) into the test center. If candidates are viewed with any such device during the exam administration, their exams will be voided and they will be asked to immediately leave the exam site.**

For further details regarding what personal belongings can (and cannot) be brought into the test site, please visit *www.isaca.org/criscbelongings*.

## Exam Guidelines: Your Exam ID

The chief examiner or designate at each test center will read aloud the instructions for entering information on the answer sheet.

It is imperative that candidates include their exam identification number as it appears on their admission ticket and any other requested information on their exam answer sheet. Failure to do so may result in a delay or errors.

## Exam Guidelines: Question Analysis

The following instructions may help the candidate to find the correct answer more efficiently:

- Read the provided instructions carefully before attempting to answer questions. Skipping over these directions or reading them too quickly could result in missing important information and possibly losing credit points.
- Mark the appropriate area when indicating responses on the answer sheet. When correcting a previously answered question, fully erase a wrong answer before writing in the new one.
- Remember to answer all questions since there is no penalty for wrong answers. Grading is based solely on the number of questions answered correctly. Do not leave any question blank.
- Identify key words or phrases in the question (**MOST**, **BEST**, **FIRST** …) before selecting and recording the answer.

## Exam Guidelines: Budget Time

The exam is administered over a four-hour period. This allows for a little over one minute per question. Therefore, it is advisable that candidates pace themselves to complete the entire exam. In order to do so, candidates should complete an average of 50 questions per hour.

Candidates are asked to sign the answer sheet to protect the security of the exam and maintain the validity of the scores.

> **Note:** Candidates are urged to record their answers on their answer sheet. No additional time will be allowed after the exam time has elapsed to transfer or record answers should candidates mark their answers in the question booklet.

## Exam Misconduct

Candidates who are discovered engaging in any kind of misconduct—such as giving or receiving help; using notes, papers or other aids; attempting to take the exam for someone else; using any type of communication device, including cell phones, during the exam administration; or removing the exam booklet, answer sheet or notes from the testing room—will be disqualified and may face legal action. Candidates who leave the testing area without authorization or accompaniment by a test proctor will not be allowed to return to the testing room and will be subject to disqualification. The testing agency will report such irregularities to ISACA's CRISC Certification Committee.

Candidates may not take the exam question booklet after completion of the exam.

The following are reasons for dismissal or disqualification:

- Unauthorized admission to the test center
- Candidate creates a disturbance or gives or receives help
- Candidate attempts to remove test materials or notes from the test center
- Candidate impersonates another candidate
- Candidate brings items into the test center that are not permitted
- Candidate possession of any communication device (i.e., cell phone, PDA, BlackBerry) during the exam administration
- Candidate unauthorized leave of the test area

## Earning the CRISC Certification

Passing the exam does not grant the CRISC designation.

To become a CRISC, each candidate must complete all requirements, including submitting an application for certification.

In order to become CRISC-certified, candidates must pass the exam and must complete and submit an application for certification (and must receive confirmation from ISACA that the application is approved). The application will be available on the ISACA web site at *www.isaca.org/criscapp*. Once the application is approved, the applicant will be sent confirmation of the approval. The candidate is not CRISC-certified, and cannot use the CRISC designation, until the candidate's application is approved.

## Exam Grading Procedure

The exam consists of 200 items:
- Candidate scores are reported as a scaled score.

A scaled score is a conversion of a candidate's raw score on an exam to a common scale:
- ISACA uses and reports scores on a common scale from 200 to 800.
- A candidate must receive a score of 450 or higher to pass the exam.

A score of 450 represents a minimum consistent standard of knowledge as established by ISACA's CRISC Certification Committee:
- A candidate receiving a score less than 450 is not successful and can retake the exam by registering and paying the appropriate exam fee for any future exam administration. To assist with future study, the result letter each candidate receives includes a score analysis by content area. There are no limits to the number of times a candidate can take the exam.

## Exam Result Distribution

**Approximately eight weeks after the test date, the official exam results will be mailed to candidates.** Additionally, with the candidate's consent on the registration form, an e-mail containing the candidates pass/fail status and score will be sent to paid candidates. This e-mail notification will only be sent to the address listed in the candidate's profile at the time of the initial release of the results. To ensure the confidentiality of scores, exam results will not be reported by telephone or fax. To prevent the e-mail notification from being sent to the candidate's spam folder, the candidate should add *exam@isaca.org* to his/her address book, whitelist or safe senders list.

## Failed Score Details

For those candidates not passing the exam, the score report contains a subscore for each job practice domain.

Subscores help identify those areas in which further study may be needed before retaking the exam.

> **Note:** Taking either a simple or weighted average of the subscores does not derive the total scaled score.

## Rescoring

As all scores are subjected to several quality control checks before they are reported, rescores are unlikely to result in a score change.

**Rescoring request process**—The rescoring procedure involves a person hand scoring the answer sheet to ensure that no stray marks, multiple responses or other conditions interfered with computer scoring.

**Rescoring request submission**—Candidates receiving a failing score on the exam may request a rescoring of their answer sheet. The requests must:

- Be submitted in writing to the ISACA certification department
- Be submitted within 90 days following the release of the exam results
- Include:
  - The candidates's name
  - The candidate's identification number
  - The candidate's mailing address
- Be accompanied by a payment of the US $75 processing fee

> **Note:** Requests for a hand score after the deadline date will not be processed.

# List of Exhibits

Starting Page No.

## Introduction



## Part I—Risk Management and Information Systems Control Theory and Concepts



## Part II—Risk Management and Information Systems Control in Practice

# Index

A slash (/) indicates that the terms are synonymous within this manual.

"See also" indicates that the terms are related or relevant to one another.

## A



## B



## C



## D



## E



## F



## G



## I

CRISC
Certified in Risk and Information Systems Control™
An ISACA® Certification

## K



## L



## M



## O



## P



## Q



## R

## S



## T

## U



## V

**Page intentionally left blank**

# Your Evaluation of the CRISC™ Review Manual

ISACA continuously monitors the swift and profound professional, technological and environmental advances affecting risk and IS control professionals. Recognizing these rapid advances, the *CRISC™ Review Manual* will be updated annually.

To assist ISACA in keeping abreast of these advances, please take a moment to evaluate the *CRISC™ Review Manual 2012*. Such feedback is valuable to fully serve the profession and future CRISC exam registrants.

To complete the evaluation on the web site, please go to *www.isaca.org/studyaidsevaluation*.

Thank you for your support and assistance.